T0139981

Marenglen Biba and Fatos Xhafa (Eds.)

Learning Structure and Schemas from Documents

# Studies in Computational Intelligence, Volume 375

## Editor-in-Chief

Prof. Janusz Kacprzyk
Systems Research Institute
Polish Academy of Sciences
ul. Newelska 6
01-447 Warsaw
Poland
*E-mail:* kacprzyk@ibspan.waw.pl

---

Marenglen Biba and Fatos Xhafa (Eds.)

# Learning Structure and Schemas from Documents

 Springer

**Editors**

Marenglen Biba, PhD
University of New York Tirana,
Rr. Komuna E Parisit,
Tirana, Albania
E-mail: marenglenbiba@unyt.edu.al

Fatos Xhafa, PhD
Technical University of Catalonia,
Campus Nord, Ed. Omega
C/Jordi Girona 1-3
08034 Barcelona, Spain
E-mail: fatos@lsi.upc.edu

ISBN 978-3-662-50671-4          ISBN 978-3-642-22913-8 (eBook)

DOI 10.1007/978-3-642-22913-8

Studies in Computational Intelligence          ISSN 1860-949X

*Typeset* & *Cover Design:* Scientific Publishing Services Pvt. Ltd., Chennai, India.

Printed on acid-free paper

9 8 7 6 5 4 3 2 1

springer.com

# Foreword

It was not long ago that database systems were revolutionized through the birth of the relational concepts and theory, which are now materialized in most commercial database management systems. It was then thought that data should be structured in a rather simple way to support the day-to-day business operations. This has certainly been made possible and supported by the mature transaction management concepts, and mathematically based query language and processes. Whilst this has been the backbone of businesses today, it cannot be ignored that most data in the world is not structured, or at least, easily and conveniently structured in a relational way.

The rise of an Internet world has undeniably contributed to the explosion of data in the digital universe. It has been reported not so long ago that there were over 1 trillion web pages – this roughly equals to almost 150 web pages per man, woman, and child on Earth, and the number of web pages increases billions per day. Another good example to illustrate data explosion in the digital universe is *facebook*. Also, it has been reported that the number of facebook users has surpassed the 500-million mark and is still growing strongly. If facebook users were citizen of a country called facebook, this country would have been the third largest country in the world, after China and India. And the kind of data is not easily categorized as structured data, since mostly, they are free from rigidly and constraining structures.

It was estimated slightly more than a couple years ago in 2008 that our digital universe grew by almost by half *zettabytes* – this is more than the data collected in the previous 5,000 years. To give readers perspective on this size, imagine gigabytes, terabytes, and then petabytes. It doesn't stop there; it continues to exabytes, and now zettabytes. It will not be surprising that in the near future, it will even surpass zettabytes to reach yottabytes. Next year in 2012, it is predicted that the annual growth rate of data will be 5 times that of 2008 – so, yottabytes, literally, are not far away. Of this data, it is estimated that 95% of them is unstructured.

Now how about the database market? It is not surprising to know that large enterprises have almost 10 thousand databases with 30-50 major data warehouses – many in tera-peta-exa byte range and double in size every 18 months. The database market is almost $30billion a couple of years ago with 4 major vendors growing at 7%. The database infrastructure is the largest, fastest growing and highest priority software category. Hence, the structured concept still governs. It is still expected that the structures and schemas still exist in the so-called unstructured world of data. It is undeniably that structures and schemas play an important role in any aspect of data – either it is in data processing, data management, data manipulation, data analysis and knowledge discovery, or even simple data or information retrieval. Every organization depends critically on the notion of structures and schemas of data.

This book addresses an important issue – that is finding, discovering, learning structures and schemas from documents and unstructured data. There is nothing more critical in the data explosion age in our digital universe than knowing that data have some degree of structures and schemas, as this is the foundation in many important applications, such as digital libraries, imaging, biomedical, medicine, and bioinformatics. This book covers the most comprehensive range of topics related to the importance of learning structures and schemas from documents – from basic techniques to advanced applications. I believe readers, from different sort of fields and background; from data oriented researchers to application practitioners, will benefits from reading a fine collection of articles compiled in this book.

Melbourne, June 2011                                                    David Taniar
                                          Editor-in-Chief *International Journal of*
                                          *Data Warehousing and Mining* (SCI-E)

# Preface by Editors

The rapidly growing volume of available digital documents of various formats and the possibility to access these through internet-based technologies, have led to the necessity to develop solid methods to properly organize and structure documents in large digital libraries and repositories. Specifically, since the extremely large volumes make it impossible to manually organize such documents and since most of the documents exist in an unstructured form and do not follow any schemas, most of the efforts in this direction are dedicated to automatically infer structure and schemas that can help to better organize hue collections. This is essential in order for these documents to be effectively and efficiently retrieved.

Dealing with unstructured information is currently a hot research topic. A growing body of work is addressing the problem of recognizing structure and schemas in documents of various types. Some areas are mainly concerned about the visual representation of documents and increasing improvements are being made in the area of pattern recognition and document layout analysis to classify documents according to structure found in their layout. On the other side, extensive research is being done in the field of machine learning to exploit attributes of documents and relationships among different documents to infer structures in large collections of documents. Important work is also being performed in the data mining and knowledge discovery community which has traditionally dealt with raw data but recently is dedicating attention to learning structure from unstructured information. In addition, Semantic Web researchers are dedicating important efforts to the problem of identifying structure and schemas in order for them to achieve ontology matching or alignment. Another related area regards the database community that has long worked with integration problems but only recently this community has started considering automatic structure and schema learning as a potential approach for schema and database integration. Finally information retrieval and extraction seek to infer structure and schemas from free text in order to build efficient information seeking models from large corpora.

## Main Contributions of This Book

This book covers the latest advances in structure inference in heterogeneous collections of documents and data. The goal is to present the state-of-the-art in the area in order to describe some lessons learned, identify new research issues and challenges as well as opportunities for further developments.

Overall, the book contributes with the following:

- Presenting state-of-the-art approaches to learning and inferring structure and schemas from documents
- Assessing whether methods and techniques of one approach can be extended or exploited by the other approaches in a multi-strategic effort to the problem
- Assisting in the identification of new and future challenges and opportunities that will raise the bar in the state-of-the-art of several research areas attacking the same problem.
- Case studies and best practices from real large scale digital libraries, repositories and corpora.

### Organization of the Book

The book consists of 19 chapters, organized as follows:

**Chapter 1:** Learning Structure and Schemas from Heterogeneous Domains in Networked Systems Surveyed

*Marenglen Biba and Fatos Xhafa*

This chapter presents a survey of the state-of-the-art methods for inferring structure from documents and schemas in networked environments. The survey is organized around the most important application domains, namely, bio-informatics, sensor networks, social networks, P2P systems, automation and control, transportation and privacy-preserving.

**Chapter 2:** Handling Hierarchically Structured Resources Addressing Interoperability Issues in Digital Libraries

*Maristella Agosti, Nicola Ferro, and Gianmaria Silvello*

This chapter presents the NEsted SeTs for Object hieRarchies (NESTOR) Framework that allows to model, manage, access and exchange hierarchically structured resources. The authors envision the framework in the context of Digital Libraries as a mean to address the complex and multiform concept of interoperability when dealing with hierarchical structures.

**Chapter 3:** Administrative Document Analysis and Structure

*Abdel Belaïd, Vincent Poulain D'Andecy, Hatem Hamza, and Yolande Belaid*

This chapter presents a novel technique about the analysis and recognition of scanned administrative documents. This technique is related to the case-based reasoning already used in data mining and various problems of machine learning.

**Chapter 4:** Automatic Document Layout Analysis through Relational Machine Learning

*Stefano Ferilli, Teresa M.A. Basile, Nicola Di Mauro, and Floriana Esposito*

This chapter presents a prototypical version of the document processing system DOMINUS, using the incremental first-order logic learner INTHELEX. The authors propose the application of supervised Machine Learning techniques to infer correction rules to be applied to forthcoming documents.

**Chapter 5:** Dataspaces: where structure and schema meet

*Maurizio Atzori and Nicoletta Dessì*

This chapter investigates the crucial problem that poses the bases to the concept of dataspaces: the need for human interaction/intervention in the process of organizing (getting the structure of) unstructured data. The authors survey the existing techniques behind dataspaces to overcome that need

**Chapter 6:** Transductive Learning of Logical Structures from Document Images

*Michelangelo Ceci, Corrado Loglisci, and Donato Malerba*

This chapter investigates the induction of a classifier for the automated recognition of semantically relevant layout components. In particular, it investigates the combination of transductive inference with principled relational classification.

**Chapter 7:** Progressive Filtering on the Web: the Press Reviews Case Study

*Andrea Addis, Giuliano Armano and Eloisa Vargiu*

This chapter presents NEWS.MAS, a multiagent system aimed at: (i) extracting information from online newspapers by using suitable wrapper agents, each associated with a specific information source, (ii) categorizing news articles according to a given taxonomy, and (iii) providing user feedback to improve the performance of the system depending on user needs and preferences.

**Chapter 8:** A Hybrid Binarization Technique for Document Images

*Vavilis Sokratis, Ergina Kavallieratou, Roberto Paredes, and Kostas Sotiropoulos*

This chapter presents a binarization technique specifically designed for historical document images. The technique is based on a hybrid approach that first applies a global thresholding technique and, then, identifies the image areas that are more likely to still contain noise.

**Chapter 9:** Digital Libraries and Document Image Retrieval Techniques: a Survey

*Simone Marinai, Beatrice Miotti, and Giovanni Soda*

This chapter presents a survey about the more recent techniques applied in the field of recognition and retrieval of text and graphical documents is presented. In particular it describes techniques related to recognition-free approaches.

**Chapter 10:** Mining Biomedical Text Towards Building a Quantitative Food-disease-gene Network

*Hui Yang, Rajesh Swaminathan, Abhishek Sharma, Vilas Ketkar, and Jason D'Silva*

This chapter presents an information extraction system that analyzes publications in the emerging discipline of Nutritional Genomics. The system is used to systematically extract relationships among bio-entities from biomedical articles.

**Chapter 11:** Mining Tinnitus Data based on Clustering and new Temporal Features

*Xin Zhang, Pamela Thompson, Zbigniew W. Raś, and Pawel Jastreboff*

This chapter presents an initial research on data mining of medical data about Tinnitus Retraining Therapy. The authors present Interesting rules about the relationship recovery against symptoms, audiological therapy parameters, and other factors revealed during the experiments.

**Chapter 12:** DTW-GO Based Microarray Time Series Data Analysis for Gene-Gene Regulation Prediction

*Andy C. Yang and Hui-Huang Hsu*

This chapter introduces a novel approach that provides an effective distance measurement for genes based on gene ontology annotations. The authors show that the proposed approach facilitates analysis for microar-ray time series data.

**Chapter 13:** Integrating Content and Structure into a Comprehensive Framework for XML Document Similarity Represented in 3D Space

*Eric Draken, Tamer N. Jarada, Keivan Kianmehr, and Reda Alhajj*

In this chapter the authors have presented a novel way to represent XML document similarity in 3D space. Their approach benefits from the characteristics of the XML documents to produce a measure to be used in clustering and classification techniques, information retrieval and searching methods for the case of XML documents.

**Chapter 14:** Modelling User Behaviour on Page Content and Layout in Recommender Systems

*Kin Fun Li and Kosuke Takano*

This chapter presents a literature survey on Web browsing behaviour, a recommender system based on user browsing behaviour, and the representation and manipulation of attributes associated with the design and layout of Web pages. A proof-of-concept system is designed with the objective to study whether there is a correlation between all these factors.

**Chapter 15:** MANENT: An Infrastructure for Integrating, Structuring and Searching Digital Libraries

*Angela Locoro, Daniele Grignani, and Viviana Mascardi*

This chapter describes MANENT, an infrastructure for integrating, structuring and searching Digital Libraries. The technologies and standards enabling the realisation of infrastructures of this kind have been reviewed, as well as the related work.

**Chapter 16:** Low-Level Document Image Analysis and Description: From Appearance to Structure.

*Emanuele Salerno, Pasquale Savino, and Anna Tonazzini*

This chapter deals with the problem of processing and analyzing digital images of ancient or degraded documents to increase the possibilities of inferring their structures.

**Chapter 17:** Model Learning from Published Aggregated Data

*Janusz Wojtusiak and Ancha Baranova*

This chapter describes the problem of machine learning of models from aggregated data as compared to traditional learning from individual examples. It presents a method of rule induction from such data as well as an application of this method.

**Chapter 18:** Data De-duplication: A Review

*Gianni Costa, Alfredo Cuzzocrea, Giuseppe Manco, and Riccardo Ortale*

This chapter presents an overview of research on data de-duplication, with the goal of providing a general understanding and useful references to fundamental concepts concerning the recognition of similarities in very large data collections.

**Chapter 19:** A Survey on Integrating Data in Bioinformatics

*Andrea Manconi and Patricia Rodriguez-Tomé*

This chapter discusses the current approaches, methods, systems, and open issues of integrating biological data.

## *Targeted Audience and Last Words*

The audiences of this book are those working in or interested in joining interdisciplinary and trans-disciplinary works in the areas of data mining, machine learning, pattern recognition, document analysis and understanding, semantic web, databases, artificial intelligence and digital libraries, whose main focus is that of learning structure and schemas from unstructured information. The book also brings contributions from application areas such as from bioinformatics, web mining, text mining, information retrieval, real-world digital libraries, data warehouses and ontology building. Specifically, audiences who are broadly involved in the domains of computer science, web technologies, applied informatics, business or management information systems are: (1) researchers or senior graduates working in academia; (2) academics, instructors and senior students in colleges and universities, and (3) business analysts from industries interested in data integration, information retrieval and enterprise search.

Finally, academic researchers, professionals and practitioners in the field can also be inspired and put in practice the ideas and experiences proposed in the book in order to evaluate them for their specific research and work.

We hope that the readers will find this book useful in their academic and professional activities!

## *Acknowledgements*

We would like to first thank all the authors of the chapters for their contributions and their efforts to produce high quality manuscripts to this volume. We are greatful to the referees who have carefully reviewed the chapters and gave useful suggestions and constructive feedback to the authors. We gratefully acknowledge the support and encouragement received from Prof. Janusz Kacprzyk, the editor in chief of Springer series "Studies in Computational Intelligence", Dr Thomas Ditzinger, Ms Heather King and the whole Springer's editorial team for their great and continuous support throughout the development of this book.

Finally, we are deeply indebted to our families for their love, patience and support!

Fatos Xhafa's work is partially done at Birkbeck, University of London, during his Sabbatical Leave from Technical University of Catalonia (Barcelona, Spain). His research is supported by a grant from the General Secretariat of Universities of the Ministry of Education, Spain.

June 2011                                                          Marenglen Biba
                                                                       Fatos Xhafa

# Contents

# List of Contributors

Marenglen Biba
Department of Computer Science,
University of New York, Tirana
Rr. "Komuna e Parisit", Tirana, Albania
marenglenbiba@unyt.edu.al

Fatos Xhafa
Department of Languages and
Information Systems
Technical University of Catalonia
Campus Nord. Ed. Omega
C/ Jordi Girona, 1-3
08034 Barcelona, Spain
fatos@lsi.upc.edu

Maristella Agosti
Department of Information Engineering
University of Padua
DEI - Via Gradenigo 6/B - 35131
Padova, Italy
agosti@dei.unipd.it

Nicola Ferro
Department of Information Engineering,
University of Padua
DEI - Via Gradenigo 6/B - 35131
Padova, Italy
ferro@dei.unipd.it

Gianmaria Silvello
Department of Information Engineering
University of Padua
DEI - Via Gradenigo 6/B - 35131
Padova, Italy
silvello@dei.unipd.it

Maurizio Atzori
University of Cagliari
Via Ospedale 72, 09124 Cagliari, Italy
atzori@unica.it

Nicoletta Dessì
University of Cagliari
Via Ospedale 72, 09124 Cagliari, Italy
dessi@unica.it

Abdel Belaïd
LORIA, University of Nancy 2
UMR 7503
Campus scientifique B.P. 239
54506 Vandoeuvre Lès Nancy Cedex,
France
abdel.belaid@loria.fr

Vincent Poulain D'Andecy
ITESOFT, France,
Vincent.PoulaindAndecy@itesoft.com

Hatem Hamza
ITESOFT, France,
Hatem.Hamza@itesoft.com

Yolande Belaïd
LORIA, University of Nancy 2
UMR 7503
Campus scientifique B.P. 239
54506 Vandoeuvre Lès Nancy Cedex,
France
Yolande.Belaid@loria.fr

Michelangelo Ceci,
Dipartimento di Informatica,
Università degli Studi di Bari "Aldo
Moro" via Orabona 4 - 70125
Bari, Italy.
ceci,@di.uniba.it

Corrado Loglisci
Dipartimento di Informatica,
Università degli Studi di Bari "Aldo
Moro" via Orabona 4 - 70125
Bari, Italy.
loglisci@di.uniba.it

Donato Malerba
Dipartimento di Informatica,
Università degli Studi di Bari
"Aldo Moro" via Orabona 4 - 70125
Bari, Italy.
malerba@di.uniba.it

Gianni Costa
ICAR-CNR, Via P. Bucci, 41C,
87036 Rende (CS)
Italy.
costa@icar.cnr.it

Alfredo Cuzzocrea
ICAR-CNR, Via P. Bucci, 41C,
87036 Rende (CS) - Italy.
cuzzocrea@icar.cnr.it

Giuseppe Manco
ICAR-CNR, Via P. Bucci, 41C,
87036 Rende (CS) Italy.
manco@icar.cnr.it

Riccardo Ortale
ICAR-CNR, Via P. Bucci, 41C,
87036 Rende (CS) - Italy.
ortale@icar.cnr.it


Vavilis Sokratis
Dept. of Information and
Communication Systems Engineering
University of the Aegean
83200 Karlovasi
Samos, Greece
sokratisvav@gmail.com

Ergina Kavallieratou
Dept. of Information and
Communication Systems Engineering
University of the Aegean
83200 Karlovasi
Samos, Greece
kavallieratou@aegean.gr

Roberto Paredes
PRHLT - Universidad Politecnica de
Valencia
Camino de Vera s/n
46071 Valencia
Spain
rparedes@dsic.upv.es

Kostas Sotiropoulos
University of Patras,
University campus, 26504, Rio, Patras,
Greece
kosotiro@upatras.gr

Stefano Ferilli
Dipartimento di Informatica,
Università degli Studi di Bari "Aldo
Moro" via Orabona 4 - 70125
Bari, Italy.
ferilli@di.uniba.it

Teresa M.A. Basile and
Dipartimento di Informatica,
Università degli Studi di Bari "Aldo
Moro" via Orabona 4 - 70125
Bari, Italy.
basile@di.uniba.it

Nicola Di Mauro
Dipartimento di Informatica,
Università degli Studi di Bari "Aldo
Moro" via Orabona 4 - 70125
Bari, Italy.
ndm@di.uniba.it

Floriana Esposito
Dipartimento di Informatica,
Università degli Studi di Bari "Aldo
Moro" via Orabona 4 - 70125
Bari, Italy.
esposito@di.uniba.it

Hui Yang,
Department of Computer Science, 1600
Holloway Avenue . San Francisco . CA
94132, USA
huiyang@cs.sfsu.edu

Rajesh Swaminathan
Department of Computer Science, 1600
Holloway Avenue . San Francisco . CA
94132, USA
rajeshs@sfsu.edu

Abhishek Sharma
Department of Computer Science, 1600
Holloway Avenue . San Francisco . CA
94132, USA
asharma1@sfsu.edu

Vilas Ketkar
Department of Computer Science, 1600
Holloway Avenue . San Francisco . CA
94132, USA
vilask@sfsu.edu

Jason D'Silva
Department of Computer Science, 1600
Holloway Avenue . San Francisco . CA
94132, USA

Andy C. Yang
Department of Computer Science &
Information Engineering, Tamkang
University, Taipei, 25137,
Taiwan R.O.C.
andyyung0215@gmail.com

Hui-Huang Hsu
Department of Computer Science &
Information Engineering, Tamkang
University, Taipei, 25137,
Taiwan R.O.C.
huihuanghsu@gmail.com

Eric Draken
Computer Science Department
University of Calgary
2500 University Dr. NW
Calgary, Alberta Canada

Tamer N. Jarada
Computer Science Department
University of Calgary
2500 University Dr. NW
Calgary, Alberta Canada
tJarada@gmail.com

Keivan Kianmehr
Computer Engineering Department,
University of Western Ontario, London,
Ontario, Canada
kkianmeh@uwo.ca

Reda Alhajj
Computer Science Department
University of Calgary
2500 University Dr. NW
Calgary, Alberta Canada
alhajj@ucalgary.ca

Kin Fun Li
Department of Electrical and Computer
Engineering, University of Victoria
P.O. Box 3055 STN CSC
Victoria, B.C. V8W 3P6
Canada
kinli@uvic.ca

Kosuke Takano
Department of Information and
Computer Sciences, Kanagawa
Institute of Technology, Japan
takano@ic.kanagawa-it.ac.jp

Angela Locoro
Computer Science Department,
University of Genova, Via Dodecaneso
35, 16146 Genova, Italy
locoro@disi.unige.it

Daniele Grignani
Department of Modern and
Contemporary History, University of
Genova, Via Balbi 6, 16126 Genova,
Italy
daniele.grignani@gmail.com

Viviana Mascardi
Computer Science Department,
University of Genova, Via Dodecaneso
35, 16146 Genova, Italy
mascardi@disi.unige.it

Andrea Manconi
Institute for Biomedical Technologies,
National Research Council, Via F.lli
Cervi 93, 20090, Seg-
rate (MI), Italy.
andrea.manconi@itb.cnr.it

Patricia Rodriguez-Tomé
CRS4 - Center for Advanced Studies,
Research and Development in Sardinia,
Loc. Piscina Manna,
Ed.1, 09010 Pula (CA), Italy
prtome@crs4.it

Simone Marinai
Dipartimento di Sistemi e Informatica.
Via di Santa Marta, 3 - 50139, Firenze,
Italy
simone.marinai@unifi.it

Beatrice Miotti
Dipartimento di Sistemi e Informatica.
Via di Santa Marta, 3 - 50139, Firenze,
Italy
miotti@dsi.unifi.it

Giovanni Soda
Dipartimento di Sistemi e Informatica.
Via di Santa Marta, 3 - 50139, Firenze,
Italy
soda@dsi.unifi.it

Xin Zhang,
University of North Carolina at
Pembroke, Dept. of Math. Comp.
Science, Pembroke, NC 28372, USA
xin.zhang@uncp.edu

Pamela Thompson,
University of North Carolina,
Dept. of Computer Science, Charlotte,
NC 28223, USA
pthompso@catawba.edu

Zbigniew W. Ras,
University of North Carolina,
Dept. of Computer Science, Charlotte,
NC 28223, USA & Warsaw
University of Technology, Institute of
Comp. Science, 00-665 Warsaw, Poland
ras@uncc.edu

Pawel Jastreboff
Emory University School of Medicine,
Dept. of Otolaryngology, Atlanta, GA
30322, USA
pjastre@emory.

Emanuele Salerno
CNR - Istituto di Scienza e Tecnologie
dell'Informazione, Via Moruzzi 1,
56124 Pisa, Italy
emanuele.salerno@isti.cnr.it

Pasquale Savino
CNR - Istituto di Scienza e Tecnologie
dell'Informazione, Via Moruzzi 1,
56124 Pisa, Italy
pasquale.savino@isti.cnr.it

Anna Tonazzini
CNR - Istituto di Scienza e Tecnologie
dell'Informazione, Via Moruzzi 1,
56124 Pisa, Italy
anna.tonazzini@isti.cnr.it

Andrea Addis
Dept. of Electrical and Electronic Engineering, University of Cagliari, Piazza
D'Armi, I-09123 Cagliari, Italy
addis@diee.unica.it

Giuliano Armano
Dept. of Electrical and Electronic
Engineering, University of Cagliari,
Piazza D'Armi, I-09123 Cagliari, Italy
armano@diee.unica.it

Eloisa Vargiu
Dept. of Electrical and Electronic
Engineering, University of Cagliari,
Piazza D'Armi, I-09123 Cagliari, Italy
vargiu@diee.unica.it

Janusz Wojtusiak
Department of Health Administration
and Policy, George Mason University
Northeast Module, Room 108
4400 University Drive, MSN 1J3
Fairfax, VA 22030, USA
jwojt@mli.gmu.edu

Ancha Baranova
The Center for Biomedical Genomics
Room 182 Discovery Hall, MSN 4D7
10900 University Blvd
Manassas VA, 20110
abaranov@gmu.edu

# Learning Structure and Schemas from Heterogeneous Domains in Networked Systems Surveyed

Marenglen Biba and Fatos Xhafa

**Abstract.** With the continuous growing amount of digital documents in many different formats and with the increasing possibility to access these through internet-based technologies in distributed environments, there is strong motivation to develop robust methods to organize documents in large and repositories. In particular, the extremely large volume of document collections makes it unfeasible to manually handle such documents. In addition, most of the documents exist in an unstructured form and do not follow any schemas. Therefore, research efforts in this direction are being dedicated to automatically infer structure and schemas. This is essential in order to properly organize huge collections as well as to effectively and efficiently retrieve documents in in . This chapter presents a survey of the state-of-the-art methods for inferring structure from documents and schemas in networked environments. The survey is organized around important application domains such as bio-informatics, sensor networks, social networks, P2P systems, automation and control, transportation and privacy-preserving for which we analyze the recent developments on dealing with unstructured data in such domains.

## 1 Introduction

For a long time, , and research areas have focused mainly on structured domains. Structured information is usually easier to process by machines, however, there are still many research issues [14]. Some research works are concerned with the visual representation of documents, while improvements are being made in the area of

Marenglen Biba
Department of Computer Science, University of New York Tirana, Albania
e-mail: marenglenbiba@unyt.edu.al

Fatos Xhafa
Technical University of Catalonia,
Barcelona, Spain
e-mail: fatos@lsi.upc.edu

M. Biba and F. Xhafa (Eds.): Learning Structure and Schemas from Documents, SCI 375, pp. 1–16.
springerlink.com

pattern recognition and to classify documents according to structure found in their layout [31]. On the other side, research works in the field of machine learning try to exploit attributes of documents and relationships among different documents to infer structures in large collections of documents taking into consideration also the uncertainty that may be present in the application domain [18].

Most of the information sources on the web contain information in the form of free text or heterogeneous documents distributed among different domains in networked environments. Recently, a growing body of research work is addressing the problem of recognizing structure and schemas in documents of various formats from heterogeneous domains. As in the case of structured information, machine learning, pattern recognition and data mining methods have the potential to uncover relevant hidden structures in heterogeneous data. However, most of the methods and techniques developed so far have focused only on the classical problem of discovering structures in documents and data, and should now re-consider their setting (and maybe their theoretical framework) in order to deal with unstructured information in heterogeneous domains. In particular new features, typical of heterogeneous distributed data, such as noise, incomplete data, missing attributes, stream data, private data, social data, etc., pose new challenges to the community of machine learning and data mining. In fact, dealing with heterogeneous data is a challenge *per se* even in non networked systems.

In this chapter we present some recent approaches in the literature that deal with inferring structure in heterogeneous collections of documents and data which are distributed in networked systems. The goal is to present the state-of-the-art in the area in order to emphasize some lessons learned, identify new research issues and challenges as well as opportunities for further developments. The growing body of research work dedicated to the problem of dealing with different types of data in heterogeneous domains is found across several application domains such as bioinformatics, sensor networks, social networks, P2P systems, automation and control, transportation and privacy-preserving. We dedicate a section to each of these areas considering recent developments and describing each of the approaches in order to outline opportunities for new research.

## 2  Learning Patterns in Sensor Networks

present large amounts of data spread over many physically distributed nodes. Machine learning and data mining techniques have the potential to deal with these kind of data. Due to the complexity of heterogeneous networked data, important challenges have arisen such as the need for run-time data aggregation, parallel computing, and distributed hypothesis formation [8]. One of the existing approaches in sensor networks is presented in [53] where the authors present an algorithm for finding distributed icebergs-elements that may have low frequency at individual nodes but high aggregate frequency (this is a problem that arises commonly in practice). The work in [7] addresses a major challenge in data mining applications where the full information about the underlying processes, such as sensor networks or large online databases, cannot be practically obtained due to physical limitations such as low bandwidth or memory, storage, or computing power. They propose a framework for detecting anomalies from these large-scale data mining applications where the full information is not practically possible to obtain.

In [43] it is presented another approach for network management in large-scale randomly-deployed sensor networks, called Energy Map, which explores the inherent relationships between the energy consumption and the sensor operation. Through nonlinear manifold learning algorithms the approach visualizes the residual energy level of each sensor in a large scale network, infers the sensor locations and the current network topology through mining the collected residual energy data in a randomly-deployed sensor network, and explores the inherent relation between sensor operation and energy consumption to find the dynamic patterns from large volumes of sensor network data for network design.

In [39] and [40] the author proposes a declarative query language and data mining techniques to discover frequent event patterns and their spatial and temporal properties. In these works, raw streams of sensor readings are collected for later offline processing and analysis and in-network data mining techniques are explored to discover frequent event patterns and their spatial and temporal properties.

The authors of [29] propose and evaluate distributed algorithms for data clustering in self-organizing ad-hoc sensor networks with computational, connectivity, and power constraints. One of the benefits of in-network data clustering algorithms is the capability of the network to transmit only relevant, high level information, namely models, instead of large amounts of raw data, also reducing drastically energy consumption. Finally, the work in [38] presents an exploration of different characteristics of sensor networks which define new requirements for knowledge discovery, with the common goal of extracting some kind of comprehension about sensor data and sensor networks, focusing on clustering techniques which provide useful information about sensor networks as it represents the interactions between sensors.

In [34] the authors propose a combination of a neural network based offline learning approach and online repuation update schemes to identify nodes reporting inconsistent data. The authors experimentally evaluate their scheme for two different network sizes and two different data patterns over the sensor field and the results show that their approach is successful in identifying multiple colluding malicious nodes without any false positives and false negatives.

## 3 Learning Structures in Biological Domains

Recent advances in computing, digital storage technologies and high throughput data acquisition technologies have led to a growing capability to gather and store huge volumes of data. For example, advances in high throughput sequencing and other data acquisition have resulted in gigabytes of DNA, protein sequence data, and gene expression data being gathered continuously. On the other side, the necessity for methods to automatically analyze large volumes of data has led to a growing effort in machine learning and data mining communities to develop robust methods that work effectively on heterogeneous domains. For example, in [20], it was presented a mixture model associative artificial neural network that integrates two heterogeneous domain knowledge (Gene Ontology (GO) annotation and gene expression profiling) for discovery of genome-wide functional patterns. The presented experiments showed that association of these domains reduces analytical noises and produces a more meaningful functional grouping. In the same direction goes also another approach presented in [49]. Since in many biomedical modeling tasks a number of different types of data may influence predictions made by the model, an established approach to pursuing supervised learning with multiple types of data is to encode these different types of data into separate kernels and use multiple kernel learning. In [49] the authors present a simple iterative approach to multiple kernel learning (MKL), focusing on multi-class classification and show that the proposed method outperforms state-of-the-art results on an important protein fold prediction dataset and gives competitive performance on a protein subcellular localization task.

Another interesting approach is presented in [47], where the authors introduced a multi-label large-margin classifier that automatically learns the underlying intercode structure and allows the controlled incorporation of prior knowledge about medical code relationships. In addition to refining and learning the code relationships, their classifier can also use this shared information to improve its performance. Experiments on a publicly available dataset containing clinical free text and their associated medical codes showed that the proposed multi-label classifier outperforms related multi-label models in this problem.

Another line of research has been that of monitoring the occurrence of topics in a stream of events, such as a stream of news articles. This has led to several models of bursts in these streams, i.e., periods of elevated occurrence of events and there are several burst definitions and detection algorithms. The authors in [21] present a topic dynamics model for the large PubMed/MEDLINE database of biomedical publications, using the MeSH (Medical Subject Heading) topic hierarchy. They show that their model is able to detect bursts for MeSH terms accurately as well as efficiently.

Another challenge in bioinformatics is that of selecting genes that are differentially expressed and critical to a particular biological process. Important developments in data gathering technologies have made various data sources available such as mRNA and miRNA expression profiles, biological pathway and gene annotation, etc. Recent works have also shown that integration of multiple data sources helps enrich knowledge for selecting genes bearing significant biological relevance. One approach is the one proposed in [52], where multiple data sources are extracted into an intrinsic global geometric pattern which is used in covariance analysis for gene selection. Another approach is presented in [36] where the authors propose a machine learning technique to identify essential genes using the experimental data of genome-wide knock-out screens from one bacterial organism to infer essential genes of another related bacterial organism. They use a broad variety of topological features, sequence characteristics and co-expression properties potentially associated with essentiality, such as flux deviations, centrality, codon frequencies of the sequences, coregulation and phyletic retention.

The analysis of trends and topics in the biomedical literature is yet another hot research direction nowadays. Often the goal in bio-informatics is to identify potential diagnostic and therapeutic bio-markers for specific diseases. In [33] the presented approach integrates several data sources to provide the user with up-to-date information on current research in the field. The BioJournalMonitor is a decision support system that deploys state-of-the-art text mining technologies to provide added value on top of the original content, including named entity detection, relation extraction, classification, clustering, ranking, summarization, and visualization. The presented results suggest that early prediction of emerging trends is possible through a probabilistic topic models that can be used to annotate recent articles with the most likely MeSH terms.

## 4  Learning in Distributed Automation and Control Systems

Machine learning and data mining provide excellent methods and techniques for dealing with automation and control in a distributed setting. Here we explore some approaches that have proven successful in important areas such as transportation, fleets and automation control.

In [22] it is presented a distributed vehicle performance data mining system designed for commercial fleets. The MineFleet system analyzes high throughput data streams onboard the vehicle, generates the analytics, sends those to the remote server over the wide-area wireless networks and offers them to the fleet managers using stand-alone and web-based user-interface. MineFleet is probably one of the first commercially successful distributed data stream mining systems. Another approach was proposed in [27] called mobility-based clustering that deals with practical research on hot spots in smart city taking into consideration unique features, such as highly mobile environments, supremely limited size of sample objects, and the non-uniform, biased samples. The authors report performance of mobility-based clustering based on real traffic situations.

The authors in [54] deal with the critical problem in a crisis situation of how to efficiently discover, collect, organize, search and disseminate real-time disaster information. The proposed system exploits the latest advances in data mining technologies to analyze the integrated input data from different sources. Another interesting approach is that in [17] where a massive quantity of complex, dynamic, and distributed location traces is handled and mined to provide effective mobile sequential recommendation.

In another recent work, a novel approach was presented based on the theory of multiple kernel learning to detect potential safety anomalies in very large data bases of discrete and continuous data from world-wide operations of commercial fleets [11]. Their results show that the proposed algorithm uncovers operationally significant events in high dimensional data streams in the aviation industry which are not detectable using state of the art methods. Another interesting approach is that of [23] where it is presented a system based on Ubiquitous Data Mining (UDM) concepts. It merges and analyses different types of information from crash data and physiological sensors to diagnose driving risks in real time.

An important feature of networked data is their uncertainty since sensors are typically expected to have considerable noise in their readings because of inaccuracies in data retrieval, transmission, and power failures. In [2] the authors propose a method for clustering uncertain data streams.

Another interesting approach is presented in [6] where the authors show that shortcomings of automating datacenters using closed-loop control can be addressed by replacing simple techniques of modeling and model management with more sophisticated techniques imported from statistical machine learning. An interesing approach is also presented in [45] where the authors present a novel framework that combines queueing networks and graphical models, allowing Markov chain Monte Carlo to be applied. The authors demonstrate the effectiveness on real-world data from a benchmark Web application.

Another challenging task has always been collective decision making. In particular, collective recognition is the problem of jointly applying multiple classifiers. The decisions are made about the class of an entity, situation, image, etc, and the joint decision is used to improve quality of the final decision by aggregation and coordination of different classifier decisions using a metalevel algorithm [19].

## 5   Learning Structures in Social Networks

Social networks have inspired a lot research in the machine learning and data mining community. This is due to the growing amount of available data that have to be analyzed. Moreover, most social networks have an outstanding marketing value and developing methods for viral marketing is a hot topic in the research community [13].

Mobile devices and wireless technologies have led to mobile social network systems which are increasingly becoming popular. In a mobile social network the spread of information and influence is in the form of word-of-mouth, therefore it is important to find a subset of influential individuals in a mobile social network such that targeting them initially (e.g. for marketing campaigns) will maximize the spread of the influence. Unfortunately, it has been shown that the problem of finding the most influential nodes is NP-hard. It has also been shown that a Greedy algorithm with provable approximation guarantees can give good approximation. However, it is computationally expensive, if not prohibitive, to run the greedy algorithm on a large mobile network. In [56] the authors propose a new algorithm called Community-based Greedy algorithm for mining top-K influential nodes. The proposed algorithm encompasses two components: 1) an algorithm for detecting communities in a social network; and 2) a dynamic programming algorithm for selecting communities to find influential nodes. Empirical experiments on a large real-world mobile social network show that their algorithm is more than an order of magnitudes faster than the state-of-the-art Greedy algorithm for finding top-K influential nodes and the error of their approximate algorithm is small.

Another line of research is the analysis of blog data. In [1] the authors present an approach that uses innovative ways to employ contextual information and collective wisdom to aggregate similar bloggers.

Social interactions that occur regularly typically correspond to significant yet often infrequent and hard to detect interaction patterns. To identify such regular behavior, the authors in [24] propose a new mining problem of finding periodic or near periodic subgraphs in dynamic social networks. They propose a practical, efficient and scalable algorithm to find such subgraphs that takes imperfect periodicity into account and demonstrate the applicability of their approach on several real-world networks and extract meaningful and interesting periodic interaction patterns.

Some recent developments regard the application of the concept of organizational structure to social network analysis which may well represent the power of members and the scope of their power in a social network. In [37], the authors propose a data structure, called Community Tree, to represent the organizational structure in the social network. They combine the PageRank algorithm and random walks on graph to derive the community tree from the social network. Experiments conducted on real data show that the methods are effective at discovering the organizational structure and representing the evolution of organizational structure in a dynamic social network.

Social networks often involve multiple relations simultaneously. People usually construct an explicit social network by adding each other as friends, but they can also build implicit social networks through daily actions like commenting on posts, or tagging photos. The authors in [15] address this problem: given a real social networking system which changes over time, do daily interactions follow any pattern? They model the formation and co-evolution of multi-modal networks proposing an approach that discovers temporal patterns in peoples social interactions. They show the effectiveness of the approach on two real datasets (Nokia FriendView and Flickr) with 100,000 and 50,000,000 records respectively, each of which corresponds to a different social service, and spans up to two years of activity.

The formation of implicit groups is an interesting related problem here. Although users of online communication tools do not usually categorize their contacts into groups such as "family", "co-workers", or "jogging buddies", they implicitly cluster contacts through their interactions with them, forming implicit groups. The authors in [41] describe the implicit social graph which is formed by users' interactions with contacts and groups of contacts, and which is distinct from explicit social graphs in which users explicitly add other individuals as their "friends". They introduce an interaction-based metric for estimating a user's affinity to his contacts and groups and propose a novel friend suggestion algorithm that exploits a user's implicit social graph to generate a friend group, given a small seed set of contacts which the user has already labeled as friends. Their experimental results prove the importance of both implicit group relationships and interaction-based affinity ranking in suggesting friends.

Also, the study of critical nodes appears to be an interesting problem related to inhibiting diffusion of complex contagions such as rumors, undesirable fads and mob behavior in social networks by removing a small number of nodes (called critical nodes). The authors in [42] develop efficient heuristics for these problems and perform empirical studies of their performance on three well known social networks, namely epinions, wikipedia and slashdot.

Another important task in an online community is to observe and track the popular events, or topics that evolve over time in the community. Existing approaches have usually focused on either the burstiness of topics or the evolution of networks, but have usually ignored the interplay between textual topics and network structures. In [25] the authors formally define the problem of popular event tracking in online communities (PET), focusing on the interplay between texts and networks. They propose a novel statistical method that models the popularity of events over time, taking into consideration the burstiness of user interest, information diffusion on the network structure, and the evolution of textual topics. The approach models the influence of historic status and the dependency relationships in the graph through a Gibbs Random Field. Empirical experiments with two different communities and datasets (i.e., Twitter and DBLP) show that their approach is effective.

Current social networks are continuously growing. This makes them hard to analyze and in some cases intractable. One solution to this is compressing social networks so that we can substantially facilitate mining and advanced analysis of large social networks. The optimal solution would be to compress social networks in a way that they still can be queried efficiently without decompression. For example, we should still be able to perform neighbor queries efficiently (which search for all neighbors of a query vertex), as these are the most essential operations on social networks. The problems has been addressed in [32] where the authors propose an social network compression approach based on a novel Eulerian data structure using multi-position linearizations of directed graphs. Their approach seems to be the first that can answer both out-neighbor and in-neighbor queries in sublinear time and they verfiy their design with an extensive empirical study on more than a dozen benchmark real data sets.

Finally, an important task often essential in some social networks is the discovery of communities. Usually, the given scenario is the one where communities need to be discovered with only reference to the input graph. However, for many interesting applications one is interested in finding the community formed by a given set of nodes. In [44] the authors study a query-dependent variant of the community-detection problem, which they call the community-search problem: given a graph G, and a set of query nodes in the graph, the goal is to find a subgraph of G that contains the query nodes and it is densely connected. A measure of density is proposed based on minimum degree and distance constraints, and an optimum greedy algorithm is developed for this measure. The authors characterize a class of monotone constraints and they generalize the algorithm to compute optimum solutions satisfying any set of monotone constraints. Finally they modify the greedy algorithm and present two heuristic algorithms that find communities of size no greater than a specified upper bound. The experimental evaluation on real datasets demonstrates the efficiency of the proposed algorithms and the quality of the solutions they obtain.

# 6 Learning Structures in Peer-to-Peer Networks

There research in machine learning and data mining on analyzing data in Peer-to-Peer (P2P) networks is attracting a lot of attention of researchers. We will briefly describe here some recent developments in this exciting area.

In [3] the authors proposed a novel P2P learning framework for concept drift classification, which includes both reactive and proactive approaches to classify the drifting concepts in a distributed manner. Their empirical study shows that the proposed technique is able to effectively detect the drifting concepts and improve the classification performance.

The authors of [5] presented an algorithm for learning parameters of Gaussian mixture models (GMM) in large P2P environments that can be used for a variety of well-known data mining tasks in distributed environments such as clustering, anomaly detection, target tracking, and density estimation, which are necessary for many emerging P2P applications in bio-informatics, web-mining and sensor networks.

Tagging information is often an important feature to exploit in analyzing test documents. In many application areas involving classification of text documents, web users participate in the tagging process and the collaborative tagging results in the formation of large scale P2P systems which can function, scale and self-organize in the presence of highly transient population of nodes and do not need a central server for co-ordination. In [16] it is presented a P2P classifier learning system for extracting patterns from text data where the end users can participate both in the task of labeling the data and building a distributed classifier on it. The approach is based on a novel distributed linear programming based classification algorithm which is asynchronous in nature. The authors provide extensive empirical results on text data obtained from the online repository of NSF Abstracts Data.

Another important challenge in data mining over P2P networks is the right data representation. In [58] the authors describe an approach to collaborative feature extraction, selection and aggregation in distributed, loosely coupled domains. The authors focus on scenarios in which a large number of loosely coupled nodes apply data mining to different, usually very small and overlapping, subsets of the entire data space. The goal is to learn a set of local concepts and not to find a global concept. The paper proposes two models for collaborative feature extraction, selection and aggregation for supervised data mining. One is based on a centralized P2P architecture, and the other on a fully distributed P2P architecture. The comparison of both models is performed on a real word data set.

An important direction for research is the self-reorganization in P2P networks. In [4] the authors employ machine learning feature selection in a novel manner: to reduce communication cost thereby providing the basis of an efficient neighbor selection scheme for P2P overlays. In addition, their method enables nodes to locate and attach to peers that are likely to answer future queries with no a priori knowledge of the queries.

Finally, an important challenge is that fully distributed data mining algorithms build global models over large amounts of data distributed over a large number of peers in a network, without moving the data itself. The difficulty of the problem stands in implementing good quality models with an affordable communication complexity, while assuming as little as possible about the communication model. In [35] the authors describe a conceptually simple, yet powerful generic approach for designing efficient, fully distributed, asynchronous, local algorithms for learning models of fully distributed data. The key idea proposed in the paper is that many models perform a random walk over the network while being gradually adjusted to fit the data they encounter, using stochastic gradient descent search. The authors demonstrate their approach by implementing the support vector machine method and by experimentally evaluating its performance in various failure scenarios over different benchmark datasets.

## 7  Learning and Privacy-Preserving in Distributed Environments

Privacy is a major issue in data mining applications and very often privacy considerations often constrain data mining projects. For this reason, a growing amount of research is being dedicated to the problem of analyzing data from distributed environments under privacy requirements. Most distributed data mining applications, such as those dealing with health care, finance, counter-terrorism and homeland security, use sensitive data from distributed databases held by different parties.

One of the recent approaches proposes mining of association rules where transactions are distributed across sources [46]. In this scenario each site holds some attributes of each transaction, and the sites wish to collaborate to identify globally valid association rules. The sites must not reveal individual transaction data. The authors present a two-party algorithm for efficiently discovering frequent patterns, without either site revealing individual transaction values.

Some research work has considered the possibility of using multiplicative random projection matrices for privacy preserving distributed data mining. The work in [26] considers the problem of computing statistical aggregates like the inner product matrix, correlation coefficient matrix, and Euclidean distance matrix from distributed privacy sensitive data possibly owned by multiple parties. The authors propose an approximate random projection-based technique to improve the level of privacy protection while still preserving certain statistical characteristics of the data.

An interesting scenario arises when different parties like businesses, governments, and others, may wish to benefit from cooperative use of their data, but privacy regulations and other privacy concerns may prevent them from sharing the data. Privacy-preserving data mining provides solutions through distributed data mining algorithms in which the underlying data is not revealed. An interesting approach is presented in [57] where a Bayesian network structure is learned for distributed heterogeneous data. In the proposed setting, two parties owning confidential databases wish to learn the structure of Bayesian network on the combination of their databases without revealing anything about their data to each other.

The authors give an efficient and privacy-preserving version of the K2 algorithm to construct the structure of a Bayesian network for the parties' joint data.

One of the approaches towards privacy-preserving data mining is to adapt existing successful knowledge discovery algorithms so that they can deal with privacy issues. One of the most widely used classification methodologies in data mining and machine learning is support vector machine classification. The work in [50] proposes privacy-preserving solution for support vector machine classification. The solution constructs the global classification model from the data distributed at multiple parties, without disclosing the data of each party to others. It is assumed that data is horizontally partitioned: each party collects the same features of information for different data objects.

Traditional research on preserving privacy in data mining focuses on time-invariant privacy issues. However with time series data mining, snapshot-based privacy solutions should take into consideration the addition of the time dimension. The work in [55] shows that current techniques to preserve privacy in data mining are not effective in preserving time-domain privacy. They show with real data, that the data flow separation attack on privacy in time series data mining, which is based on blind source separation techniques from statistical signal processing, is effective. The authors propose possible countermeasures to the data flow separation attack in the paper.

Sharing data among multiple parties, without disclosing the data, is an important issue. In [51] it is presented an approach for sharing private or confidential data where multiple parties, each with a private data set, want to collaboratively conduct association rule mining without disclosing their private data. The approach is based on homomorphic encryption techniques to exchange the data while keeping it private. The proposed solution is distributed, i.e., there is no central, trusted party accessing all the data. Another similar problem regarding distributed association rule mining is presented in [48] where the authors come up with a protocol based on a new semi-trusted mixer model. Their protocol can protect the privacy of each distributed database against the coalition up to n-2 other data sites or even the mixer if the mixer does not collude with any data site.

It should be noted that many of the existing techniques have strict assumptions on the involved parties which need to be relaxed in order to reflect the real-world requirements. In [12] the authors present a distributed scenario where the data is partitioned vertically over multiple sites and the involved sites would like to perform clustering without revealing their local databases. They propose a new protocol for privacy preserving k-means clustering based on additive secret sharing and show that the new protocol is more secure than the state of the art.

Interesting research has been dedicated to distributed environments where the participants in the system may also be mutually mistrustful. The work in [30] discusses the design and security requirements for large-scale privacy-preserving data mining (PPDM) systems in a fully distributed setting, where each client possesses its own records of private data. The authors argue in favor of using some well-known cryptographic primitives, borrowed from the literature on Internet elections. They also show how their approach can be used as a building block to obtain Random Forests classification with enhanced prediction performance.

A powerful approach is extending the privacy preservation notion to original learning algorithms. An interesting effort that addresses this problem is presented in [9] where the authors focus on preserving the privacy in an important learning model, multilayer neural networks. They present a privacy-preserving two-party distributed algorithm of backpropagation which allows a neural network to be trained without requiring either party to reveal her data to the other.

Finally, a large body of research has been devoted to address corporate-scale privacy concerns related to social networks. The main concern has been on how to share social networks without revealing the identities or sensitive relationships of users. An interesting work is proposed in [28] that addresses privacy concerns arising in online social networks from the individual users viewpoint. The authors propose a framework to compute the privacy score of a user, which indicates the potential privacy risk caused by his participation in the network. The definition of privacy score satisfies the following: the more sensitive the information revealed by a user, the higher his privacy risk. Also, the more visible the disclosed information becomes in the network, the higher the privacy risk. The authors develop mathematical models to estimate both sensitivity and visibility of the information.

## 8  Conclusion

In this chapter we presented a survey of the current state-of-the-art methods for inferring structure and schemas from documents in heterogeneous networked environments. The rapidly growing volume of available digital documents of various formats and the possibility to access these through internet-based technologies in distributed environments, have led to the necessity to develop solid methods to properly organize and structure documents in large digital libraries and repositories. Specifically, since the extremely large volumes make it impossible to manually organize such documents and since most of the documents exist in an unstructured form and do not follow any schemas, most of the efforts in this direction are dedicated to automatically infer structure and schemas that can help to better organize huge collections. This is essential in order for these documents to be effectively and efficiently retrieved in heterogeneous domains in networked system.

**Acknowledgment.**The authors would like to thank the Brain Gain program of the Albanian Government for the support.

# References

1. Agarwal, N., Liu, H., Subramanya, S., Salerno, J.J., Yu, P.S.: Connecting Sparsely Distributed Similar Bloggers. In: Proc. of Ninth IEEE International Conference on Data Mining, pp. 11–20 (2009)
2. Aggarwal, C., Yu, P.: A framework for clustering uncertain data streams. In: Proc. of 24th International Conference on Data Engineering, Cancún, México (2008)
3. Ang, H.H., Gopalkrishnan, V., Ng, W.K., Hoi, C. H.: On classifying drifting concepts in P2P networks. In: Balcázar, J.L., Bonchi, F., Gionis, A., Sebag, M. (eds.) ECML PKDD 2010. LNCS, vol. 6321, pp. 24–39. Springer, Heidelberg (2010)
4. Beverly, R., Afergan, M.: Proceedings of USENIX Tackling Computer Systems Problems with Machine Learning Techniques (SysML 2007) Workshop, Cambridge, MA (April 2007)
5. Bhaduri, K., Srivastava, A.N.: A Local Scalable Distributed Expectation Maximization Algorithm for Large Peer-to-Peer Networks. In: Proc. of Ninth IEEE International Conference on Data Mining, pp. 31–40 (2009)
6. Bodik, P., Griffith, R., Sutton, C., Fox, A., Jordan, M.I., Patterson, D. A.: Statistical Machine Learning Makes Automatic Control Practical for Internet Datacenters. In: Workshop on Hot Topics in Cloud Computing, HotCloud 2009 (2009)
7. Budhaditya, S., Pham, D., Lazarescu, M., Venkatesh, S.: Effective Anomaly Detection in Sensor Networks Data Streams. In: Proc. of Ninth IEEE International Conference on Data Mining, pp. 722–727 (2009)
8. Cantoni, V., Lombardi, L., Lombardi, P.: Challenges for Data Mining in Distributed Sensor Networks. In: Proc. of 18th International Conference on Pattern Recognition (ICPR 2006), vol. 1, pp. 1000–1007 (2006)
9. Chen, T., Zhong, S.: Privacy-preserving backpropagation neural network learning. IEEE Transactions on Neural Networks 20(10), 1554–1564 (2009)
10. Das, S., Egecioglu, O., Abbadi, A.E.: Anonymizing weighted social network graphs. In: Proc. of IEEE 26th International Conference on Data Engineering (ICDE), pp. 904–907 (2010)
11. Das, S., Matthews, B.L., Srivastava, A.N., Oza, N.C.: Multiple kernel learning for heterogeneous anomaly detection: algorithm and aviation safety case study. In: Proc. of the 16th ACM SIGKDD International Conference on Knowledge Discovery and Data Mining, July 25-28. ACM, USA (2010)
12. Doganay, M.C., Pedersen, T.B., Saygin, Y., Savas, E., Levi, A.: Distributed privacy preserving k-means clustering with additive secret sharing. In: Proc. of the 2008 International Workshop on Privacy and Anonymity in Information Society, Nantes, France, March 29-29 (2008)
13. Domingos, P.: Mining Social Networks for Viral Marketing. IEEE Intelligent Systems 20(1), 80–82 (2005)
14. Domingos, P.: Structured Machine Learning: Ten Problems for the Next Ten Years. Machine Learning 73, 3–23 (2008)
15. Du, N., Wang, H., Faloutsos, C.: Analysis of large multi-modal social networks: Patterns and a generator. In: Balcázar, J.L., Bonchi, F., Gionis, A., Sebag, M. (eds.) ECML PKDD 2010. LNCS, vol. 6321, pp. 393–408. Springer, Heidelberg (2010)
16. Dutta, H., Zhu, X., Mahule, T., Kargupta, H., Borne, K., Lauth, C., Holz, F., Heyer, G.: TagLearner: A P2P Classifier Learning System from Collaboratively Tagged Text Documents. In: Proc. of Ninth IEEE International Conference on Data Mining Workshops, pp. 495–500 (2009)

17. Ge, Y., Xiong, H., Tuzhilin, A., Xiao, K., Gruteser, M., Pazzani, M.: An energy-efficient mobile recommender system. In: Proc. of the 16th ACM SIGKDD International Conference on Knowledge Discovery and Data Mining, Washington, DC, USA (2010)

18. Getoor, L., Taskar, B.: Introduction to statistical relational learning. MIT Press, Cambridge (2007)

19. Gorodetskiy, V.I., Serebryakov, S.V.: Methods and algorithms of collective recognition. Automation and Remote Control 69(11), 1821–1851 (2008)

20. He, J., Dai, X., Zhao, P.X.: Mixture Model Adaptive Neural Network for Mining Gene Functional Patterns From Heterogenous Knowledge Domains. International Journal of Information Technology and Intelligent Computing (2007)

21. He, D., Parker, D.S.: Topic dynamics: an alternative model of bursts in streams of topics. In: Proc. of the 16th ACM SIGKDD International Conference on Knowledge Discovery and Data Mining, July 25-28. ACM, Washington (2010)

22. Kargupta, H., Sarkar, K., Gilligan, M.: MineFleet: an overview of a widely adopted distributed vehicle performance data mining system. In: Proc. of the 16th ACM SIGKDD International Conference on Knowledge Discovery and Data Mining, Washington, DC, USA (2010)

23. Krishnaswamy, S., Loke, S.W., Rakotonirainy, A., Horovitz, O., Gaber, M. M.: Towards Situation-awareness and Ubiquitous Data Mining for Road Safety: Rationale and Architecture for a Compelling Application. In: Proc. of Conference on Intelligent Vehicles and Road Infrastructure (IVRI 2005), February 16-17, University of Melbourne (2005)

24. Lahiri, M., Berger-Wolf, T.Y.: Mining Periodic Behavior in Dynamic Social Networks. In: Proc. of the 8th IEEE International Conference on Data Mining (ICDM 2008), December 15-19. IEEE Computer Society, Pisa (2008)

25. Lin, C.X., Zhao, B., Mei, Q., Han, J.: PET: a statistical model for popular events tracking in social communities. In: Proc. of the 16th ACM SIGKDD International Conference on Knowledge Discovery and Data Mining, Washington, DC, USA, July 25-28, ACM, New York (2010)

26. Liu, K., Kargupta, H., Ryan, J.: Random Projection-Based Multiplicative Data Perturbation for Privacy Preserving Distributed Data Mining. IEEE Transactions on Knowledge and Data Engineering 18(1), 92–106 (2006)

27. Liu, S., Liu, Y., Ni, L.M., Fan, J., Li, M.: Towards mobility-based clustering. In: Proceedings of the 16th ACM SIGKDD International Conference on Knowledge Discovery and Data Mining, Washington, DC, USA (2010)

28. Liu, K., Terzi, E.: A Framework for Computing the Privacy Scores of Users in Online Social Networks. In: Proc. of the Ninth IEEE International Conference on Data Mining, pp. 288–297, 932–937 (2009)

29. Lodi, S., Monti, G., Moro, G., Sartori, C.: Peer-to-Peer Data Clustering in Self-Organizing Sensor Networks. In: Intelligent Techniques for Warehousing and Mining Sensor Network Data, pp. 179–212. IGI Global (2010)

30. Magkos, E., Maragoudakis, M., Chrissikopoulos, V., Gritzalis, S.: Accurate and large-scale privacy-preserving data mining using the election paradigm. Data and Knowledge Engineering 68(11), 1224–1236 (2009)

31. Marinai, S., Fujisawa, H. (eds.): Machine Learning in Document Analysis and Recognition. SCI, vol. 90. Springer, Heidelberg (2008)

32. Maserrat, H., Pei, J.: Neighbor query friendly compression of social networks. In: Proc. of the 16th ACM SIGKDD International Conference on Knowledge Discovery and Data Mining, Washington, DC, USA, July 25-28. ACM, New York (2010)

33. Morchen, F., Dejori, M., Fradkin, D., Etienne, J., Wachmann, B., Bundschus, M.: Anticipating annotations and emerging trends in biomedical literature. In: Proc. of the 14th ACM SIGKDD International Conference on Knowledge Discovery and Data Mining, Las Vegas, Nevada, USA, August 24-27. ACM, New York (2008)

34. Mukherjee, P., Sen, S.: Using learned data patterns to detect malicious nodes in sensor networks. In: Rao, S., Chatterjee, M., Jayanti, P., Murthy, C.S.R., Saha, S.K. (eds.) ICDCN 2008. LNCS, vol. 4904, pp. 339–344. Springer, Heidelberg (2008)

35. Ormandi, R., Hegedu, I., Jelasity, M.: Asynchronous Peer-to-peer Data Mining with Stochastic Gradient Descent. In: Proceedings of 17th International European Conference on Parallel and Distributed Computing, EuroPar 2011, Bordeux, France (2011)

36. Plaimas, K., Eils, R., Konig, R.: Identifying essential genes in bacterial metabolic networks with machine learning methods. In: BMC Systems Biology 2010, vol. 4, p. 56 (2010)

37. Qiu, J., Lin, Z., Tang, C., Qiao, S.: Discovering Organizational Structure in Dynamic Social Network. In: Proc. of the Ninth IEEE International Conference on Data Mining, pp. 932–937 (2009)

38. Rodrigues, P.P., Gama, J., Lopes, L.: Knowledge Discovery for Sensor Network Comprehension. In: Intelligent Techniques for Warehousing and Mining Sensor Network Data, pp. 179–212. IGI Global (2010)

39. Römer, K.: Discovery of frequent distributed event patterns in sensor networks. In: Verdone, R. (ed.) EWSN 2008. LNCS, vol. 4913, pp. 106–124. Springer, Heidelberg (2008)

40. Romer, K.: Distributed Mining of Spatio-Temporal Event Patterns in Sensor Networks. In: EAWMS / DCOSS 2006, San Francisco, USA, pp. 103–116 (June 2006)

41. Roth, M., Ben-David, A., Deutscher, D., Flysher, G., Horn, I., Leichtberg, A., Leiser, N., Matias, Y., Merom, R.: Suggesting friends using the implicit social graph. In: Proc. of the 16th ACM SIGKDD International Conference on Knowledge Discovery and Data Mining, Washington, DC, USA, July 25-28. ACM, New York (2010)

42. Saito, K., Kimura, M., Ohara, K., Motoda, H.: Selecting information diffusion models over social networks for behavioral analysis. In: Balcázar, J.L., Bonchi, F., Gionis, A., Sebag, M. (eds.) ECML PKDD 2010. LNCS, vol. 6323, pp. 180–195. Springer, Heidelberg (2010)

43. Song, C.: Mining and visualising wireless sensor network data Source. International Journal of Sensor Networks archive 2(5/6), 350–357 (2007)

44. Sozio, M., Gionis, A.: The community-search problem and how to plan a successful cocktail party. In: Proc. of the 16th ACM SIGKDD International Conference on Knowledge Discovery and Data Mining, Washington, DC, USA, July 25-28. ACM, New York (2010)

45. Sutton, C., Jordan, M.I.: Learning and Inference in Queueing Networks. In: Conference on Artificial Intelligence and Statistics, AISTATS (2010)

46. Vaidya, J., Clifton, C.: Privacy preserving association rule mining in vertically partitioned data. In: Proc. of 8th ACM SIGKDD International Conference on Knowledge Discovery and Data Mining (2002)

47. Yan, Y., Fung, G., Dy, J.G., Rosales, R.: Medical coding classification by leveraging inter-code relationships. In: Proc. of the 16th ACM SIGKDD International Conference on Knowledge Discovery and Data Mining, Washington, DC, USA, July 25-28. ACM, New York (2010)

48. Yi, X., Zhang, Y.: Privacy-preserving distributed association rule mining via semi-trusted mixer. Data and Knowledge Engineering 63(2), 550–567 (2007)

49. Ying, Y., Campbell, C., Damoulas, T., Girolami, M.: Class Prediction from Disparate Biological Data Sources Using an Iterative Multi-kernel Algorithm. In: 4th IAPR International Conference on Pattern Recognition in Bioinformatics, Sheffield (2009)
50. Yu, H., Jianga, X., Vaidya, J.: Privacy-preserving SVM using nonlinear kernels on horizontally partitioned data. In: Proc. of the 2006 ACM Symposium on Applied computing, April 23-27, Dijon, France (2006)
51. Zhan, J., Matwin, S., Chang, L.: Privacy-preserving collaborative association rule mining. Journal of Network and Computer Applications 30(3), 1216–1227 (2007)
52. Zhao, Z., Wang, J., Liu, H., Ye, J., Chang, Y.: Identifying biologically relevant genes via multiple heterogeneous data sources. In: Proc. of the 14th ACM SIGKDD International Conference on Knowledge Discovery and Data Mining, Las Vegas, Nevada, USA, August 24-27. ACM, New York (2008)
53. Zhao, H., Lall, A., Ogihara, M., Jun, X.: Global iceberg detection over distributed data streams. In: Proc. of IEEE 26th International Conference on Data Engineering, ICDE (2010)
54. Zheng, L., Shen, C., Tang, L., Li, T., Luis, S., Chen, S., Hristidis, V.: Using data mining techniques to address critical information exchange needs in disaster affected public-private networks. In: Proc. of the 16th ACM SIGKDD International Conference on Knowledge Discovery and Data Mining, Washington, DC, USA (2010)
55. Zhu, Y., Fu, Y., Fu, H.: On privacy in time series data mining. In: Proc. of the 12th Pacific-Asia Conference on Advances in Knowledge Discovery and Data Mining, Osaka, Japan, May 20-23 (2008)
56. Wang, Y., Cong, G., Song, G., Xie, K.: Community-based greedy algorithm for mining top-K influential nodes in mobile social networks. In: Proc. of the 16th ACM SIGKDD International Conference on Knowledge Discovery and Data Mining, Washington, DC, USA (2010)
57. Wright, R., Yang, Z.: Privacy-preserving Bayesian network structure computation on distributed heterogeneous data. In: Proc. of the 10th ACM SIGKDD International Conference on Knowledge Discovery and Data Mining (2004)
58. Wurst, M., Morik, K.: Distributed feature extraction in a p2p setting: a case study. Future Generation Computer Systems 23(1), 69–75 (2007)

# Handling Hierarchically Structured Resources Addressing Interoperability Issues in Digital Libraries

Maristella Agosti, Nicola Ferro, and Gianmaria Silvello

**Abstract.** We present and describe the NEsted SeTs for Object hieRarchies (NESTOR) Framework that allows us to model, manage, access and exchange hierarchically structured resources. We envision this framework in the context of Digital Libraries and using it as a mean to address the complex and multiform concept of interoperability when dealing with hierarchical structures. The NESTOR Framework is based on three main components: The Model, the Algebra and a Prototype. We detail all these components and present a concrete use case based on archives that are collections of historical documents or records providing information about a place, institution, or group of people, because the archives are fundamental and challenging entities in the digital libraries panorama. Within the archives we show how an archive can be represented through set data models and how these models can be instantiated. We compare two instantiations of the NESTOR Model and show how interoperability issues can be addressed by exploiting the NESTOR Framework.

## 1 Introduction and Motivation

An important challenge in the research work on Digital Libraries is to transform them in a new type of information infrastructures that can be user-centered, able to support content management tasks together with tasks devoted to communication and cooperation. That is information infrastructures that become common vehicle by which every user can access, discuss, evaluate, and enhance information of all forms. Although they are still places where information resources can be stored and made available to end users, the current design and development efforts are moving

Maristella Agosti · Nicola Ferro · Gianmaria Silvello
Department of Information Engineering, University of Padua, Italy
Tel.: +39 049 827 7650; Fax: +39 049 827 7799
e-mail: agosti@dei.unipd.it
Tel.: +39 049 827 7939; Fax: +39 049 827 7799
e-mail: ferro@dei.unipd.it
Tel.: +39 049 827 7929; Fax: +39 049 827 7799
e-mail: silvello@dei.unipd.it

M. Biba and F. Xhafa (Eds.): Learning Structure and Schemas from Documents, SCI 375, pp. 17–49.
springerlink.com                                        © Springer-Verlag Berlin Heidelberg 2011

in the direction of transforming them into infrastructures able to support the user in different information centric activities. In the context of digital libraries we need to take into account several distributed and heterogeneous information sources with different community backgrounds such as libraries, archives and museums and different information objects ranging from full content of digital information objects to the metadata describing them. These objects can be exchanged between distributed systems or they can be aggregated and accessed by users with distinct information needs and living in different countries. Digital Libraries are heterogeneous systems with peculiarities and functionalities that range from data representation to data exchange passing through data management. Furthermore, Digital Libraries are meaningful parts of a global information network which includes scientific repositories, curated databases and commercial providers. All these aspects need to be taken into account and balanced to support final users with effective and interoperable information systems.

A common goal in the design and development of Digital Libraries is to build systems which rely as much as possible on existing building blocks, thus maximizing the exploitation of Web and Internet standards. This trend is evident if we consider the standard technologies and protocols adopted over the years by Digital Libraries. Two of the main technologies of choice in Digital Libraries are: the eXtensible Markup Language (XML)[1] and the Open Archives Initiative Protocol for Metadata Harvesting (OAI-PMH)[2]. These technologies are very strongly interlinked with the Internet and the Web: XML is the technology of choice representing and encoding metadata in Digital Libraries. It was originally designed to meet the challenges of large-scale electronic publishing, XML is also playing an increasingly fundamental role in the exchange of a wide variety of data on the Web. OAI-PMH is the standard *de-facto* for metadata exchange in distributed environments and its basic functioning is based on the Internet infrastructure (e.g. OAI-PMH requests are based on HyperText Transfer Protocol (HTTP) requests) [30].

These technologies are designed and shaped to be used within the Digital Libraries but at the same time their scope is fairly broad and they can be used within a large corpus of resources and systems. For these reasons they are either the means for addressing interoperability or the possible sources of interoperability issues. Indeed, on the one hand, we can build on these technologies and exploit them to constitute an integrated framework which handles the heterogeneity of Digital Library resources and functionalities. On the other hand, they can constitute a barrier towards the very interoperability they are aiming to foster. When it comes to modeling, managing, accessing and exchanging the resources of interest, often the design of the models and systems is driven by the technology-of-choice characteristics, thus they are bound to the technologies. A fundamental step is to define an organic and general framework free from specific technologies; in this way the technologies of choice can be

---

[1] http://www.w3.org/XML/
[2] http://www.openarchives.org/pmh/

conveyed into a well-defined path where their characteristics can be exploited to the maximum to address interoperability and to meet users requirements.

The difficulty in accomplishing this goal is due to the very nature of interoperability as a complex and multiform concept, which can be defined - as by the "ISO/IEC 2382-01, Information Technology Vocabulary, Fundamental Terms" - as follows: *"The capability to communicate, execute programs, or transfer data among various functional units in a manner that requires the user to have little or no knowledge of the unique characteristics of those units"*. When the concept of interoperability is considered in the context of digital libraries it takes on different dimensions as it has been evidenced by the *European Commission Working Group on Digital Library Interoperability* [13] which has identified six dimensions that can be distinguished and taken into account:

**Interoperating entities.**   These can be assumed to be the traditional cultural heritage institutions, such as libraries, museums, archives, and other institutions in charge of preservation of artifacts or that offer digital services.

**Information objects.**   The entities that actually need to be processed in interoperability scenarios. Choices range from the full content of digital information objects to the metadata describing them.

**Functional perspective.**   This may simply be the exchange and/or propagation of digital content. Other functional goals are aggregating digital objects into a common content layer.

**Multilingualism.**   Linguistic interoperability can be thought of in two different ways: as multilingual user interfaces to digital library systems or as dynamic multilingual techniques for exploring the digital library systems object space.

**User perspective.**   Interoperability concepts of a digital library system manager differ substantially from those of a content consuming end user.

**Interoperability technology.**   Enabling different kinds of interoperability constitutes a major dimension and several technologies designed in the context of Digital Libraries such as OAI-PMH and the Dublin Core[3].

The dimensions of interoperability are often analyzed and addressed focusing on a specific one; e.g. we can consider interoperability between the Digital Libraries entities with their information objects, we can consider the functional perspective of Digital Libraries by focusing on the exchange of objects or we can take into account only the cross-language access of information objects. Our aim is to consider all six dimensions of interoperability and define a common framework that can address all of them; to do so, we consider one of the most diffuse and important resources – i.e. hierarchically structured resources; in particular we employ a meaningful and challenging reality: archives. Archives are one of the main organizations of interest for Digital Libraries; they are a meaningful example of the need to support document management and access, as well as interoperability among the systems that manage different co-operating and related archives. The fundamental characteristic of archives resides in their internal hierarchical organization that constitutes both a

---

[3] http://www.dublincore.org/

challenge for their representation, managing and, exchange and a relevant feature
for addressing interoperability.

In this chapter we present a conceptual and logical framework called "NEsted
SeTs for Object hieRarchies (NESTOR) Framework" and show how it can be
adopted to model, manage, access and exchange hierarchically structured resources.
The presentation of this framework also shows how it can be used to address interoperability in Digital Libraries. As a guide use case we make use of archives both
because of their peculiar characteristics and because of the challenges we have to
accomplish to model, manage, access, and exchange their resources. The results that
are here presented are rooted on those presented in [11].

The presentation is organized as follows: Section 2 presents the objectives and the
key contributions of this chapter and introduces the composition of the NESTOR
Framework which is defined by the NESTOR Model, the NESTOR Algebra and
the NESTOR Prototype. Section 3 describes the context background of the work;
it introduces the main concepts about archives and related technologies as well as
a panoramic on the Digital Library technologies we are going to employ in this
chapter. Section 4 describes the NESTOR Model based on two set data models and
Section 5 depicts the main features of the NESTOR Algebra. In Section 6 we show
how the NESTOR Prototype can be modeled and used to achieve the presented requirements. Lastly, we draw some conclusions and give indications for future work
in Section 7.

## 2   Objectives and Contributions

The main target of this chapter is to provide a framework allowing us to model,
manage, access and exchange hierarchically structured resources. A key aspect is to
envision this framework in the context of Digital Libraries and to use it as a mean to
address the complex and multiform concept of interoperability when dealing with
hierarchical structures. A significant goal is to understand which is the best option
for modeling hierarchies in order to meet the higher number of user and system requirements. The modeling and representation of data (or more in general resources)
is a fundamental step towards building automatic and effective systems and providing services and functionalities to the users. In this context we take into account
one of the most important data structures in computer science – the tree data structure [19]; this is widely adopted in many scientific fields to model and represent
hierarchies. The tree data structure is often regarded as the only way to model hierarchies; our aim is to investigate alternative data models that can do the work and
then compare their effectiveness in specific application contexts.

We propose the NESTOR Framework as a conceptual and logical mean for modeling and representing hierarchically structured data and specifically to overcome
some of the difficulties that we encounter when we have to address interoperability
in Digital Libraries. There is the lack of a general framework satisfying these requirements which often need to be addressed on a case-by-case basis. The NESTOR

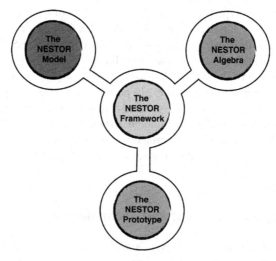

**Fig. 1** The graphical outline of the composition of the NESTOR Framework: Model, Algebra and Prototype

Framework is constituted by three parts – pointed out in Figure 1: The *NESTOR Model*, the *NESTOR Algebra* and the *NESTOR Prototype*.

The NESTOR Model is the heart of the Framework; it is based on two set data models called *Nested Set Model* (NS-M) and *Inverse Nested Set Model* (INS-M) which are based on an organization of nested sets. The foundational idea underlying these set data models is that an opportune set organization can maintain all the features of a tree data structure with the addition of some new relevant functionalities. We define these functionalities in terms of flexibility of the model, rapid selection and isolation of easily specified subsets of data and extraction of only those data necessary for satisfying specific needs. We can use these set data models to represent hierarchical structures disclosing a variety of properties which can be related to the properties of the tree data structure and which are also peculiar of these models. The representation of hierarchies by one of these models lays the ground for an environment leading to new ways of modeling and consequently accessing, managing and querying hierarchical data.

Each data model has to specify a set of operations to manipulate and query the data represented by a specific model; for instance, in the relational model this operation set is defined by the relational algebra [6]. A formal bulk algebra is essential to a data model first of all because it provides a formal basis for the operations on the data sets and second of all because it is used as a basis to implement and optimize the queries written in some query language against these data sets. We develop an algebra, called the NESTOR Algebra, for the manipulation and query of data represented throughout the set data models defined in the NESTOR Model.

The NESTOR Prototype gives an actual instantiation of the model and of the algebra allowing the application of the formal concepts defined in the NESTOR Model and Algebra. The prototype is presented by the use case of the archives describing how a hierarchy can be modeled by means of the NESTOR Model and specifically how the archival records can be represented through it. The NESTOR Prototype takes into account the dimensions of interoperability in Digital Libraries [13] showing how the adoption of the NESTOR Model and the exploitation of the Algebra addresses several interoperability issues. In particular, we face the problem of access and exchange of hierarchically organized resources in distributed environment and discuss the relationships between the NESTOR Prototype and Digital Library technologies such as the OAI-PMH protocol. Furthermore, we illustrate how we can address multilingualism in the archival context by exploiting widely-adopted techniques and technologies.

## 3   The Background Context and Technologies

The background of this chapter relies on the archives environment, thus it is fundamental to describe the nature of archival practice and the peculiarities of archival resources. In the beginning we present the nature of archives and then we talk about digital archives. In this context we present the standard XML metadata format adopted by the archives, i.e. is the Encoded Archival Description (EAD).

Two important Digital Library technologies are the OAI-PMH protocol and the Dublin Core. We will exploit these technologies in the context of the NESTOR Prototype, thus it is worthwhile to describe their characteristics and functionalities.

### 3.1   Archives and Archival Descriptions

An archive is not simply constituted by a series of objects that have been accumulated and filed with the passing of time. Instead, it represents the trace of the activities of a physical or juridical person in the course of their business which is preserved because of their continued value. Archives have to keep the context in which their records[4] have been created and the network of relationships between them in order to preserve their informative content and provide understandable and useful information over time.

Archival description is defined in [23] as "the process of analyzing, organizing, and recording details about the formal elements of a record or collection of records, to facilitate the work's identification, management, and understanding"; archival descriptions have to reflect the peculiarities of the archive, retain all the informative power of a record, and keep trace of the provenance and original order in which resources have been collected and filed by archival institutions [12]. This is emphasized by the central concept of *fonds*[5], which should be viewed primarily as an

---

[4] In [21] a record is defined as: "Any document made or received and set aside in the course of a practical activity".

[5] The term *fonds* is not a commonly used English word but it is derived from the French and it is used both for the singular and plural form of the noun.

"intellectual construct", the conceptual "whole" that reflects an organic process in which a records creator produces or accumulates series of records [8]. In this context, provenance becomes a fundamental principle of archives; the principle of the "*respect des fonds*" which dictates that resources of different origins be kept separate to preserve their context; the "*respect des fonds*" is often regarded as the principle of *provenance* [12, 10].

[10] highlights that maintaining provenance leads archivists to evaluate records on the basis of the importance of the creator's mandate and functions, and fosters the use of a hierarchical method. The hierarchical structure of the archive expresses the relationships and dependency links between the records of the archive by using what is called the archival bond defined as "the interrelationships between a record and other records resulting from the same activity" [23]. Archival bonds, and thus relations, are constitutive parts of an archival record: if a record is taken out from its context and has lost its relations, its informative power would also be considerably affected. Therefore, archival descriptions need to be able to express and maintain such structure and relationships in order to preserve the context of a record. To this end, the International Council on Archives (ICA)[6] has developed a general standard for archival description called International Standard for Archival Description (General) (ISAD(G)) [15]. According to ISAD(G), archival description proceeds from general to specific as a consequence of the provenance principle and has to show, for every unit of description, its relationships and links with other units and to the general fonds. Therefore, archival descriptions produced according to the ISAD(G) standard take the form of a tree which represents the relationships between more general and more specific archive units going from the root to the leaves of the tree.

## 3.2 EAD: Encoded Archival Description

EAD is an archival description metadata standard that reflects and emphasizes the hierarchical nature of ISAD(G) [24]. EAD fully enables the expression of multiple description levels central to most archival descriptions and reflects hierarchy levels present in the resources being described. EAD cannot be considered a one-to-one ISAD(G) implementation, although it does respect ISAD(G) principles and is useful for representing archival hierarchical structures. EAD is composed of three high-level components: `<eadheader>`, `<frontmatter>`, and `<archdesc>`.

The `<eadheader>` contains metadata about the archive descriptions and includes information about them such as title, author, and date of creation. The `<frontmatter>` supplies publishing information and is an optional element, while the `<archdesc>` contains the archival description itself and constitutes the core of EAD. The `<archdesc>` may include many high-level sub-elements, most of which are repeatable. The most important element is the `<did>` or descriptive identification which describes the collection as a whole. The `<did>` element is composed of numerous sub-elements that are intended for brief, clearly designated

---

[6] http://www.ica.org/

statements of information and they are available at every level of description. Finally, the `<archdesc>` contains an element that facilitates a detailed analysis of the components of a fonds, the `<dsc>` or description subordinate components. The `<dsc>` contains a repeatable recursive element, called `<c>` or component. A component may be an easily recognizable archival entity such as series, subseries or items. Components not only are nested under the `<archdesc>` element, they usually are nested inside one another. Components usually are indicated with `<cN>` tag, where $N \in \{01, 02, \ldots, 12\}$.

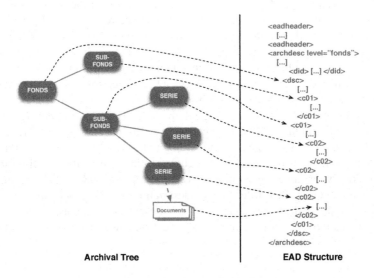

**Fig. 2** How an archive represented as a tree is mapped into an EAD XML file

EAD reflects the archival structure and holds relations between entities in an archive. In addition, EAD encourages archivists to use collective and multilevel description, and because of its flexible structure and broad applicability, it has been embraced by many repositories [17].

On the other hand, EAD allows for several degrees of freedom in tagging practice, which may turn out to be problematic in the automatic processing of EAD files, since it is difficult to know in advance how an institution will use the hierarchical elements. The EAD permissive data model may undermine the very interoperability it is intended to foster. Indeed, it has been underlined that only EAD files meeting stringent best practice guidelines are shareable and searchable [26]. Moreover, there is also a second relevant problem related to the level of material being described. Unfortunately, the EAD schema rarely requires a standardized description of the level of the materials being described, since the `<level>` attribute is required only in the `<archdesc>` tag, while it is optional in `<cN>` components and in very few EAD files this possibility is used, as pointed out by [25]. As a consequence, the level of description of the lower components in the hierarchy needs to be inferred by navigating the upper components, maybe up to the <archdesc>, where the presence of the <level> attribute is mandatory. Therefore, access to individual items might be difficult without taking into consideration the whole hierarchy.

We highlight this fact in Figure 2 where we present the structure of an EAD file. In this example we can see the top-level components <eadheader> and <archdesc> and the hierarchical part represented by the <dsc> component; the <level> attribute is specified only in the <archdesc> component. Therefore, the archival levels described by the components of the <dsc> can be inferred only by navigating the whole hierarchy.

## 3.3   OAI-PMH and Dublin Core

OAI-PMH is based on the distinction between two main components that are Data Provider and Service Provider. Data Providers are repositories that export records in response to requests from a software service called harvester. On the other hand, Service Providers are those services that harvest records form Data Providers and provide services built on top of aggregated harvest metadata.

The protocol defines two kinds of harvesting procedures: incremental and selective harvesting. Incremental harvesting permits users to query a Data Provider and ask it to return just the new, changed or deleted records from a certain date or between two dates. Selective harvesting is based on the concept of *OAI set*, which enables logical data partitioning by defining groups of records. Selective harvesting is the procedure that permits the harvesting only of metadata owned by a specified OAI set. [30] states that in OAI-PMH a set is defined by three components: setSpec which is mandatory and a unique identifier for the set within the repository, setName which is a mandatory short human-readable string naming the set, and setDesc which may hold community-specific XML-encoded data about the set.

OAI set organization may be flat or hierarchical, where hierarchy is expressed in setSpec field by the use of a colon [:] separated list indicating the path from the root of the set hierarchy to the respective node. For example if we define an OAI set for whose setSpec is *"A"*, its sub-set *"B"* would have *"A:B"* as setSpec. In this case "B" is a proper sub-set of "A": $B \subset A$. When a repository defines a set organization it must include set membership information in the headers of items returned to the harvester requests. Harvesting from a set which has sub-sets will cause the repository to return metadata in the specified set and recursively to return metadata from all the sub-sets. In our example, if we harvest set A, we also obtain the items in sub-set B [29].

The Dublin Core (DC) metadata format is tiny, easy-to-move, shareable and remarkably suitable for a distributed environment. Thanks to these characteristics it is required as the lowest common denominator in OAI-PMH. Thus, DC metadata are very useful in information sharing but are not broadly used by archivists. Indeed, the use of DC seems to flatten out archive structure and lose context and hierarchy information. For this reason, even though DC is used in several contexts ranging from Web to digital libraries, it is less used in the archival domain. Nevertheless, we can apply it to the archival domain and meet the three requirements discussed above, if we use it in combination with OAI-PMH: in this way, the OAI set provides us with context and hierarchy requirements compliance, while the DC metadata format gives us the expected variable granularity support.

## 4   The NESTOR Model

The NESTOR Framework is the composition of three main parts: the Model, the Algebra and the Prototype. The NESTOR Model is the core of the framework because it defines the set data models on which every component of the framework relies. In Figure 3 we can see a graphical representation of the principal components composing the NESTOR Model. In the upper part we have the set data models – i.e. the Nested Set Model (NS-M) and the Inverse Nested Set Model (INS-M). The second component represents the properties of the set data models such as: the mapping function to go from the NS-M to the INS-M and vice versa, the meaning of the union or intersection of two sets or the definition of distance measures. The latter component represents the relationships between the set data models and the tree data structure; this component contains the functions for mapping a tree into one of the two models and it compares the properties of the tree with the properties of the set data models.

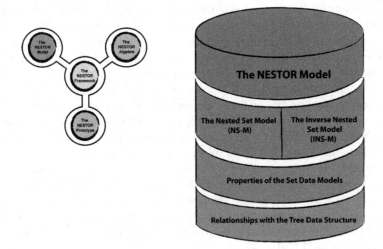

**Fig. 3** The main components of the NESTOR Model

The formal definition of the NESTOR Model within all its components relies on set theory, and particularly, on the basic concept of *family of subsets*. It is not in the scope of this chapter to give a complete mathematical definition of all the components of the NESTOR Model. For this reason, in order to properly understand how the set data models are defined it is worthwhile to get an intuitive idea of their main characteristics. After this intuitive presentation we explain some basic concepts of set theory which allow us to understand the formal definition of the models and to introduce the minimum set of notation and terminology indispensable to understand the properties of the models, the relationships with the tree data structure and, afterwards, the NESTOR Algebra.

Now, we informally present the two data models with examples of mapping between them and a sample tree, with a clear understanding that these models are independent from the tree data structure. The first model we present is the **Nested Set Model** (NS-M). The intuitive graphic representation of a tree as an organization of nested sets was used in [19] to show different ways of representing tree data structure and in [5] to explain an alternative way for solving recursive queries over trees in SQL language. An organization of sets in the NS-M is a collection of sets in which any pair of sets is either disjoint or one contains the other. In Figure 4 (b) we can see how a sample tree is mapped into an organization of nested sets based on the NS-M.

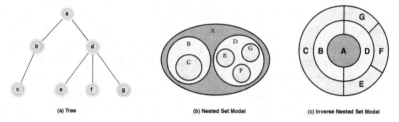

(a) Tree     (b) Nested Set Model     (c) Inverse Nested Set Model

**Fig. 4** (a) A tree. (b) Euler-Venn Diagram of a NS-M. (c) Doc-Ball representation of a INS-M.

From Figure 4 (b) we can see that each node of the tree is mapped into a set, where child nodes become *proper subsets* of the set created from the parent node. Every set is subset of at least of one set; the set corresponding to the tree root is the only set without any supersets and every set in the hierarchy is subset of the root set. The external nodes are sets with no subsets. The tree structure is maintained thanks to the nested organization and the relationships between the sets are expressed by the set inclusion order. Even the disjunction between two sets brings information; indeed, the disjunction of two sets means that these belong to two different branches of the same tree.

The second data model is the **Inverse Nested Set Model** (INS-M). We can say that a tree is mapped into the INS-M by transforming each node into a set, where each parent node becomes a subset of the sets created from its children. The set created from the tree's root is the only set with no subsets and the root set is a proper subset of all the sets in the hierarchy. The leaves are the sets with no supersets and they are sets containing all the sets created from the nodes composing the tree path from a leaf to the root. An important aspect of INS-M is that the intersection of every couple of sets obtained from two nodes is always a set representing a node in the tree. The intersection of all the sets in the INS-M is the set mapped from the root of the tree.

Unlike the NS-M, representing the INS-M with the Euler-Venn diagrams is not very expressive and can be confusing for the reader [1]. We can represent the INS-M in a straightforward way by means of the *"DocBall representation"* [9]. The DocBall representation is used in [9] to depict the structural components of the documents and can be considered as the representation of a tree structure. We exploit the DocBall ability to show the structure of an object and to represent the *"inclusion order of one or more elements in another one"* [31]. The DocBall is composed of a set of circular sectors arranged in concentric rings as shown in Figure 4 (c). In a DocBall each ring represents a level of the hierarchy with the center (level 0) representing the root. In a ring, the circular sectors represent the nodes in the corresponding level. We use the DocBall to represent the INS-M, thus for us each circular sector corresponds to a set.

In Figure 4 (c) we can see the INS-M mapping of a sample tree by means of the DocBall representation. The root *"a"* of the tree is mapped into set *"A"* represented by the inner ring at level 0 of the DocBall; at level 1 we find the children of the root and so on. With this representation a subset is presented in a ring within the set including it. Indeed, we can see that set $A$ is included in all the other sets. If the intersection of two or more sets is empty then these sets have no common circular sector in the inner rings of the DocBall; in the INS-M this is not possible because the set representing the root $(A)$ is common to all the sets in the INS-M. For instance, we can see that the circular sectors $C$ and $E$ have in common only $A$, indeed $C \cap E = A$; instead, $G$ and $E$ have in common sectors $D$ and $A$, thus $G \cap E = \{D, A\}$.

Both the NS-M and the INS-M have been presented as "organizations of nested sets"; two sets are nested if one contains the other and thus, if one is the subset of the other one. The nesting between two sets determines an order inclusion between them. Let us consider a set, call it $A$, that contains all the elements organized in the hierarchy we want to represent throughout the NS-M or the INS-M. Now, let us consider two sets, call them $A_1$ and $A_2$, which are subsets of $A$ such that: $A_1 \subset A$ and $A_2 \subset A$. The collection $\mathscr{C}$ composed by the two sets $A_1, A_2$ is called the collection of subsets of $A$; a family of subsets of $A$ is just the collection $\mathscr{C}$ indexed by an "index set". The following definition formally states the very concepts we have just described.

**Definition 1.** *Let $A$ be a set, $I$ a non-empty set and $\mathscr{C}$ a collection of subsets of $A$. Then a bijective function $\mathscr{A} : I \longrightarrow \mathscr{C}$ is a **family** of subsets of $A$. We call $I$ the **index set** and we say that the collection $\mathscr{C}$ is **indexed** by $I$.*

We use the extended notation $\{A_i\}_{i \in I}$ to indicate the family $\mathscr{A}$; the notation $A_i \in \{A_i\}_{i \in I}$ means that $\exists i \in I \mid \mathscr{A}(i) = A_i$. In the rest of the chapter we will use the shorthand notation $\mathscr{A}$ when there is no risk of ambiguity and when it is not necessary to indicate the index set. We call **subfamily** of $\{A_i\}_{i \in I}$ the **restriction** of $\mathscr{A}$ to $J \subseteq I$ and we denote this with $\{B_j\}_{j \in J} \subseteq \{A_i\}_{i \in I}$.

From this definition we can see that an organization of nested sets in set theory is defined as a family of subsets or just "family" if in the context in which it is used there is no risk of ambiguity. Thus, in the context of the NS-M we have Nested Set Families (NS-F) and in INS-M we have Inverse Nested Set Families (INS-F). The differences between these two models are expressed in the constraints we impose respectively on a NS-F and a INS-F. Let us consider the formal definition of NS-F.

**Definition 2.** *Let A be a set and let $\{A_i\}_{i\in I}$ be a family. Then $\{A_i\}_{i\in I}$ is a **Nested Set Family** if:*

$$A \in \{A_i\}_{i\in I}, \tag{1}$$

$$\emptyset \notin \{A_i\}_{i\in I}, \tag{2}$$

$$\forall A_h, A_k \in \{A_i\}_{i\in I}, h \neq k \mid A_h \cap A_k \neq \emptyset \Rightarrow A_h \subset A_k \vee A_k \subset A_h. \tag{3}$$

Thus, we define a Nested Set Family (NS-F) as a family where three conditions must hold. The first condition (1) states that set $A$ which contains all the sets in the family must belong to the NS-F. The second condition states that the empty-set does not belong to the NS-F and the last condition (3) states that the intersection of every couple of distinct sets in the NS-F is not the empty-set only if one set is a proper subset of the other one [14, 3].

In the same way we define the Inverse Nested Set Model (INS-M):

**Definition 3.** *Let A be a set and let $\{A_i\}_{i\in I}$ be a family. Then $\{A_i\}_{i\in I}$ is an **Inverse Nested Set Family** if:*

$$\emptyset \notin \{A_i\}_{i\in I}, \tag{4}$$

$$\forall \{B_j\}_{j\in J} \subseteq \{A_i\}_{i\in I} \Rightarrow \bigcap_{j\in J} B_j \in \{A_i\}_{i\in I}. \tag{5}$$

$$\forall \{B_j\}_{j\in J} \subseteq \{A_i\}_{i\in I}$$
$$\Rightarrow \exists B_k \in \{B_j\}_{j\in J} \mid \forall B_h \in \{B_j\}_{j\in J}, B_h \subseteq B_k \tag{6}$$
$$\Rightarrow \forall B_h, B_g \in \{B_j\}_{j\in J}, B_h \subseteq B_g \vee B_g \subseteq B_h.$$

Thus, we define an Inverse Nested Set Family (INS-F) as a family where three conditions must hold. The first condition (4) states that the empty-set does not belong to the INS-F. The second condition (5) states that the intersection of every subfamily of the INS-F belongs to the INS-F itself. Condition 6 states that for every possible subfamily of a INS-F there cannot exist a set in the subfamily which is a superset of all the other sets in the subfamily, unless all the sets in the subfamily form a *chain*[7].

In a family of subsets the sets establish a hierarchical relationship one with the other as well as in the tree data structure the nodes are in a parent-child or ancestor-descendant relationship. Also in a family we can have different kind of relationships between the sets; let us consider a family $\{A_i\}_{i\in I}$ where $A_1, A_2 \in \{A_i\}_{i\in I}$ are two

---

[7] A family of subsets $\{A_i\}_{i\in I}$ forms a chain (or it is linearly ordered) if each set in $\{A_i\}_{i\in I}$ is nested inside the next: the family $\{A_i\}_{i\in I}$ is defined a **chain** if $\forall A_j, A_k \in \{A_i\}_{i\in I}, A_j \subseteq A_k \vee A_k \subseteq A_j$.

sets, $A_2$ is a direct subset of the set $A_1$ if and only if it does not exists a third set $A_3 \in \{A_i\}_{i \in I}$ such that $A_2 \subset A_3 \subset A_1$. In the same way we say that $A_1$ is a direct superset of $A_2$ if and only if it does not exists a third set $A_3 \in \{A_i\}_{i \in I}$ such that $A_2 \subset A_3 \subset A_1$. The following definition presents the concept of collection of proper subsets and supersets; afterwards we present the definition of collection of proper direct subsets and supersets.

**Definition 4.** *Let* $\{A_i\}_{i \in I}$ *be a family and* $A_j \in \{A_i\}_{i \in I}$ *be a set. We define* $\mathscr{S}_{\mathscr{A}}^{+}(A_j) = \{A_k : A_k \in \{A_i\}_{i \in I} \wedge A_k \subset A_j\}$ *to be the* **collection of proper subsets** *of* $A_j$ *in the family* $\mathscr{A}$. *In the same way we define the* **collection of proper supersets** *of* $A_j$ *in the family* $\mathscr{A}$ *as* $\mathscr{S}_{\mathscr{A}}^{-}(A_j) = \{A_k : A_k \in \{A_i\}_{i \in I} \wedge A_j \subset A_k\}$.

It is worthwhile for the rest of the work to introduce the definition of *collection of direct super/sub-sets* as a restriction of the above defined collection of proper super/sub-sets.

**Definition 5.** *Let* $\{A_i\}_{i \in I}$ *be a family and* $A_j \in \{A_i\}_{i \in I}$ *be a set. We define* $\mathscr{D}_{\mathscr{A}}^{+}(A_j) = \{A_k : A_k \in \{A_i\}_{i \in I} \wedge A_k \subset A_j \wedge \nexists A_t \in \{A_i\}_{i \in I} \mid A_k \subset A_t \subset A_j\}$ *to be the* **collection of direct subsets** *of* $A_j$ *in the family* $\mathscr{A}$. *In the same way we define the* **collection of direct supersets** *of* $A_j$ *in the family* $\mathscr{A}$ *as* $\mathscr{D}_{\mathscr{A}}^{-}(A_j) = \{A_k : A_k \in \{A_i\}_{i \in I} \wedge A_j \subset A_k \wedge \nexists A_t \in \{A_i\}_{i \in I} \mid A_j \subset A_t \subset A_k\}$.

Now that we have at our disposal the fundamental concepts of set theory necessary for understanding the work and that we have formally defined the NS-M and the INS-M, we can examine the relationships between them and their properties. From the collection of properties of these models we choose to introduce only those used in this chapter to show how the NESTOR Framework addresses the interoperability issues we have presented. As we have done for the definition of the models, we present some of their properties and their relationships with the tree data structure in an informal way by means of some examples. In [11, 2] the reader can find the formal definitions and theorems proving the claims that we present in the following. In order to explain the characteristics of the set data models we have already shown how a tree can be mapped into a NS-F or an INS-F; in the same way it is important to show how a NS-F can be mapped into a INS-F and vice versa, thus establishing a bijective relation between the set data models.

**Example 1.** *Let* $\{A_i\}_{i \in I}$ *be a NS-F and let* $\{A_i\}_{i \in I} = \{A_1, A_2, A_3, A_4, A_5\}$ *where* $A_1 = \{a, b, c, d, e, f, g\}$, $A_2 = \{b, g\}$, $A_3 = \{c, d, e\}$, $A_4 = \{d\}$ *and* $A_5 = \{e\}$. *Then we can map the NS-F* $\{A_i\}_{i \in I}$ *into a correspondent INS-F* $\{B_j\}_{j \in J} = \{B_1, B_2, B_3, B_4, B_5\}$, *mapping each set of* $\{A_i\}_{i \in I}$ *into a set of* $\{B_j\}_{j \in J}$*):*

$B_1 = \bigcup_{A_t \in \{A_1 \cup \mathscr{S}_{\mathscr{A}}^{-}(A_1)\}} (A_t \setminus \bigcup \mathscr{S}_{\mathscr{A}}^{+}(A_t)) =$
$A_1 \setminus \bigcup \{A_2, A_3, A_4, A_5\} = \{a, b, c, d, e, f, g\} \setminus \{b, c, d, e, g\} = \{a, f\}$.

$B_2 = \bigcup_{A_t \in \{A_2 \cup \mathscr{S}_{\mathscr{A}}^{-}(A_2)\}} (A_t \setminus \bigcup \mathscr{S}_{\mathscr{A}}^{-}(A_t)) =$
$A_2 \setminus \{\emptyset\} \cup A_1 \setminus \bigcup \{A_2, A_3, A_4, A_5\} = \{b, g\} \cup \{a, f\} = \{a, f, b, g\}$.

$B_3 = \bigcup_{A_t \in \{A_3 \cup \mathscr{S}_{\mathscr{A}}^{-}(A_3)\}} (A_t \setminus \bigcup \mathscr{S}_{\mathscr{A}}^{+}(A_t)) =$
$A_3 \setminus \{A_4, A_5\} \cup A_1 \setminus \bigcup \{A_2, A_3, A_4, A_5\} = \{c\} \cup \{a, f\} = \{c, a, f\}$.

$B_4 = \bigcup_{A_t \in \{A_4 \cup \mathscr{S}_{\mathscr{A}}^-(A_4)\}} (A_t \setminus \bigcup \mathscr{S}_{\mathscr{A}}^+(A_t)) =$
$A_4 \setminus \{\emptyset\} \cup A_3 \setminus \{A_4, A_5\} \cup A_1 \setminus \bigcup \{A_2, A_3, A_4, A_5\} = \{d\} \cup \{c\} \cup \{a, f\} = \{d, c, a, f\}.$
$B_5 = \bigcup_{A_t \in \{A_5 \cup \mathscr{S}_{\mathscr{A}}^-(A_5)\}} (A_t \setminus \bigcup \mathscr{S}_{\mathscr{A}}^+(A_t)) =$
$A_5 \setminus \{\emptyset\} \cup A_3 \setminus \{A_4, A_5\} \cup A_1 \setminus \bigcup \{A_2, A_3, A_4, A_5\} = \{e\} \cup \{c\} \cup \{a, f\} = \{e, c, a, f\}.$

*In Figure 5 we can see a graphical representation of the NS-F mapped into a INS-F. The mapping between $\{A_i\}_{i \in I}$) and $\{B_j\}_{j \in J}$ can be defined by means of a function; we define $\zeta : \{A_i\}_{i \in I} \to \{B_j\}_{j \in J}$ to be a function such that $\forall A_k \in \{A_i\}_{i \in I}$, $\exists B_k \in \{B_j\}_{j \in J} \mid B_k = \bigcup_{A_t \in \{A_k \cup \mathscr{S}_{\mathscr{A}}^-(A_k)\}} (A_t \setminus \bigcup \mathscr{S}_{\mathscr{A}}^+(A_t)).$*

*In the same way we can define the function $\xi : \{B_j\}_{j \in J} \to \{A_i\}_{i \in I}$ such that $\forall B_k \in \{B_j\}_{j \in J}, \exists A_k \in \{A_i\}_{i \in I} \mid A_k = \bigcup (B_k \cup \mathscr{S}_{\mathscr{B}}^-(B_k)) \setminus \bigcup \mathscr{S}_{\mathscr{B}}^+(B_k).$ The function $\xi$ maps $\{B_j\}_{j \in J}$ into $\{A_i\}_{i \in I}$ thus a NS-F into a INS-F.*

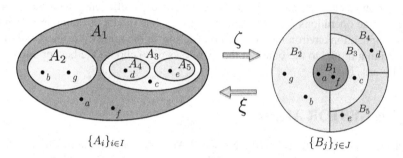

**Fig. 5** A NS-F $\{A_i\}_{i \in I}$ and its correspondent INS-F $\{B_j\}_{j \in J}$

This example showed us that if we model a hierarchy by means of a model we can map it into the other model and go from one to the other whenever it is necessary. We will see that this possibility is very useful in a concrete application because it allows us to be free to change the representation of a hierarchy and thus not to be bound to a single representation. The possibility of going from one model to the other is particularly useful when we have to exploit a property of a specific set data model; as an example we present the problem of how to find the **lowest common ancestor** in a tree $T(V, E)$ where $V$ is the set of vertexes and $E$ is the set of edges of the tree. We already know that $T(V, E)$ can be mapped in a NS-F or in a INS-F; let us see how the concept of lowest common ancestor is handled in the set data models. We can say that the lowest common ancestor, call it $v_t \in V$, of nodes $v_j \in V$ and $v_k \in V$ in a tree $T(V, E)$ is the ancestor of $v_j$ and $v_k$ that is located farthest from the root. Many algorithms have been proposed in the literature [4] to efficiently determine the lowest common ancestor of a tree. The same operation can be done both in the NS-M and the INS-M. If we map the tree $T(V, E)$ in a NS-F $\{A_i\}_{i \in I}$ each node $v_i, v_k, v_t \in V$ is mapped in a correspondent set $A_i, A_k, A_t \in \{A_i\}_{i \in I}$; the same operation can be done by mapping the same tree into an INS-F $\{B_j\}_{j \in J}$ where each node $v_i, v_k, v_t \in V$ is mapped in a correspondent set $B_i, B_k, B_t \in \{B_j\}_{j \in J}$.

In $\{A_i\}_{i \in I}$ set $A_t$ represents the lowest common ancestor between the sets $A_i$ and $A_k$. In order to determine $A_t$ we have to consider a collection $\mathscr{C}_i = A_i \cup \mathscr{S}_{\mathscr{A}}^+(A_i)$

containing $A_i$ and all its supersets in the family and a collection $\mathscr{C}_k = A_k \cup \mathscr{S}_{\mathscr{A}}^+(A_k)$ containing $A_k$ and all its supersets in the family; then, we have to intersect these two collections $C_x = C_i \cap C_k$. The collection $C_x$ contains all the sets which are common supersets of $A_j$ and $A_k$ and the set $A_t$ is the set with smaller cardinality in $C_x$. So, to determine the correspondent of the lowest common ancestor $v_t \in V$ in the NS-F $\{A_i\}_{i \in I}$ we have to do several operations: determine all the supersets of $A_i$ and $A_k$, intersect the collections containing these sets, calculate the cardinality of all the sets in the intersection and take the set with the smaller one.

On the other hand, in $\{B_j\}_{j \in J}$ set $B_t$ represents the lowest common ancestor between sets $B_i$ and $B_k$. In order to determine $B_t$ we have to intersect $B_i$ and $B_k$ and the resulting set is the correspondent of the lower common ancestor. So, in INS-M the problem of finding the lowest common ancestor is reduced to a single intersection between two sets; we can see that the choice of the set data model to adopt to represent a hierarchy is going to influence the efficiency with which we can do some operations. The best choice between one model or the other depends on the application environment we are considering; at the same time, the possibility to go from a model to the other on the fly allows us to choose the best option on a case-by-case basis.

## 5   The NESTOR Algebra

We designed an algebra, called the NESTOR Algebra, for the manipulation of data represented throughout the set data models defined in the NESTOR Model. In order to clarify the functioning and the fundamental principles on which the NESTOR Algebra is based we will relate them to the widely known relational algebra basics. Most of the relational operations such as selection, projection, product and set operations are important operations we want to perform on the data represented as families of sets. The NESTOR Algebra can be uniformly adopted by the NS-F and the INS-F. In Figure 6 we can see a graphical representation of the basic components composing the NESTOR Algebra.

The first component contains the data model on which the algebra is based, the predicates of the algebra and the fundamental concept of pattern family on which the whole algebra relies. The second component contains the definition of the operators of the algebra which are both manipulation and query ones: renaming, value update, insertion, deletion, selection, projection, union, intersection, set difference, product, join, grouping and aggregation. The third component contains the relationships of the NESTOR Algebra with the relational algebra and the Tree Algebra for XML [16, 22]. In this chapter we do not describe each one of the algebra operators and characteristics, nor do we show its completeness for the relational algebra and the Tree Algebra for XML [16]; we present the main features of the algebra in order to understand its functioning and its possible uses to address the interoperability issues we have presented.

A central feature of the relational algebra is the declarative expression of queries over the collection of tuples; an important thing we have to take into consideration is

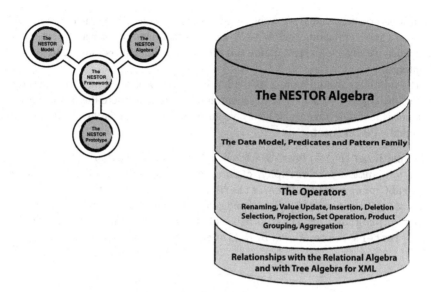

**Fig. 6** The main components of the NESTOR Algebra

that the manipulation of data in the NESTOR Framework often uses structural constructs, and element inclusion (i.e. the determination of set/superset relationships between two sets). We need to maintain this information when we manipulate the sets organized into families in the set data models. Thus, in the NESTOR Algebra a whole family of subsets is the fundamental unit, similar to a relational tuple. We do not present an algebra where the families are transformed in a collection of tuples to be processed and then re-transformed in families following a relational construction/deconstruction paradigm as is done in XML shredding [27]. Instead, we follow the approach adopted in Tree Algebra for XML [16] developed for the manipulation of XML data modeled as forests of labeled, ordered trees; thus, we choose to manage collections of families of sets directly.

A relation in the relational model is a collection of tuples and the counterpart in the nested sets models is a collection of families. Each relational algebra operator takes one or more relations as input and produces a relation as output. Correspondingly, each operator of the NESTOR Algebra takes a collection of one or more families as input and produces a collection of one or more families as output. This means that the NESTOR Algebra considers a whole family as the basic unit and not, for instance, a single set belonging to a family or the elements belonging to the sets in a family.

Predicates are central to much of querying. While the choice of the specific set of allowable predicates is orthogonal to the NESTOR Algebra, any given implementation will have to make a choice in this matter. In the NESTOR Algebra we point out structural and content predicates; the structural predicates allow us to express conditions on the structure of the families of subsets, instead the content predicates allow us to define conditions on the elements belonging to the sets. For instance, we use a structural predicate if we want to express a condition saying that a set $A_j \in \mathscr{A}_l$ [8] must be a subset of another set $A_k \in \mathscr{A}_l$. On the other hand, we use a content predicate when we need to express conditions on specific elements belonging to the sets of interest.

The following example presents the NS-F we will use as a basis in the description of the algebra; it is a toy representation of an archive where the label of every element is a simple string that also give an indication of the archival information brought by that element.

**Example 2.** *Let* $\{B_j\}_{j \in J}$ *be a NS-F where* $\{B_j\}_{j \in J} = \{B_1, B_2, B_3, B_4, B_5, B_6\}$.

$B_1 = \{$ summary, biography, chronology, programA, letterG, letterF, programC, programD, letterA, letterB, letterC, testamentA, letterD, testamentB, testamentC$\}$, $B_2 = \{$programA, letterG, letterF$\}$.

$B_3 = \{$programC, programD$\}$, $B_4 = \{$letterD, testamentB, testamentC$\}$, $B_5 = \{$letterA, letterB$\}$ *and* $B_6 = \{$letterC, testamentA$\}$.

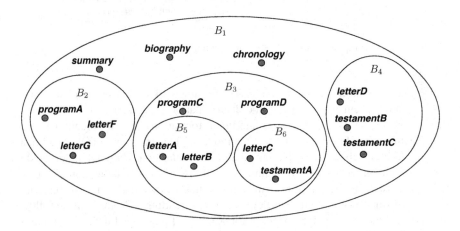

**Fig. 7** Venn-diagram of the NS-F described in Example 2

A basic syntactic requirement of any algebra is the ability to specify the attributes of interest. In relational algebra this is accomplished straightforwardly by considering domain-ordered relations or using names instead of position numbers for "identifying the domains" [7]; doing so for a collection of families is tricky for several reasons. First-of-all merely specifying elements is ambiguous: the elements of which set and belonging to which family? Furthermore, we also need to define structural constraints; we have to specify the characteristics of elements belonging to some sets as well as the relationships between the sets in a family. In a family we cannot use information such as the position of the elements in a set or the linear order between the subsets of some sets. If we consider a homogeneous collection of families,

---

[8] In the following, a family of subsets $\{A_i\}_{i \in I}$ will be indicated usiing the shorthand notation $\mathscr{A}_l$.

we could define a family with an identical structure to those in the collection, label its sets, and use these labels to specify elements. But in a collection of families, data families are heterogeneous and often we do not care about the complete structure of a family but we wish only to reference some portion of the family we care about. The NESTOR Algebra needs a mechanism to:

- Identify and manipulate an element on the basis of its value.
- Identify a set on the basis of the elements it contains.
- Define the relationships of a set with the other sets in a family.

In order to address these issues we define the concept of *pattern triple* which provides a specification of the sets and elements of interest. The pattern family is fixed for a given operation, and hence provides the needed standardization over a heterogeneous collection of families. All algebraic operations manipulate sets and elements identified by means of a pattern family.

A pattern triple, that we can indicate as $P = \langle \mathscr{A}_l, F_s, F_e \rangle$, constraints each sets in two ways: The formula $F_s$ imposes structural predicates on sets requiring each set to have structural relatives (subset, superset, direct superset, etc.) satisfying other content predicates defined in the formula $F_e$ imposed on any set. In order to understand how the pattern triple works, we propose an example presenting a pattern triple; the pattern triple we present is drawn from the NS-F presented in the Example 2.

**Example 3.** *Let* $P = \langle \mathscr{A}_l, F_s, F_e \rangle$ *be a pattern triple where* $\mathscr{A}_l = \{A_1, A_2\}$ *is a NS-F.*

*P is a pattern triple defining a NS-F composed by two sets* $A_1, A_2$ *where* $A_1$ *is required to be the direct superset of* $A_2$ *and* $A_1$ *must not have any superset – i.e.* $\mathscr{S}^-_{\mathscr{A}}(A_1) = \{\emptyset\}$; *these conditions are expressed by means of structural predicates. On the other hand, by means of content predicates we require* $A_2$ *to contain an element whose label starts with the string:* letter. *We can see a graphical representation of the pattern family described by the pattern triple in this example in Figure 8.*

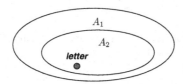

**Fig. 8** A graphical representation of the pattern triple P of Example 3

Now, we can introduce another important concept of the NESTOR Algebra: the order embedding of a pattern triple P into a collection of families $\mathscr{C}$. It is useful to point out that $\mathscr{C}$ can be composed by one or more families of sets. In the following we indicate with $B_k \in \mathscr{C}$ a generic set in the collection meaning that $\exists \{B_i\}_{i \in I} \in \mathscr{C} \mid B_k \in \{B_i\}_{i \in I}$.

Let $\mathscr{C}$ be a collection of families of sets, and $P = \langle \mathscr{A}_I, F_s, F_e \rangle$ be a pattern triple. An **order-embedding** [14] of P into $\mathscr{C}$ produces as output a collection of families of sets where each family has the same structure of $\mathscr{A}_I$ and satisfies the structural and content predicates defined in the pattern triple.

The families of sets outputted by the order-embedding of a pattern family into a collection of families are called *witness families*. There may be no, one or more than one embedding of a pattern triple into a collection of families and each embedding induces a *witness family* of the embedding.

A witness family is composed of each set in the input collection $\mathscr{C}$ that matches a set in the pattern triple and if there exists a set $B_t \in \mathscr{C}$ such that it is both a subset of a set that matches a pattern set and superset of another set matching a pattern set, then $B_t$ belongs to the witness family even if it does not match any pattern set.

The meaning of "witness family" is that the sets in an instance that satisfies the pattern triple are retained and the original family structure is restricted to the retained sets to yield a witness family. If a given pattern triple can be embedded in an input family instance in multiple ways, then multiple witness families are obtained, one for each order-embedding as we show in the next example.

**Example 4.** *In this example we show some order-embeddings of the pattern triple presented in the Example 3 into the NS-F presented in the Example 2. This pattern triple can be embedded in three ways into the input family and then it returns three different witness families. We can see a graphical representation of the witness NS-families in Figure 9.*

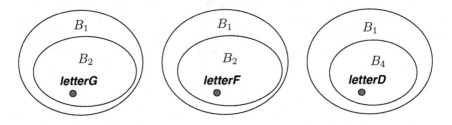

**Fig. 9** A graphical representation of the witness families resulting from the pattern triple and the collection of families in Example 4

Most-of-all the operators in the NESTOR Algebra are based on the concepts of pattern triple and witness family. As an example we present the selection operator that exploits the concept of pattern triple and witness family; furthermore, we introduce the concept of *adornment list* which is a list containing pattern sets: let $\mathscr{A}_I$ be a pattern family then the adornment list is SL $= \{A_j\}$ for some $A_j \in \mathscr{A}_I$. The **selection operator** takes a family of subsets (or a collection of families) as input, a pattern triple and an adornment list as parameters and it outputs a witness family for each embedding of the pattern triple in the input family; the witness families

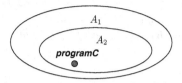

**Fig. 10** A graphical representation of the pattern triple P of Example 5

produced as output are then augmented with all the subsets of each set in the adornment list even if these subsets do not match any set in the pattern triple. This means that for each set $A_k$ in the adornment list we have to insert all the sets in $\mathscr{S}_{\mathscr{A}}^{+}(A_k)$ in each witness family embedded by the pattern triple. If the adornment list is empty, the selection operator straightforwardly returns the witness families. The selection operator works in the same way for the NS-M and the INS-M.

Figure 9 presents the output of the selection operator when the adornment list is empty. Let us see another example about how the selection operator can be used.

**Example 5.** *Let us consider the NS-F $\mathscr{B}_J$ presented in Example 2 and shown in Figure 7. Let us consider the pattern triple* $P = \langle \mathscr{A}_I, F_s, F_e \rangle$ *where* $\mathscr{A}_I = \{A_1, A_2\}$ *is a NS-F. The structural predicates in $F_s$ say that $A_1 \subseteq A_2$ and that $A_1$ must not have any superset. The content predicates in $F_e$ say that $A_2$ must contain an element which label is "programC". This pattern is represented in Figure 10.*

*If we match the presented pattern triple with the input family $\mathscr{B}_J$ we obtain a single witness family as output because there is only one configuration in the input family for which we have a set with no superset which contains a set at which belongs an element with label "programC". This witness family is presented in Figure 11.*

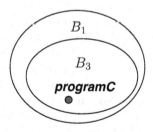

**Fig. 11** A graphical representation of the witness family resulting from the pattern triple and the collection of families in Example 5

*If we consider a selection operator which takes the family $\mathscr{B}_J$ as input and the just described pattern triple and an empty adornment list as parameters, the output will be exactly the witness family in Figure 11. Let us consider the same selection operator with a non-empty adornment list – i.e.* SL $= \{A_2\}$. *This means that all the* *subsets of the set matching $A_2$ in the input collection will be added to the resulting witness family. So, in this example we have to augment the witness family in Figure 11 with all the subsets of set $B_3$; we can see the output of the selection in Figure 12.*

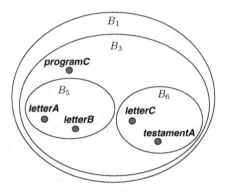

**Fig. 12** A graphical representation of the output of the selection operator presented in Example 5

The other operators in the NESTOR Algebra are defined on the same theoretical basis of the selection.

## 6  The NESTOR Prototype: Addressing Interoperability for Digital Archives

The NESTOR Prototype is the actual instantiation of the NESTOR Model; in Figure 13 we can see the main components of the NESTOR Prototype. The first component details how the entities and the information objects we are considering are represented through the NESTOR Model; the second describes the possible instantiations of the model in an actual environment and the third examines the relationships between the instantiations and the technologies of choice.

The NESTOR Prototype presents two possible applications of the NESTOR Model. The first application shows how an archive modeled through the NESTOR Model can be instantiated by means of the EAD metadata format. We show how the archival hierarchy and the context of archival descriptions can be retained by means of an XML file. The second application shows how we can access and share archival descriptions with variable granularity by means of the OAI-PMH while retaining the fundamental characteristics of the archives. First of all we will show how an archive can be modeled through a NS-F and a INS-F; then, we analyze the requirements of interoperability in the archival context and afterwards we present the two applications of the NESTOR Model and describe how they can or cannot address the interoperability aspects for Digital Libraries.

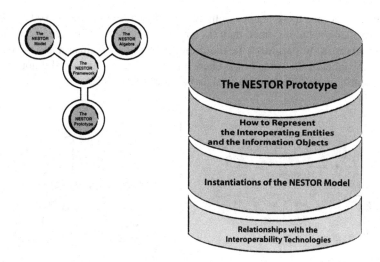

**Fig. 13** The main components of the NESTOR Prototype

## 6.1 How to Represent an Archive through the NESTOR Model

At this stage it is quite straightforward to model a digital archive throughout the set data models defined in the NESTOR Model. Let us consider an archive composed of several divisions – e.g. fonds, sub-fonds, series – each division contains a bunch of records – e.g. letters, registers, testaments; a representation of such an archive is given in Figure 7 where we represented an archive through a NS-F. In order to model this archive we have to represent the hierarchical relationships between its divisions and the records belonging to them. By adopting the NESTOR Model we represent each division as a set maintaining the hierarchical relationships by means of the inclusion order defined between the nested sets. Let us consider another example: if a fonds contains three sub-fonds, then in the NS-M we will have a set representing the fonds containing three subsets representing the subfonds. Vice versa in the INS-M we will have a set representing the fonds which is the common subset of the three sets representing the sub-fonds. Each record belonging to a division is represented as an element belonging to the set corresponding to that division. In this context we consider each element is a metadata – defined in whatever format – describing the archival resource; please note that the model does not necessarily require that the elements be metadata, they can also be full content digital objects represented as well as elements belonging to sets; we have seen in the NESTOR Algebra that the value of an element can be of whatever domain we want. At this level of definition there are no constraints on the nature of the elements belonging to the sets.

In Figure 2 we have seen a sample archive represented by a tree mapped into an EAD XML file; in Figure 14 below, we can see the same archive represented with the Nested Set Model and the Inverse Nested Set Model. Please note that with these models the records belonging to each archival division are properly represented as elements. In the tree there is not a defined way of representing the records; indeed, as an expedient, we represented them as a bunch of documents linked to a division by a dotted arrow. In Figure 14 we focus on the sets order inclusion and we do not indicate the label or the value of the elements just as in Figure 2 we did not specify the elements contained by the archival divisions encoded in EAD.

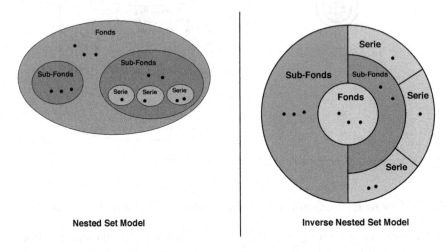

**Fig. 14** A sample archive represented by means of the Nested Set Model and the Inverse Nested Set Model

## 6.2 Analysis of the Requirements

In the definition of a data model it is important to define the requirements that the instantiations of the data model have to fulfill. For the purposes of this chapter we point out the requirements that if fulfilled allow interoperability issues for the digital archives to be addressed.

> **R1.** The archival descriptions have to be accessible from multiple entry points and at the same time they have to disclose their relationships allowing the user to consult contextual information. Furthermore, the users must have a means at their disposal for manipulating both the archival structure and the archival descriptions and for defining and performing queries on-the-fly.

This requirement is important because it says that we have to be able to consult an archive starting from the required description without having to navigate the whole archival hierarchy from a unique entry point to find the information of interest. At the same time from each description we have to be able to reconstruct its relationships with the other elements of the archive – i.e. preserving and exploiting the archival bonds. When we manipulate and query an archive we have to be able to express constraints on the structure and on the content of an archive; to do so we need to have a well-defined mechanism that allows us to express our needs in a standard way. This requirement if fulfilled addresses the *"user perspective"* aspect of interoperability.

> **R2.** The archival descriptions have to be shareable in a distributed environment with a variable granularity and have to provide a mechanism for reconstructing the hierarchy when necessary.

This requirement states that we have to be able to exchange archival descriptions with different degrees of coarseness and belonging to whatever level of the archival hierarchy without having to exchange the whole archive. Furthermore, a mechanism needs to be available for reconstructing the archival relationships of an exchanged description whenever it is necessary. In the current state of development of DL an important technological requirement for such a data model is compliance with OAI-PMH. Through the fulfillment of this requirement we can address the *"functional perspective"* and the *"interoperability technology"* interoperability technology aspects of interoperability.

> **R3.** Advanced language techniques allowing cross-language access to the resources have to be straightforwardly applicable to the archival descriptions.

Cross-language access to information leads to problems of both semantic and syntactic interoperability [20]. Two "metadata-related challanges" need to be addressed that usually are faced by involving the specification of the language of the metadata fields" [20]: false friends and term ambiguity. Another important issue is "name resolution" which regards the necessity to disambiguate between words that are proper names that do not require a translation or nouns that have to be translated for multilingual purposes. For instance, the term "Bush" can be seen as the surname of a former president of the United States or as a noun indicating a shrub. On the other hand, we may need to translate some proper names; for instance, the proper name"Kepler" has to be translated as "Keplero" in Italian.

To address these issues we can point out three main solutions. In the **Translation** technique a query formulated into the user language is automatically translated in the other languages supported and then submitted to the system. This solution is not free from the false friends, name resolution and term ambiguity issues. The **Enrichment of Metadata** is understood as making the intended meaning of information resources explicit and machine-processable, thus allowing machines and humans to better identify and access the resources. The language would be thus provided in the metadata itself. Lastly, the **Association to a Class** is the association of terms to a fairly broad class in a library classification system such as the Dewey Decimal Classification. This is a common solution for the term ambiguity problem and it is similar to synsets used in WordNet[9]. More advanced language techniques such as semantic annotation and tagging may also be taken into account and related to this solution.

The specification of the language of metadata field permits us to fully exploit metadata for cross-language purposes. If metadata do not come with or cannot be

---

[9] http://wordnet.princeton.edu/

enriched with the language of the field, it is useful to rely on the association to a class technique. This useful technique relies on the use of the subject field of metadata; it is not always possible to determine the subject of a metadata or of a term. This is particularly true for archival metadata where determining the subject can be very difficult. This requirement is directly related to addressing the "*multilingualism*" aspect of interoperability.

## 6.3   Retaining Archival Hierarchy and Context throughout an XML Tree

The EAD metadata format is the standard means for representing and encoding an archive. In Figure 2 we have seen how a tree is mapped into an EAD file; in Figure 15 we can see how the same sample archive modeled by means of the Nested Set Model (see Figure 14) can be mapped in the same EAD file. The order inclusion between the sets defining the hierarchical relationships between the archival divisions is retained in the EAD by means of nested tags in the XML file. The elements representing the archival descriptions are encoded by a sub portion of XML nested inside each tag representing the corresponding archival division.

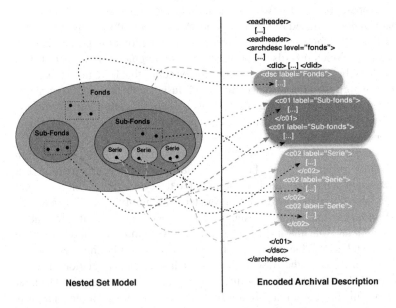

**Fig. 15**  A sample archive represented throughout the NSM and mapped into an EAD file

The main feature of this instantiation of the model is that both the structural and the content elements are represented by means of XML elements (i.e. tags). The EAD metadata allows us to encode the description part defined by means of the data models and thus it is a proper means for representing an archive. Let us see how it behaves with the three requirements we pointed out in the previous section.

The first requirement (**R1**, Section 6.2) states that we have to be able to access the archival description – i.e. descriptive metadata – at different degrees of coarseness. EAD is encoded as a unique XML file which mixes structural and content information while the entry point to access the information is the root of the XML file. From the root we have to navigate the hierarchy to access the information of interest. In order to overcome this issue we can define some superstructures to the EAD; for instance, we can settle some predefined entry points by the use of XPointers[10] pointing to specific elements of the XML or by using predefined paths driving the user through the hierarchical structure. These solutions are palliatives because they can only adequately match a well-defined reality with limited and specific needs; moreover, they have to be revised and adapted when the user needs or the requirements change. Furthermore, for each instantiation of the EAD, we have to know in advance how the XML elements are used; this is not a problem in general because we can make use of the EAD schema, but to do so each instantiation of EAD has to meet stringent best practice guidelines [26] otherwise the use of tags may be inconsistent, leading to wrong interpretations of the information as has happened in practice [18]. This peculiar aspect is problematic also from the manipulation and query point of view. We can adopt the Tree Algebra for XML [16] as a natural way to manipulate and query the EAD file and consider both structure and content of the encoded archive. The users can express their need throughout algebra operators, but they have to know in advance how the EAD elements are used otherwise the algebra operators are ineffective.

The second requirement (**R2**, Section 6.2) states that we have to be able to exchange archival descriptions in a distributed environment. The same issues affecting EAD for the access requirement can be found here for metadata exchange; indeed, the encoding of all the archival descriptions as a unique XML file forces us to exchange the archive as a whole. We cannot share a specific piece of information – e.g. the descriptions of the documents belonging to a specific *serie* – without extracting it from the XML file and losing in this way the structural information retained thanks to the nested tags in the XML itself.

The third requirement (**R3**, Section 6.2) regards the possibility of using language techniques with the archival metadata. The first language technique (e.g. "translation") is not affected by the choice of EAD and it can be directly applied. Furthermore, when we consider the translation of an EAD file we have the advantage of a big file with a large amount of contextual information which can be used to disambiguate the terms. On the other hand, the "enrichment of metadata" technique requires the metadata to be machine-readable in order to be automatically processed and enriched. The very flexibility of EAD leading to a not always consistent use of structure and content elements precludes the possibility of adopting this technique with many EAD files. Lastly, we know that a single EAD metadata is used to describe an entire archive, thus in a single metadata we can find very different subjects.

---

[10] http://www.w3.org/TR/xptr-framework/

With this organization it is very difficult to disambiguate terms or to identify the subject of metadata; with the EAD metadata the "association to a class" technique is essentially unworkable.

We can see that in this case we are not able to meet the interoperability requirements for a digital archive. We can think of different solutions that address one specific aspect at a time. These solutions must be designed on a case-by-case basis and they do not constitute a general environment that can be applied to hierarchically structured resources for addressing interoperability in Digital Libraries.

## 6.4  Encoding, Accessing and Sharing an Archive through Sets

This application of the NESTOR Framework follows a different approach and it is based on the joint use of some basic features of OAI-PMH and the Dublin Core metadata.

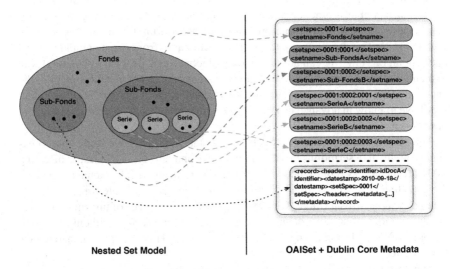

**Fig. 16** A sample archive represented throughout the NSM and mapped into OAI sets and DC metadata

Now we informally discuss how the set data models are mapped into a structure of OAI sets and Dublin Core metadata; a formal definition of the ideas behind this approach can be found in [1, 11]. First-of-all we present the mapping of an archive represented by means of a NS-F into an organization of OAI sets and DC metadata; the mapping of the INS-F is symmetrical to this procedure but it leads to a slightly different outcome [1]. Let us consider the sample archive represented by the Euler-Venn diagram in Figure 14. As we can see in Figure 16, each set composing this nested set structure is mapped into an OAI set with a proper setSpec; the set called "fonds" is mapped into an OAI set with $<$ setspec $> 0001 </$setspec$>$. This set has two subsets that are mapped into two OAI sets: $<$ setspec $> 0001 : 0002$ $</$ setspec $>$ and $<$ setspec $> 0001 : 0003 </$ setspec $>$ and so on for the other sets. We can see that the hierarchical relationships and thus the inclusion order between the sets is maintained by the identifiers of the OAI sets which are defined as materialized paths from the root to the identified set. Each single archival description is mapped into a DC metadata belonging to an OAI set; the membership information is added to the header of these metadata that are seen as OAI-records. In this way each archival description can be encoded by a single metadata without any constraints on its format; indeed, an OAI set can contain different kinds of metadata formats. With this model we do not impose any conditions on the archival descriptions, thus allowing the possibility of changing the metadata, updating the information or adding a new metadata format without affecting the structure of the archive and without changing the data model. The choice of the DC metadata format is lead by its widespread use in libraries and the possibility of defining Dublin Core *application profiles* which allow us to make it domain-specific; indeed, DC application profiles allow the definition of DC metadata formats well-suited for the reality we intend to represent.

This instantiation of the set data models has two main differences from the EAD one: it clearly divides the structural elements (i.e. the sets) from the content elements (i.e. the archival descriptions) and it does not bind the archival descriptions to a unique, fixed and predefined metadata format. These differences have a major impact on the fulfillment of the three presented requirements.

The **R1** requirement is fulfilled because each OAI set is individually accessible as well as each single metadata. From a set we can easily reconstruct the relationships with the other sets by exploiting the setspec; from a metadata we can reconstruct the relationships with the other metadata thanks to the membership information contained in their header. At the same time we can straightforwardly adopt the NESTOR Algebra to manipulate and query the archival descriptions. Indeed, each set is uniquely identified by a setspec value and the name of the set is a mandatory requirement; the metadata are encoded by means of the Dublin Core and thus the use of the tags is simple and consistent. By means of the pattern triple we can express requirements on the structure of the archive (i.e. the nested sets by means of the pattern family and the boolean formula $F_s$) and on the archival descriptions (i.e. the metadata by means of the boolean formula $F_e$). The NESTOR Algebra gives users a standard way to manipulate and query hierarchical structure that can be applied to different *interoperating entities*; users do not have to change the manipulation and query language and thus they can perform a query both in the library and the archival context in the same way. The NESTOR Algebra defines a natural way to modify the structure and the content of an archive represented with the NESTOR Model; at the same time we can query an archive naturally in the context of the NESTOR Framework.

Let us consider the **R2** requirement; throughout the OAI sets and DC metadata approach we can easily use OAI-PMH to exchange a single set or a single metadata, thus allowing a variable granularity exchange. Furthermore, from the identifier of an OAI set we can reconstruct the hierarchy through the ancestors to the root. By means of OAI-PMH it is possible to exchange a specific part of the archive while at the same time maintaining the relationships with the other parts of it. The NS-M fosters the reconstruction of the lower levels of a hierarchy; thus, with the couple NS-M and OAI-PMH applied to the archive, if a harvester asks for an OAI set representing for instance a sub-fonds it recursively obtains all the OAI subsets and items in the subtree rooted in the selected sub-fonds.

It is worthwhile highlighting that this approach can also be applied with the INS-M, then if a harvester asks for an OAI set representing for instance an archival serie, it recursively obtains all the OAI-subsets and records in the path from the archival serie to the principal fonds that is the root of the archival hierarchy. The choice between a NS-M or INS-M should be made on the basis of the application context. In the archival context the application of the INS-M would be more significant than the NS-M. Indeed, often the information required by a user is stored in the external nodes of the archival tree [28]. If we represent the archival tree by means of the INS-M, when a harvester requires an external node of the tree it will receive all the archival information contained in the nodes up to the root of the archive. This means that a Service Provider can offer a potential user the required information stored in the external node and also all the information stored in its ancestor nodes. This information is very useful for inferring the context of an archival metadata which is contained in the required external node; indeed, the ancestor nodes represent and contain the information related to the series, sub-fonds and fonds in which the archival metadata are classified. The INS-M fosters the reconstruction of the upper levels of a hierarchy which in the archival case often contain contextual information which permit the relationships of the archival documents to be inferred with the other documents in the archive and with the production and preservation environment. We can see how the possibility of changing from one set data model to the other by means of the defined mapping functions is very useful in the archival context; we can address the user requirements in the most effective way without being bound to the properties of the model of choice.

With regard to the **R3** requirement, we can see that this approach is particularly well-suited for use in conjunction with the presented language techniques. Indeed, the representation of an archive as an organization of sets and DC metadata makes it easier to determine the subject of each single metadata and thus to apply the "association to a class" solution; in the same way the metadata enrichment can be adopted because the DC metadata are well-suited to automatic processing. In this way the solutions proposed to enable cross-language access to digital contents can be applied also with the archival metadata, thus opening up these valuable resources to a significant service offered by the Digital Library technology.

## 7   Conclusions and Future Work

The NESTOR Framework has been introduced as a conceptual and logical environment that can be exploited to address interoperability issues in Digital Libraries. The NESTOR Framework focuses on hierarchical structured resources by proposing two set data models alternative to the tree data structure which, as we have seen in some application contexts are well-suited to addressing interoperability issues. Furthermore, we presented the NESTOR Algebra that allows us to manipulate and query the hierarchies represented through the NESTOR Model in a natural way. We presented a concrete use case based on archives, which are a fundamental and challenging entity in the Digital Libraries panorama. Within the archives we showed how an archive can be represented through set data models and how these models can be instantiated. We compared two instantiations of the NESTOR Model dealing with issues of interoperability for Digital Libraries. We showed that the use of sets to express archives open them up to new important functions that meet interoperability issues.

An archival information management system has been envisioned and it is under design and development to implement the NESTOR Model for the modeling and managing of archival resources. In the design of the system, it has been possible to separate the structural representation from the content representation of the archival resources thanks to the features of the NESTOR Model. The design of a system that allows for the separation of the two levels of representation makes a step forward in the area where the available management systems do not allow such a level of abstraction in the possible application implementations.

The archival information management system, under development, is going to be adopted in the next future in the context of a project of the Italian Veneto Region[11]. Main aim of the project is to make available a regional archival information system, which allows the management of the resources of archives that are present on the region territory.

**Acknowledgments.** The work reported has been envisaged in the context of an agreement between the Italian Veneto Region and the University of Padua that aims at drawing up innovative information management solutions for improving the final user access to archives.

EuropeanaConnect[12] (Contract ECP-2008-DILI-52800) and the PROMISE network of excellence[13] (contract n. 258191) projects, as part of the 7th Framework Program of the European Commission, have partially supported the reported work.

# References

1. Agosti, M., Ferro, N., Silvello, G.: Access and Exchange of Hierarchically Structured Resources on the Web with the NESTOR Framework. In: IEEE/WIC/ACM International Conference Web Intelligence and Intelligent Agent Technology, pp. 659–662 (2009)
2. Agosti, M., Ferro, N., Silvello, G.: The NESTOR Framework: Manage, Access and Exchange Hierarchical data Structures. In: Proceedings of the 18th Italian Symposium on Advanced Database Systems, Società Editrice Esculapio, Bologna, Italy, pp. 242–253 (2010)

---

[11] http://www.regione.veneto.it/

[12] http://www.europeanaconnect.eu/

[13] http://www.promise-noe.eu/

3. Anderson, K.W., Hall, D.W.: Sets, Sequences, and Mappings: The Basic Concepts of Analysis. John Wiley & Sons, Inc., New York (1963)

4. Bender, M.A., Farach-Colton, M., Pemmasani, G., Skiena, S., Sumazin, P.: Lowest Common Ancestors in Trees and Directed Acyclic Graphs. J. Algorithms 57, 75–94 (2005)

5. Celko, J.: Joe Celko's SQL for Smarties: Advanced SQL Programming. Morgan Kaufmann, San Francisco (2000)

6. Codd, E.F.: A Relational Data Model of Data for Large Shared Data Banks. Communications of the ACM 13(6), 377–387 (1970)

7. Codd, E.F.: Relational Completeness of Data Base Sublanguages. In: Rustin, R. (ed.) Database Systems, pp. 65–98 (1972)

8. Cook, T.: The Concept of Archival Fonds and the Post-Custodial Era: Theory, Problems and Solutions. Archiviaria 35, 24–37 (1993)

9. Crestani, F., Vegas, J., de la Fuente, P.: A Graphical User Interface for the Retrieval of Hierarchically Structured Documents. Inf. Process. Management 40(2), 269–289 (2004)

10. Duranti, L.: Diplomatics: New Uses for an Old Science. Society of American Archivists and Association of Canadian Archivists in association with Scarecrow Press (1998)

11. Ferro, N., Silvello, G.: The NESTOR framework: How to handle hierarchical data structures. In: Agosti, M., Borbinha, J., Kapidakis, S., Papatheodorou, C., Tsakonas, G. (eds.) ECDL 2009. LNCS, vol. 5714, pp. 215–226. Springer, Heidelberg (2009)

12. Gilliland-Swetland, A.J.: Enduring Paradigm, New Opportunities: The Value of the Archival Perspective in the Digital Environment. Council on Library and Information Resources (2000)

13. Gradmann, S.: Interoperability of Digital Libraries: Report on the work of the EC working group on DL interoperability. In: Seminar on Disclosure and Preservation: Fostering European Culture in The Digital Landscape, National Library of Portugal, Directorate-General of the Portuguese Archives, Lisbon, Portugal (2007)

14. Halmos, P.R.: Naive Set Theory. D. Van Nostrand Company, Inc., New York (1960)

15. International Council on Archives. ISAD(G): General International Standard Archival Description, 2nd edn. International Council on Archives, Ottawa (1999)

16. Jagadish, H.V., Lakshmanan, L.V.S., Srivastava, D., Thompson, K.: TAX: A tree algebra for XML. In: Ghelli, G., Grahne, G. (eds.) DBPL 2001. LNCS, vol. 2397, pp. 149–164. Springer, Heidelberg (2002)

17. Kiesling, K.: Metadata, Metadata, Everywhere - But Where Is the Hook? OCLC Systems & Services 17(2), 84–88 (2001)

18. Kim, J.: EAD Encoding and Display: A Content Analysis. Journal of Archival Organization 2(3), 41–55 (2004)

19. Knuth, D.E.: The Art of Computer Programming, 3rd edn., vol. 1. Addison-Wesley, Reading (1997)

20. Levergood, B., Farrenkopf, S., Frasnelli, E.: The Specification of the Language of the Field and Interoperability: Cross-Language Access to Catalogues and Online Libraries (CACAO). In: Greenberg, J., Klasore, W. (eds.) DC 2008, Proc. of the Int'l Conf. on Dublin Core and Metadata Applications 2008, pp. 191–196. Universitätsverlag Göttingen, Germany (2008)

21. MacNeil, H., Wei, C., Duranti, L., Gilliland-Swetland, A., Guercio, M., Hackett, Y., Hamidzadeh, B., Iacovino, L., Lee, B., McKemmish, S., Roeder, J., Ross, S., Wan, W., Zhon Xiu, Z.: Authenticity Task Force Report. InterPARES Project: Vancouver, Canada (2001)
22. Paparizos, S., Jagadish, H.V.: The importance of algebra for XML query processing. In: Grust, T., Höpfner, H., Illarramendi, A., Jablonski, S., Fischer, F., Müller, S., Patranjan, P.-L., Sattler, K.-U., Spiliopoulou, M., Wijsen, J. (eds.) EDBT 2006. LNCS, vol. 4254, pp. 126–135. Springer, Heidelberg (2006)
23. Pearce-Moses, R.: Glossary of Archival And Records Terminology. Society of American Archivists (2005)
24. Pitti, D.V.: Encoded Archival Description. An Introduction and Overview. D-Lib Magazine 5(11) (1999)
25. Prom, C.J.: Does EAD Play Well with Other Metadata Standards? Searching and Retrieving EAD Using the OAI Protocols. Journal of Archival Organization 1(3), 51–72 (2002)
26. Prom, C.J., Rishel, C.A., Schwartz, S.W., Fox, K.J.: A Unified Platform for Archival Description and Access. In: Rasmussen, E.M., Larson, R.R., Toms, E., Sugimoto, S. (eds.) Proc. 7th ACM/IEEE Joint Conference on Digital Libraries (JCDL 2007), pp. 157–166. ACM Press, New York (2007)
27. Rys, M., Chamberlin, D.D., Florescu, D.: XML and Relational Database Management Systems: The Inside Story. In: Özcan, F. (ed.) Proc. of the ACM SIGMOD International Conference on Management of Data (SIGMOD 2005), pp. 945–947. ACM Press, New York (2005)
28. Shreeves, S.L., Kaczmarek, J.S., Cole, T.W.: Harvesting Cultural Heritage Metadata Using the OAI Protocol. Library Hi Tech 21(2), 159–169 (2003)
29. Van de Sompel, H., Lagoze, C., Nelson, M., Warner, S.: Implementation Guidelines for the Open Archive Initiative Protocol for Metadata Harvesting - Guidelines for Harvester Implementers. Technical report, Open Archive Initiative, p. 6 (2002)
30. Van de Sompel, H., Lagoze, C., Nelson, M., Warner, S.: The Open Archives Initiative Protocol for Metadata Harvesting, 2nd edn., Technical report, Open Archive Initiative, p. 24 (2003)
31. Vegas, J., Crestani, F., de la Fuente, P.: Context Representation for Web Search Results. Journal of Information Science 33(1), 77–94 (2007)

# Administrative Document Analysis and Structure

Abdel Belaïd, Vincent Poulain D'Andecy, Hatem Hamza, and Yolande Belaïd

**Abstract.** This chapter reports our knowledge about the analysis and recognition of scanned administrative documents. Regarding essentially the administrative paper flow with new and continuous arrivals, all the conventional techniques reserved to static databases modeling and recognition are doomed to failure. For this purpose, a new technique based on the experience was investigated giving very promising results. This technique is related to the case-based reasoning already used in data mining and various problems of machine learning. After the presentation of the context related to the administrative document flow and its requirements in a real time processing, we present a case based reasonning for invoice processing. The case corresponds to the co-existence of a problem and its solution. The problem in an invoice corresponds to a local structure such as the keywords of an address or the line patterns in the amounts table, while the solution is related to their content. This problem is then compared to a document case base using graph probing. For this purpose, we proposed an improvement of an already existing neural network called Incremental Growing Neural Gas.

## 1 Introduction

One could think that paper is something from the past, that documents will all be "electronic" and that our offices and world is becoming paperless. This will be true one day, but it will take several decades or centuries because it is not only a

Abdel Belaïd
LORIA, University of Nancy 2
e-mail: abdel.belaid@loria.fl

Vincent Poulain D'Andecy
e-mail: Vincent.PoulaindAndecy@itesoft.com

Hatem Hamza
e-mail: Hatem.Hamza@itesoft.com

Yolande Belaïd
e-mail: yolande.belaid@loria.fl

M. Biba and F. Xhafa (Eds.): Learning Structure and Schemas from Documents, SCI 375, pp. 51–71.
springerlink.com © Springer-Verlag Berlin Heidelberg 2011

technological mater but a real cultural issue. Even though they all surf and chat over the Internet, our kids still learn how to read with paper books and how to write with pen and paper. Whatever the future, if we just have a look today at our administration services such as public organisations and services (city hall), social security services (health insurance, family benefits, hospitals, pension funds ), large companies (bank, insurance) etc., they handle each day a large volume of documents from their citizens and customers. On the other hand they face several challenges in order to reduce the charges and increase the customer satisfaction. They do look for processing times reduction; process error reduction and delivery of new services that can share quickly electronic information in an internet or a mobile context.

One solution is to turn the administrative paper flow into an electronic document flow. That means digitize the documents, extract automatically by document analysis techniques some information from the content of these digital images, push the extracted indexes into a business system like an ERP (Enterprise Resource Planning) and finally archive the indexed images into a DMS (Document Management System).

For the last 20 years, several software editors have proposed intelligent systems to automatically read the administrative documents. The current most famous companies are EMC-Captiva, ReadSoft, Top Image System, Kofax, Itesoft, A2IA, AB-BYY, Nuancet, Pegasus, Mitek, Parascript,

The performance of these systems is a good enough fit for some business objectives on few documents such as cheques, invoices, orders or forms but the performance deteriorates quickly when the system faces the complexity and the large variability of administrative document types. Basically an administrative document is a textual document, usually a form. The issue is that the information to extract is very heterogeneous:

- The extracted information can be high level information like a document class (this is an invoice, a complain, a form) or low level information such as fields (invoice number, social security ID, cheque amount) or tables (list of ordered items).
- The information can be machine printed (barcodes, characters, symbols, logos), hand-printed (script, marks) in a constraint frame or handwritten (cursive) with no constraint.
- The information can be very structured like forms. For one form each field is always located at the same physical position. Notice that there is a vast amount of different administrative document structures. Then if the automation of one form can be quite simple, the automation of hundreds of forms in a single flow becomes a bottleneck.
- The information can be semi-structured like bank cheques or invoices. The document looks like a structured document but the field location is floating, respecting only a logical structure. Each supplier prints the document in their own way but respecting (almost) for instance the rule <invoice number is closed to the top of the page beside the keyword "invoice no">.
- The information can be unstructured like a free text in a natural language (handwritten mails).

- The information can be explicit, directly readable within the document, or implicit, i.e. not directly readable but inferred from readable information, eventually combined with customer context data.
- The information can be isolated and easy to locate (printed or written on a white background) or with overlapping and then hard to segment (printed on a textured background or written over a comb or a printed text, ).
- The information can be in color while most of the industrial capture systems produce binary images.

All the proposed systems rely on 2 important kinds of technologies:

- Classifiers to convert the pixels into symbols: OCR (Optical Character Recognition) for machine printed symbols, ICR (Intelligent Character Recognition) for hand-printed characters [1], IWR (Intelligent Word Recognition) for handwritten cursive words [3] [2], Pattern Matching [4] for some graphic elements, Logo Recognition [5] [6], etc.
- Segmentation and extraction strategies to locate and interpret the field to extract. We identify two different techniques: the template matching approach [7] [8] and full-text approach [26] [31].

In the template matching approach, the extraction is driven by a mask. The mask defines an accurate physical location (bounding box) for each field to extract within the image. The system interprets the document in applying the classifier (OCR, ICR) to the snippet framed by the bounding box. In the industrial systems, the mask description is usually done by manual settings on a graphical interface. This is easy to handle, very efficient for fixed-structure forms, but is time-consuming when the user has to define many templates. Some researchers have proposed automatic model methods. The system builds itself the template by detecting graphical features: Duygulu [9] uses horizontal and vertical lines; Mao [10], Ting [12] use lines, logo, text lines; Hroux [13] uses pixel density; Sako [14] uses keywords ; Cesarini [15] uses both keywords and graphical features, Belaïd [16] uses connected component sequences. These features are exploited by a physical structure analysis algorithm which can perform top-down analysis as RLSA based techniques [17]; bottom-up analysis for example, X-Y cut [18] X-Y tree [9] [7]; and generally said more efficient, hybrid approaches like Pavlidis Split & Merge [19].

When the document flow contains heterogeneous documents that can not be described by a single template, then a classification step is required to select the adapted model. Some classification techniques rely on the same graphical features as described above. For instance Mao [10] uses also the lines to discriminate the document classes; Heroux [13] uses a vector of pixel density pyramid; Sako [20] describes a document class with keywords. By extension, text classification strategy with vector of words [21] can be used. For instance EMC, Itesoft and A2IA products integrate both graphical and textual classifiers. With these features many categorization algorithms have been explored like decision trees, nearest neighbour [22] [12], neural networks [23] [13], prototype-based methods, support vector machines [24].

The automatic model is an interesting method but limited to one or few similar document classes. In fact the features and the document structure analysis hardcode the application domain: invoices [15], orders [16], cheques, forms or tables [10]. Therefore if the method allows a good flexibility in the domain scope, it can not manage heterogeneous flows without an important development invest. The full-text approach answers to this issue.

In the full-text approach, the extraction is driven by the data itself and a set of rules (knowledge base) to find logical labels [25]. The rules can be parametrized by a user or trained on labeled documents like Cesarini [26] who trains an M-X-Y tree to learn the table structures. Most of the approaches are formal grammar based as in Couasnon [28] or Conway [29]. Niyogi [30] presented a system called DeLoS for document logical structure derivation. Dengel [31] described a system (DAVOS) that is capable of both learning and extracting document logical structure. Grammar and tree-representation seem to be a major area of research, the technique can even work for specialized tasks such as table form recognition [20]. A few systems try to learn the relationship between physical and logical structures [37] that do not focus especially on logical labeling. These systems can outperform classical rule base systems because they use more flexible models, which can tolerate variations of the structure [34]. In most recent works, new human brain based approaches have been developed: Rangoni [35] combines learning and knowledge integration by using a transparent neural network;

However all these systems face the same problems:

- Any segmentation and extraction strategies need an expert to describe the documents and "teach" the system. These settings or training are still time consuming and not often final-user friendly.
- Only few solutions address the full scope of heterogeneous information.
- The solutions depend on the intrinsic performance of the character classifiers. These classifiers are very efficient in standard contexts but usually collapse in noisy context.

We proposed in [38] a case-based reasoning system to model and train the knowledge for the recognition of administrative documents corresponding to different kind of invoices. They have a typical format that changes depending on the issuing company or administration. It may vary and evolve even within the same company. To cope with these changes, we seek to develop a system able to rely on experience to recognize new samples. The case-based reasoning seems to be a methodology capable of meeting this challenge. We will first describe the methodology, then we will show how we exploit it in the case of invoices.

This chapter will be organized as follows. Section 2 provides a general definition of CBR, explaining its various components. Then, Section 3 shows the use of CBR for document analysis. The proposed approach is described in Section 4. Experiments on invoices are detailed in Section 5. Finally, we conclude and give some perspectives of this work in Section 6.

## 2 Case-Based Reasoning

The case-based reasoning (CBR) is a reasoning paradigm that uses past experiences to solve new problems [39]. It is applied today in all areas where we need to use or synthesize past experiences to propose new solutions. Early work in CBR has been proposed in 1983 by Kolodner [41], which created a system of questions and answers using a knowledge base already established.

CBR is based on several steps we will detail in this paper. Several models have been proposed to define the CBR (see Fig. 1), but the most common is that of Aamodt and Plaza [39]. It consists of four steps: research, reuse, revision, retention (or learning). A preliminary step, also needed, is the formulation of the problem. In this work, we combined stages for reuse and review a step conventionally called "adaptation". This is also done in most researches in CBR.

### 2.1 CBR Terminology

In CBR, a problem is posed by the user or the system. That is the part to resolve. A case is the set formed by the problem and its solution.

$$case = \{Problem, Solution\}$$

We call "target case" the case to solve and "source case" the case from the case base used to solve the target case. The base case is the core of the system cases. These are either given by the user, or by the system automatically enriched.

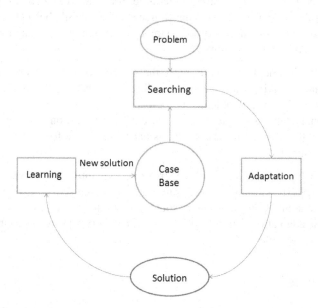

**Fig. 1** CBR paradigm

## 2.2   Problem Elaboration

This phase involves the indices and problem descriptors extraction, and their representation. It can be done either by the user who, based on his data, constitutes the problem to solve, or by the system which knows itself how to elaborate a problem based on available information. For example, in the system FAQFinder [40], the system input is a question given by the user. The question is then processed by the system which extracts the problem (extraction of the most significant terms, removing the less informative terms). The final problem in FAQFinder is a vector representing the query.

## 2.3   Similar Case Search

Similar case search is one of the most important parts of CBR. It involves taking into account the representation of cases, a measure of similarity between cases and the choice of one or more close cases. The similarity measure between cases is often limited to a similarity measure between problems. It depends directly on the data representation and complexity. Once the similarity measure appropriate to the problem is chosen, it becomes easy for each new problem, to find similar cases in the case base.

## 2.4   Adaptation

The adaptation is to find a solution to the target problem from a solution given to the source problem . The solution to the source problem should be kept or modified to allow the proposal of a new solution to the target problem. Two major categories of adaptation exist in the literature:

- Structural adjustment: it tries to apply the solution of the source case through some changes, taking into account the differences between source and target problems, on the target problem.
- Derivational adaptation which traces how the source solution has been proposed to produce the solution to target. This is particularly true for planning applications for example.

These two broad categories can be broken down into several types of other adaptations[42]. The simplest, and it's one we'll use is the zero adjustment. It consists, from the solution of the problem source, to "paste" the solution on the target problem. This adaptation, although it is very simple, solves problems as the case of documents for example.

## 2.5   Learning

The role of the learning is to integrate new solved and revised cases in the case base. This process must be designed to avoid filling the database with each new case resolved (redundancy can be dangerous for the system as far as it can slow it down considerably). It allows the system to manage its knowledge. We propose here an interesting solution for classifying the case base. We stand where the case base is constantly fed with new cases resolved as different, and we try to build a system to classify these cases so as to always have quick access to the database.

## 3  CBR for Document Image Analysis: CBRDIA

In an administrative Document Image Analysis (DIA) system, it often happens that documents are processed in batches. We call batch a set of similar documents, from the same supplier, the same company or administration. Documents in a batch all have the same characteristics, they are represented in the same manner and only the specific content of each document is different (for example, the cell contents in a form). Currently, in a company, a batch requires a user intervention to model the material on this batch. From the model, it becomes easier for the system to extract the desired information in the document.

The first problem is to try to model documents automatically in a batch and use this model later on the batch. The second problem is more complex. The system must also be able to handle heterogeneous documents. In this case, two configurations are possible:

- The first, and it is the easiest, is the case of the document for which you can find a similar model. This means that documents similar to this document have already been treated before, and it can speed up the processing. Documents in Figure 2 have exactly the same structures. As well the tables as the address structures and payment are similar. It is clear then that knowing the general model of this batch, it becomes very easy to treat all documents belonging to this batch.
- The second is more complex. This is the case of a completely new document, for which no existing model can be associated, and must of course be analyzed. The easiest solution is to appeal to a user who will help to extract the necessary information.

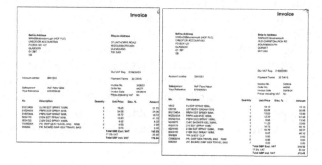

**Fig. 2** Two invoices from the same batch. The model of these documents is the same

These two problems raised other issues as important:

- how to model documents, what information must be represented in this model,
- how to take advantage of existing templates,
- which approach to take for that user intervention is minimal.

In our application, the system tries to analyze the documents individually, without resorting models of existing documents. The example of invoices in Figure 3 shows two invoices from two different batches. It is clear that both invoices should not (and can in fact not) be treated in the same way. Special treatment should be performed on each of them.

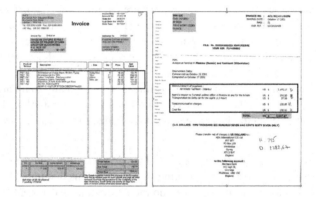

**Fig. 3** Two invoices from different batches

The ideal solution would be to try to enjoy the experience of the system, instead of the user. As to treatment, the system accumulates the experience of document analysis. After a while, it is certain that the system has treated so many different documents that the knowledge it gained could be used to analyze any type of documents. We must therefore take advantage of that knowledge. This is the main idea of the solution to the problems mentioned above. We will therefore introduce a system that not only deals with documents, but also learns as you go.

For this reason we chose to use CBR for the proposed system. First, the existing mode of reasoning in CBR is well suited to the needs of a system for document analysis. Indeed, CBR can benefit from previous experiences to offer new solutions to new problems. In the case of our application, a new problem is simply a new document to be analyzed. The fact that this document be in a batch or completely new does not change the fact that we must be able to analyze and interpret. Previous experiences are therefore other than documents previously modeled, analyzed and interpreted.

## 4   The Proposed Approach

Figure 4 shows the block diagram of our approach. It consists of two CBR cycles. For each new document, the problem is first extracted. This is compared with document problems of the case base.

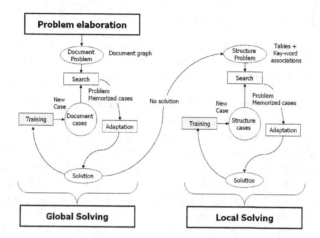

**Fig. 4** CBRDIA flow

- If a similar problem exists in the database, then the solution of the nearest (source problem) is applied to the problem of the target document.
- If no similar problem exists in the case base, then we proceed to the next CBR cycle. The structure problems are used. Each structure issue is compared with the problems of the structure case base. The solutions of the closest problems are then applied to the problem of target structure. Through this process we can obtain a complete solution for the document problem. So we can inject this resolved case in the document case base.

### 4.1   Document Structures

The document structures constitute in our approach the problems in CBR terminology. The documents that are used in our approach are real world documents taken from a document processing chain. First, they are OCRed thanks to some commercial softwares combined with other local built OCRs. The OCR output is then all the document words with their coordinates in the document. These words have to be organized in a more logical way. For this, we create groups of information given below:

1. Words: which are returned by the OCR. They are given the following attributes: position (top, left, right, bottom), tags (alphabetical, numerical, alphanumerical).

2. Fields: which are groups of neighbor words aligned horizontally. They are given the following attributes: positions and tags. The field tags correspond to the concatenation of the tags of the words composing the field.

3. Horizontal lines: which are groups of neighboring fields. They are given the following attributes: position, pattern. A pattern is the concatenation of the tags of the fields composing the horizontal lines.

4. Vertical blocks: which are groups of fields aligned vertically. They are given the following attributes: position coordinates.

From this physical re-structuring of the document, we extract now the logical structure. We have noticed through our observations on thousands of documents that two kinds of structures exist in invoices:

- The first one is the keywords structures (KWS) that are based on the extraction of some keywords. This extraction uses semantic dictionaries related to the domain of invoice (example: words like Amount, VAT, Total...).

- The second type of structures is tables which are very important structures in administrative documents. As tables correspond generally to a repetition of a pattern, these structures are called pattern structures (PS). Once these two types of structures are extracted, the document model takes them into account as well as their relative positions (top, left, bottom, right). Further details about our document model extraction, and especially on table extraction can be found in [43].

## 4.2 Problem Representation

The final document model is a graph of the different structures of the documents. This representation allows us not only to describe a document, but also a whole class of documents. In this way, whenever a document from the class 'X' is presented to the system, it can be directly recognized as belonging to this class. The example in Figure 5 shows clearly how two documents can be represented with the same model.

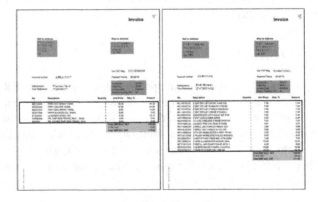

**Fig. 5** Two documents from the same class

Figure 6 shows a document model. This document is composed of two KWS. It can be clearly seen that the nodes are the document keywords, or the labels of the structures, whereas the edges describe the relative positions between the elements.

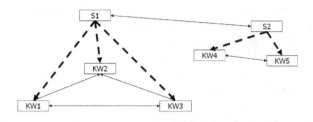

**Fig. 6** A document model

## 4.3   Problem Solving

Once the document model is extracted, global problem solving starts. It consists of checking if a similar document is available in the document database (using graph probing or edit distance). If it is the case, then the solution of the nearest document is applied on the current document. Otherwise, local solving is used. It consists of finding a solution for every structure (KWS or PS) in the document independent of others. Similarly, as in global solving, the system looks for similar structures in the structure database and tries to apply the solution of the nearest ones on the current structures. Figure 4 shows the flow of this approach. Further details can be found in [43].

### 4.3.1   Graph Matching Using Graph Probing

In CBRDIA, every document is to be matched with the documents of the database. Since our documents are represented by graphs, graph matching techniques have to be used. Many graph similarity measures exist. Edit distance as well as the maximum common subgraph distance can be employed, but time and complexity are factors that lead us to think about a faster similarity measure.

Graph probing distance is a graph dissimilarity measure that was first presented in [48]. It is a fast and fairly accurate technique of graph comparison. It has also a direct relation with graph edit distance. Let G1 and G2 be two graphs, then $d_{graphprobing}(G1, G2) \leq 4.d_{editdistance}(G1, G2)$. Quite often graph probing is a good approximation of graph edit distance. Its principal drawback is the fact that if $d_{graphprobing}(G1, G2) = 0$, then G1 and G2 are not necessarily isomorphic.

Graph probing is based on the computation of the frequency of each vertex and each edge in its graph. Let A, B, C be the nodes of G1, and B, C, D be the nodes of G2. First, we compute the frequency of A, B, C in G1, and do the same with the nodes of G2. The probe on nodes is then: $Pb1 = \sum |freq(N_{G1}) - freq(N_{G2})|$, where $freq(N_{G1})$ and $freq(N_{G2})$ are respectively the frequencies of a node $N$ in G1 and G2.

On the other hand, we have to calculate the probe on edges. For every node, we extract its edge structure. If a node N has an edge with the tag (top, left) and another one with the tag (top, right), then the edge structure of N is (top:2, left:1, bottom:0, right:1). This is done for every edge in the graph. The probe over edges is then $Pb2 = \sum |freq(E_{G1}) - freq(E_{G2})|$, where $freq(E_{G1})$ and $freq(E_{G2})$ are respectively the frequencies of an edge structure $E$ in G1 and G2.

The total probe is then: $Pb = Pb1 + Pb2$.

Graph probing is applied in this way to every new document in order to find the most similar document in the database.

### 4.3.2 Incremental Learning in CBRDIA

Now that new cases of documents have to be learnt, incremental learning is adopted. It allows the system to use the solved cases in the future and to avoid the same processing for every similar document. Incremental learning is used as the following: every new case has to be learnt and retained by the case base. Moreover, similar cases have to be grouped together in order to make the comparison between an incoming case and the case base more accurate.

To find an approach of incremental learning, we focused on incremental neural networks. The earliest incremental neural networks were the Growing Cell Structures (GCS [44]), followed by the Growing neural Gas (GNG). Then, many other variations were built on these two networks. The Hierarchical GNG (TreeGNG) is a network that builds classes over the classes given by the GNG. Similarly, the Hierarchical GCS (TreeGCS [45]) uses the same principle. Other types of incremental neural networks are those which use self organizing maps. One has to make the difference between incremental neural networks which perform incremental learning (GNG, IGNG) and incremental neural networks which are just incremental because they can add or remove neurons. We will just introduce IGNG in this paper as it is the method we are using.

### 4.3.3 Incremental Growing Neural Gas

The IGNG is an improvement of the GNG in some aspects (algorithm explained below). As shown in [47], IGNG gives better results for online and in incremental learning. This can be explained by the fact that IGNG creates neurons only when a new datum is very far from the already created neurons, contrary to the GNG which creates neurons periodically. IGNG has also better memory properties. This means that when a new class of data having different properties appears, the IGNG can really adapt its topology without loss of the previous information, whereas the GNG can lose some of its already created neurons.

This neural network IGNG suffers however from the choice of the threshold S. In their original paper, Prudent et al. proposed to initialize S at the standard deviation of the whole database for which classification is done. This is in our opinion contrary to the principles of incremental learning as we do not know *apriori* which kind data is coming later.

### 4.3.4 IGNG Improvement

The first point on which we worked was to try to be free from the choice of the threshold S. The first constraint is that the only information available at a time T is the information about the already processed data. Moreover, we cannot use the whole previous data to determine the class of the new data. The solution is to use some local information related to each neuron. Let:

- N be the number of IGNG neurons at T.
- E be an entry.
- $m_i$ be the average distance between every element in a class $i$ and its representative neuron $n_i$. $\sigma_i$ the standard deviation of these distances.

It is logical to say that E belongs to a class $i$ if $d(E, n_i) < m_i$. In order to be more flexible, we propose that the threshold S becomes: $S = m_i + \alpha.\sigma_i$, where $n_i$ is the nearest neuron to E. By taking into account the mean and standard deviation of this class, we are using intrinsic parameters related to this class, not to the whole data. Two cases typically occur:

- the new data is close enough to the nearest class (meaning $d(E, n_i) < m_i + \alpha.\sigma_i$), this data will belong to the class $i$ and the neuron $n_i$ is updated.
- the new data is too far from its nearest class. In this case, a new neuron is created (embryon neuron), and becomes effective in classification only if its age exceeds $a_{neuron_{max}}$.

Figure 7 shows a typical case where the nearest class is too far from the new data. This new data will then create an embryon neuron.

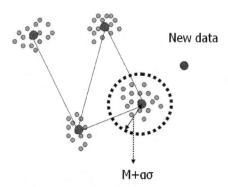

**Fig. 7** A new data creates an embryon neuron

### 4.3.5 Node Deletion

The second point of interest is the condition of node deletion. In GNG and IGNG, an edge is removed whenever its age exceeds a threshold $a_{edge}$. Then, if a neuron is not connected to any other neuron, it is removed as well. However, depending on the application, one may not want to remove neurons that are not connected to others, as these single neurons may represent an important information (eventhough it can be very rare). For example, in our application, some rare invoices can be processed from time to time, and can in this way form neurons that are very far from the other neurons. A simple example is shown on Figure 8. The neuron representing the rare data (X) is connected to two other neurons (Y and Z) which have many more data compared to X. Every time a data is attributed to Y or Z, the edges (X-Y) and (X-Z) are incremented. When $age(X-Y) > a_{edge}$ and $age(X-Z) > a_{edge}$, these edges are removed. In the classical scheme, X will also be removed. Moreover, its data will be assigned to its nearest neurons in the network (Y and Z). Then, even if Y and Z are not representative of the data associated with X, their data will include new data very different from their original data.

We propose in this case to examine the distance between X and its nearest neuron, let this distance be d(X,Y). If $d(X,Y) > \beta \cdot ((m_y + \alpha.\sigma_y) + (m_x + \alpha.\sigma_x))$, then this X has to be kept in the network, even if it has no connection with the other neurons. Otherwise, this neuron can be removed and its data can be assigned to the other neurons. The $\beta$ factor can be chosen by the user, depending on the application.

From now on, the Improved Incremental Growing Neural Gas will be noted I2GNG.

**Fig. 8** A neuron that should not be deleted, even if it is not connected to any other neuron

### 4.3.6 Adaptation to the Case of Graph Classification

In [46], Gunter and al. proposed a method of adapting Self Organizing Maps (SOM) to graph classification. This method was used to digit classification and was based on the computation of the edit distance between graphs. Every formula in the SOM algorithm was then adapted based on the edit path between any two graphs. Here is a simple explanation of the idea:

let G1 and G2 be two graphs. Let d(G1,G2) be the edit distance between G1 and G2. This distance corresponds to the cost of some additions, deletions or substitutions of nodes or/and edges which transform G1 into G2.

$$d(G1,G2) = \sum cost(editing)$$

In the vector domain, when the distance between a vector X and vector Y is d(X,Y), it is easy to transform X by $\varepsilon(X,Y)$, $\varepsilon$ being a real quantity. The same operation in the graph domain means that G1 has to be modified by $\beta = \varepsilon \cdot d(G1,G2)$ (equivalent to $Neuron = \varepsilon_b.(Neuron_{nearest} - entry)$). Modifying G1 by $\beta$ means that we have to apply only $\beta$ edit operations on G1. As we already know the edit path that allowed us to compute the distance between G1 and G2, $\beta$ corresponds just to a part of this edit path. Modifying G1 becomes in this way an easy task as we just have to find the edit operations which cost approaches $\beta$ as much as possible. More elaborate details can be found in [46].

Adapting the I2GNG formula using the principles cited above allows us now to classify graphs or trees using I2GNG.

## 5 Experiments

The first part of these experiments shows the results of the whole CBRDIA system. The second experiments are only related to incremental learning using administrative documents.

### 5.1 Experiments on CBRDIA

CBRDIA was tested on 950 invoice documents taken from invoice processing chain of ITESOFT. They are divided in 2 groups:

- the first one contains 150 documents where each one has a similar case in the document database: this is used to test global solving. The document database contains 10 different cases;
- the second one contains 800 documents for which no associated case exists in the document database. Hence, local solving will be applied on these documents.

The results are described thanks to three different measures as in 1. In this equation, X can be a document, a KWS or a PS.

$$R_X = \frac{|found\ solutions|}{|solutions\ in\ ground\ truth\ X|}.\qquad(1)$$

Global solving produced 85.29% of good results whereas local solving produced 76.33%. More detailed results can be found in [43].

## 5.2 Experiments on Administrative Documents

Our experiments were performed on a dataset of real documents (invoices) taken from a real invoice processing chain. Every invoice was modeled with its graph and then submitted to the I2GNG. The complexity of these documents is variable. Whereas some documents are very clean and have almost no OCR errors, others can be degraded and have very few key-words to be identified.

The dataset was divided in two parts: a learning set (324 documents) and a testing set (169 documents). 8 classes of invoices were used for this purpose. We chose this strategy of I2GNG evaluation as the learning procedure helps in knowing about the incremental capabilities of the modified I2GNG applied to graphs, whereas the testing phase helps knowing about its classification properties.

We performed two different series of tests. The first one is concerned with the influence of $\alpha$, the second one concerns the influence of the threshold age of neurons (above which neurons become mature). Table 1 gives these results.

**Table 1** Influence of $\alpha$ and $a_{edge}$

| $a_{edge}$ | neurons | rec | $\alpha$ | neurons | rec |
|---|---|---|---|---|---|
| 10 | 14 | 99.40% | 0.5 | 10 | 98.22% |
| 20 | 18 | 97.63% | 1 | 15 | 98.22% |
| 30 | 18 | 97.63% | 1.5 | 12 | 98.81% |
| 40 | 16 | 98.22% | 2 | 14 | 98.81% |
| 50 | 16 | 98.22% | 2.5 | 12 | 99.40% |
| 60 | 16 | 98.22% | 3 | 18 | 97.63% |

The results show that the number of neurons is always greater than the number of classes (8). This is due to the variations found in these classes. Representing one class with several neurons is not a problem as far as a neuron is not shared by two classes. In the training process, we tag manually the obtained neurons (by giving them the name of the class they represent).

The obtained results are very encouraging. They mean that the I2GNG with our improvements is working well. As shown in table 1, the bigger $\alpha$ is, the more neurons we obtain. This can be explained as the following:

- when $\alpha$ is big, the threshold $(m + \alpha.\sigma)$ for each neuron is also big. A new class is created if and only if it is outside the range of an existing neuron. Neurons are then not close to each others. Their ages increase quickly and they become mature quickly too.

- on the other hand, when $\alpha$ is small, neurons are very close to each other. One class can be represented by high number of neurons, few of which become mature because of the high competition among classes.

## 6  Conclusion

In this paper, we presented the different steps composing the system CBRDIA. This system and its incremental learning part have been implemented and tested on real data.

An existing neural network was improved and extended to graph classification. The obtained results are satisfying, but some work still needs to be done in order to improve the performance of the I2GNG. Two studies are being done. The first one concerns the automation of the choice of the maximum ages of edges and neurons. These parameters have an influence on the final results. The second study concerns $\alpha$ the choice of which can be done using some characteristics of the studied neuron. For example, we can use the density of a class, its entropy or other descriptors to get an adaptive $\alpha$.

We have demonstrated that the Case-Based Reasoning approach is natively adapted to heterogeneous document flows because each document is a case. The aim of the system is to organize and retrieve the different cases from an incremental knowledge base and moreover to adapt the previous knowledge to cope with a new document. The major concept of the adaptation approach comes from the observation that, specifically in the invoicing domain, the data (address, invoice date, amount) are often presented in few similar ways, in spite of each supplier editing and printing the invoices by its own feeling. Therefore, the 2-scales reasoning Global and Local approach demonstrates how it is possible to benefit from the redundancy of local structures to automatically model a whole new document.

This adaptation feature should reduce the time spent on modeling, which is one of the most important issues of Administration Documents Processing systems in the face of a large heterogeneous document flow.

In addition to invoices, this system can be applied directly to a very large majority of administrative documents in all domains: insurance, health, banking. We should recall that document processing targets the extraction of a finite set of indexes. These indexes are framed by business definitions obviously shared by the writer and the reader, otherwise there is no communication. Each writer can be free to organize the information set within a document and then introduce a large variety of documents. But the various ways to represent an index may converge, depending the accuracy of the business definition. Very formalized indexes will be very redundant (social security number for instance) and, on the other hand, indexes explained in a natural language text will support a batch of representation (that cannot be modeled by a structure).

Another perspective is to generalize the multi-scale CBR approach to manage the document flow structure. So far we have dealt only with documents, and supposed that the system input is the document. This supposition is true from a functional point of view. The end user handles each document separately and may define a data-extraction scope for each document. This is why all document analysis systems use the document scale as the system input scale. In fact the document is not the input scale but the output scale. The real input of the system is the result of the capture process: single page images. Using a fax machine or a desktop scanner, the user can digitize documents one by one. This makes sense when the user digitizes one or two documents, it makes no sense when production scanners which capture 1 000 to 10 000 pages per hour are used. Here the paper documents are turned into a flat sequence of page images. The challenge is then to segment the flat sequence into a structured document flow. The easiest solution is to introduce document separators (white sheets for instance) when preparing the batches to capture. This takes time and is paper-consuming. The other solution is document analysis to identify certain master documents in order to trigger the flow segmentation. For instance, a document starts at each page nb. 1 of an invoice or at each cheque. This solution supposes modeling of the flow sequence and the ability to identify before segmenting. To complete the complexity we should note here that the flow structure can be a multi-level tree: for instance the end users payment business document is a complex document composed of two more natural documents that are one (multi-page) invoice plus one cheque. We believe that we can expand the multi-scale CBR system in adding new CBR cycles to model the whole flow. We expect in this very global system to take advantage of redundancy in the documents and sub-structures at each level of the tree to enlarge the system scope and minimize the end users conception efforts.

Introducing the Case-Based Reasoning combined with Optical Character Recognition and Document Structure representation is a new step from Vision to Artificial Intelligence. For the last 40 years, the biggest challenge was to interpret the pixel into more symbolic information: characters, words, tables introducing more and more language models and semantic information to drive recognition. This pure recognition challenge is far from being solved, but the solution has reached a good enough level of quality to move slowly away from the document analysis scope towards a more ambitious document understanding scope. We now do not just ask the computer to read information and follow a fixed analysis process, defined by a grammar or a technical template. We can now ask the future system to simply learn from samples, to build its own rules, to infer new knowledge and to memorize enough of it so that human assistance is no longer needed.

If you will allow us the following comparison: the document reader system was yesterday a little boy able to recognize some patterns. He can today re-use part of his experience and learn himself from read documents in order to infer all this information onto new documents. On our side we will continue to raise our little boy and we wish you to conceive many brothers.

# References

1. Grosicki, E., Carr, M., Brodin, J.-M., Geoffrois, E.: Results of the RIMES Evaluation Campaign for Handwritten Mail Processing. In: Int. Conf. on Document Analysis and Recognition (ICDAR) (2009)
2. Grosicki, E., El Abed, H.: ICDAR 2009 Handwriting Recognition Competition. In: 10th Int. Conf. on Document Analysis and Recognition, ICDAR (2009)
3. Bunke, H.: Recognition of cursive Roman handwriting - past, present and future. In: Int. Conf. on Document Analysis and Recognition (ICDAR 2003), vol. 1, pp. 448–459 (2003)
4. Lladós, J., Valveny, E., Sánchez, G., Martí, E.: Symbol recognition: Current advances and perspectives. In: Blostein, D., Kwon, Y.-B. (eds.) GREC 2001. LNCS, vol. 2390, pp. 104–127. Springer, Heidelberg (2002)
5. Chang, M., Chen, S.: Deformed trademark retrieval based on 2d pseudo-hidden markov model. Pattern Recognition 34, 953–967 (2001)
6. Tombre, K., Tabbone, S., Dosch, P.: Musings on symbol recognition. In: Liu, W., Lladós, J. (eds.) GREC 2005. LNCS, vol. 3926, pp. 23–34. Springer, Heidelberg (2006)
7. Krishnamoorthy: Syntactic segmentation and labeling of digitalized pages from technical journals. PAMI (1993)
8. Yamashita, A., Amano, T., Takahashi, I., Toyokawa, K.: A model based layout understanding method for the document recognition system. In: Int. Conf. on Document Analysis and Recognition 2003, ICDAR (1991)
9. Duygulu, P., Atalay, V.: A hierarchical representation of form documents for identification and retrieval. Int. Journal on Document Analysis and Recognition (IJDAR) 5(1), 17–27 (2002)
10. Mao, J., Abayan, M., Mohiuddin, K.: A model-based form processing sub-system. In: Int. Conf. on Pattern Recognition, ICPR (1996)
11. Sako, H., Seki, M., Furukawa, N., Ikeda, H., Imaizumi, A.: Form reading based on form-type identification and form-data recognition. In: Int. Conf. on Document Analysis and Recognition 2003 (ICDAR), Scotland (2003)
12. Ting, A., Leung, M.K.H.: Business form classification using strings. In: 13th International Conference on Pattern Recognition (ICPR), p. 690. IEEE Computer Society, Washington, DC, USA (1996)
13. Hroux, P., Diana, S., Ribert, A., Trupin, E.: Etude de methodes de classification pour l'identification automatique de classes de formulaires. In: Int. Francophone Conferenceon Writing and Document Analysis, CIFED (1998)
14. Ishitani, Y.: Model based information extraction and its application to document Images. In: Int. Workshop on Digital Library and Image Analysis, DLIA (2001)
15. Cesarini, F., Gori, M., Marinai, S., Soda, G.: Informys: A flexible invoice-like form-reader system. IEEE Trans. Pattern Anal. Mach. Intell. 20(7), 730–745 (1998)
16. Belad, A., Belad, Y., Valverde, L.N., Kebairi, S.: Adaptive Technology for Mail-Order Form Segmentation. In: Int. Conf. on Document Analysis and Recognition (ICDAR), Seattle, USA, pp. 689–693 (2001)
17. Wahl, F., Wong, K., Casey, R.: Block segmentation and text extraction in mixed text/image documents. Graphical Models and Image Processing 20 (1982)
18. Nagy, G., Seth, S., Viswanathan, M.: A prototype document image analysis system for technical journals. Computer 25 (1992)
19. Pavlidis, T., Zhou, J.: Page segmentation and classification. Graphical Models and Image Processing 54 (1992)

20. Sako, H., Seki, M., Furukawa, N., Ikeda, H., Imaizumi, A.: Form reading based on form-type identification and form-data recognition. In: Int. Conf. on Document Analysis and Recognition (ICDAR), Scotland (2003)

21. Laroum, S., Bchet, N., Roche, M., Hamza, H.: Hybred: An OCR document representation for classification tasks. International Journal on Data Engineering and Management (2009)

22. Zhong, S.: Efficient online spherical k-means clustering. In: Proceedings IEEE of the International Joint Conference on Neural Networks, IJCNN 2005, Montreal, Canada, July 30 - August 4, pp. 3180–3185 (2005)

23. Vapnik, V., Chervonenkis, A.: A note on one class of perceptrons. Automation and Remote Control 25 (1964); SVM & Boosting

24. Bartlett, P., Shawe-Taylor, J.: Generalization performance of support vector machines and other pattern classifiers. In: Scholkopf, B., Burges, C.J.C., Smola, A.J. (eds.) Advances in Kernel Methods Support Vector Learning, pp. 43–54. MIT Press, Cambridge (1999)

25. Kim, J., Le, D.X., Thoma, G.R.: Automated labeling in document images. In: Document Recognition and Retrieval VIII (2001)

26. Cesarini, F., Marinai, S., Sarti, L., Soda, G.: Trainable table location in document images. In: Int. Conf. on Pattern Recognition (ICPR), vol. 3, pp. 236–240 (2002)

27. Coüasnon, B.: Dealing with noise in DMOS, a generic method for structured document recognition: An example on a complete grammar. In: Lladós, J., Kwon, Y.-B. (eds.) GREC 2003. LNCS, vol. 3088, pp. 38–49. Springer, Heidelberg (2004)

28. Coasnon, B.: Dmos, "a generic document recognition method: application to table structure analysis in a general and in a specific way. Int. Journal on Document Analysis and Recognition 8(2-3), 111–122 (2006)

29. Conway, A.: Page grammars and page parsing: A syntatic approach to document layout recognition. In: Int. Conf. on Document Analysis and Recognition, ICDAR (1993)

30. Niyogi, D., Srihari, S.N.: Knowledge-based derivation of document logical structure. In: Int. Conf. on Document Analysis and Recognition, ICDAR (1995)

31. Dengel, A., Dubiel, F.: Computer understanding of document structure. IJIST (1996)

32. Amano, A., Asada, N.: Graph Grammar Based Analysis System of Complex Table Form Document. In: Int. Conf. on Document Analysis and Recognition (ICDAR) (2003)

33. Sainz Palmero, G.I., Cano Izquierdo, J.M., Dimitriadis, Y.A., Lopez, J.: A new neuro-fuzzy system for logical labeling of documents. Pattern Recognition (1996)

34. LeBourgeois, F., Souafi-Bensafi, S., Duong, J., Parizeau, M., Cotc, M., Emptoz, H.: Using statistical models in document images understanding. In: DLIA (2001)

35. Rangoni, Y., Belad, A.: Data Categorization for a Context Return Applied to Logical Document Structure Recognition. In: Int. Conf. on Document Analysis and Recognition (ICDAR) (2005)

36. Rangoni, Y., Belad, A.: Data Categorization for a Context Return Applied to Logical Document Structure Recognition. In: Int. Conf. on Document Analysis and Recognition, ICDAR (2005)

37. Sainz Palmero, G.I., Cano Izquierdo, J.M., Dimitriadis, Y.A., Lopez, J.: A new neuro-fuzzy system for logical labeling of documents. Pattern Recognition (1996)

38. Hamza, H., Belaïd, Y., Belaïd, A.: Case-based reasoning for invoice analysis and recognition. In: Weber, R.O., Richter, M.M. (eds.) ICCBR 2007. LNCS (LNAI), vol. 4626, pp. 404–418. Springer, Heidelberg (2007)

39. Aamodt, A., Plaza, E.: Case-based reasoning: Foundational issues, methodological variations, and system approaches. IOS Press, Amsterdam (1994)

40. Burke, R., Hammond, K., Kozlovsky, J.: Knowledge-based information retrieval from semistructured text (1995)

41. Kolodner, J.: Maintaining organization in a dynamic long-term memory. Cognitive Science (1983)

42. Watson, I., Marir, F.: Case-based reasoning: A review 9, 355–381 (1994)

43. Hamza, H., Belaïd, Y., Belaïd, A.: Case-based reasoning for invoice analysis and recognition. In: Weber, R.O., Richter, M.M. (eds.) ICCBR 2007. LNCS (LNAI), vol. 4626, pp. 404–418. Springer, Heidelberg (2007)

44. Fritzke, B.: Growing cell structuresa self-organizing network for unsupervised and supervised learning. Neural Networks 7(9), 1441–1460 (1994)

45. Hodge, V.J., Austin, J.: Hierarchical growing cell structures: Treegcs. Knowledge and Data Engineering 13(2), 207–218 (2001)

46. Gunter, S., Bunke, H.: Self-organizing map for clustering in the graph domain. Pattern Recognition Letters 23(4), 405–417 (2002)

47. Prudent, Y., Ennaji, A.: A new learning algorithm for incremental self-organizing maps. In: ESANN, p. 712 (2005)

48. Lopresti, D.P., Wilfong, G.T.: A fast technique for comparing graph representations with applications to performance evaluation. IJDAR 6(4), 219–229 (2003)

# Automatic Document Layout Analysis through Relational Machine Learning

Stefano Ferilli, Teresa M.A. Basile, Nicola Di Mauro, and Floriana Esposito

**Abstract.** The current spread of digital documents raised the need of effective content-based retrieval techniques. Since manual indexing is infeasible and subjective, automatic techniques are the obvious solution. In particular, the ability of properly identifying and understanding a document's structure is crucial, in order to focus on the most significant components only. At a geometrical level, this task is known as Layout Analysis, and thoroughly studied in the literature. On suitable descriptions of the document layout, Machine Learning techniques can be applied to automatically infer models of classes of documents and of their components. Indeed, organizing the documents on the grounds of the knowledge they contain is fundamental for being able to correctly access them according to the user's needs.

Thus, the quality of the layout analysis outcome biases the next understanding steps. Unfortunately, due to the variety of document styles and formats, the automatically found structure often needs to be manually adjusted. We propose the application of supervised Machine Learning techniques to infer correction rules to be applied to forthcoming documents. A first-order logic representation is suggested, because corrections often depend on the relationships of the wrong components with the surrounding ones. Moreover, as a consequence of the continuous flow of documents, the learned models often need to be updated and refined, which calls for incremental abilities. The proposed technique, embedded in a prototypical version of the document processing system DOMINUS, using the incremental first-order logic learner INTHELEX, revealed good performance in real-world experiments.

Stefano Ferilli · Teresa M.A. Basile · Nicola Di Mauro · Floriana Esposito
Dipartimento di Informatica – University of Bari (Italy)
e-mail: {ferilli,basile,ndm,esposito}@di.uniba.it

M. Biba and F. Xhafa (Eds.): Learning Structure and Schemas from Documents, SCI 375, pp. 73–96.

# 1   Introduction

The current spread of documents available in digital format raised the need of effective retrieval techniques based on their content. Since manual indexing is infeasible due to the amount of documents to be handled and to the subjectivity of experts' judgements, automatic techniques are the obvious solution. In particular, a key factor for such techniques to be successful is represented by the ability of properly identifying and understanding the structure of documents, in order to focus on the most significant components only. The task aimed at identifying the geometrical structure of a document is known as Layout Analysis, and represents a wide area of research in document processing, for which several solutions have been proposed in the literature. However, they are mostly based on statistical and numerical approaches that may fail in classification and learning, being not able to deal with the lack of a strict layout regularity in the variety of documents available. On the other hand, on suitable descriptions of the document layout, Machine Learning techniques can be applied to automatically infer models of classes of documents and of their components. Indeed, organizing the documents on the grounds of the knowledge they contain is fundamental for being able to correctly access them according to the user's particular needs. For instance, in the scientific papers domain, in order to identify the subject of a paper and its scientific context, an important role is played by the information available in components such as Title, Authors, Abstract and Bibliographic references.

Thus, the quality of the layout analysis outcome is crucial, because it determines and biases the quality of the next understanding steps. Unfortunately, the variety of document styles and formats to be processed makes the layout analysis task a non-trivial one, so that the automatically found structure often needs to be manually fixed by domain experts. To this aim, in this work we propose the application of Machine Learning techniques to infer models of correction rules for wrong document layouts from sample corrections performed by expert users, in order to automatically apply them to future incoming documents. This task requires a first-order logic representation, because the structure of a document is not fixed, and the corrections typically depend on the relationships of the wrong components with the surrounding ones. Moreover, as a consequence of the continuous flow of new and different documents, the learned models often need to be updated and refined, which calls for incremental abilities of the system. Indeed, the layout correction setting prevents the possibility of defining and fixing the number of correction typologies since the beginning of layout analysis step, as it is not possible to foresee how many and what kinds of corrections have to be performed in order to obtain a satisfactory document layout. Hence the need of incremental abilities of the approach in order to be able to deal with totally new layout correction instances.

This chapter proposes a technique that was actually embedded in a prototypical version of the document processing system DOMINUS, using the incremental first-order logic learner INTHELEX. Experiments in a real-world task confirmed the good performance of the proposed solution. The chapter is organized as follows: Sections 2 and 3 introduce the background in which the proposed solution is intended to work; Sections 4 and 5 present the details of the proposal; Section 6 discusses the experimental evaluation of the technique; lastly, Section 7 concludes the work.

## 2 Related Work

Document Image Understanding (*DIU*) is the process aimed at transforming the informative content of a (paper or digital) document into an electronic format outlining its logical content. DIU strongly relies on a preliminary Document Image Analysis process, whose goal is to discover the geometrical and logical layout structure of the document. Specifically, the geometric layout analysis aims at producing a description of the geometric structure of the document, while the logical layout analysis aims at identifying the different logical roles of the detected regions (titles, paragraphs, captions, headings) and the relationships among them. In this work we focus our attention on the geometric layout analysis step, with a particular emphasis on the ability for a document analysis system to automatically fix the wrong behaviour of this phase.

The geometric layout analysis phase involves several processes, among which page decomposition. Several works concerning the page decomposition step are present in the literature, exploiting different approaches and having different objectives. Specifically, it is possible to distinguish text segmentation approaches, which analyse the document in order to extract and segment text. The textual part is divided into columns, paragraphs, lines, and words in order to reveal the hierarchical structure of the document. Page segmentation approaches aim at partitioning the document into homogeneous regions. They can be grouped into different typologies according to the technique they adopt: smoothing/smearing [1], projection profile analysis [2, 3, 4, 5, 6, 7], texture-based or local analysis [8, 9], and analysis of the background structure [10, 11]. On the other hand, there exist the Segmentation/Classification mixed approaches. These hybrid approaches do not clearly separate the segmentation step from the classification one. The techniques they exploit are based on connected component analysis [12, 13] and texture or local analysis [14, 15]. Finally, the block classification approaches aim at labelling regions previously extracted in a block segmentation phase. Most of them are based on feature extraction and linear discriminant classifiers [1, 3, 7, 16].

All these approaches can be viewed as relying on two basic operations: split and merge. Indeed, they all exploit the features extracted from an elementary block to decide whether splitting or merging two or more of the identified basic blocks in a top-down, bottom-up or hybrid approach to the page decomposition step.

One of the most representative top-down approaches is the recursive X–Y cut method [5] that relies on projection profiles to cut a textual region into several sub-regions. However, this approach has some difficulty in dealing with text regions that lack a fully extended horizontal or vertical cut. In the same way, approaches that exploit maximal white rectangles [17] or white streams [7] may also fail to find white margins that are large enough. For this reason, a multi-scale analysis method that examines a document at various scales was proposed [18].

Significant examples of the bottom-up behavior are the document spectrum method [19], the minimal-cost spanning tree method [13], and the component-based algorithm [20]. The underlying idea of these methods is to build basic blocks based on the distances between connected components. In particular, [21] used the spanning tree as a pre-classifier to gather connected components into sub-graphs. In this approach, one crucial step involves cutting away a vertical sub-graph that has been wrongly merged into a horizontal text line, or vice versa. Some empirically estimated heuristics are exploited to solve this problem. Other authors developed a rule-based bottom-up method [22] in which a set of rules is encoded, that allows to merge connected components in text lines. Specifically, the rules are based on the concept of nearest-neighbour connect-strength, which varies according to the size similarity, distance, and offset of the components. In [23] a hybrid segmentation method is proposed, that partitions a document into blocks based on field separators and white streams, and then merges the components of each block into text lines.

Since all methods split or merge blocks/components based on certain parameters, parameter estimation is crucial in layout analysis. All these methods exploit parameters that are able to model the split or merge operations in specific classes of the document domain. Few adaptive methods, where split or merge operations are performed using estimated parameter values, are present in the literature [24, 25]. A step forward is represented by the exploitation of Machine Learning techniques in order to automatically assess the parameters/rules able to perform the document page decomposition, and hence the eventual correction of the performed split/merge operations, without requiring an empirical evaluation on the specific document domain at hand. To this regard, learning methods have been used to separate textual areas from graphical areas [26] and to classify text regions as headline, main text, etc. [27, 28] or even to learn split/merge rules in order to carry out the corresponding operations and/or correction [29, 30].

However, a common limit of the above reported methods regards the consideration that they are all designed with the aim of working on scanned documents, and in some cases on documents of a specified typology, thus lacking any generality of the proposal with respect to the online available documents that can be of different digital formats. On the other hand, methods that work on natively digital documents assume that the segmentation phase can be carried out by simply performing a matching of the document itself with a standard template, even in this case, of a specified format.

## 3   Preliminaries

Based on the ODA/ODIF standard, any document can be progressively partitioned into a hierarchy of abstract representations, called its *layout structure*. Here we describe an approach, named DOC (Document Organization Composer) [31], for discovering a full layout hierarchy in digital documents based primarily on layout

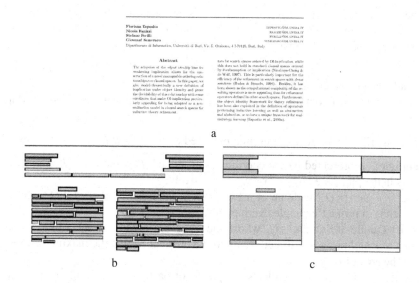

a

b          c

**Fig. 1** Layout analysis steps on the original document (a): preprocessing for world/line blocks aggregation (b) and final high-level layout structure (c).

information. The layout analysis process, whose main steps are shown in Figure 1, starts with a preprocessing step performed by a module that takes as input a generic digital document and extracts the set of its elementary layout components (*basic-blocks*), that will be exploited to identify increasingly complex aggregations of basic components.

The first step in the document layout analysis concerns the identification of rules to automatically shift from the basic digital document description to a higher level one. Indeed, the basic-blocks often correspond just to fragments of words (e.g., in PS/PDF documents), thus a preliminary aggregation based on their overlapping or adjacency is needed in order to obtain blocks surrounding whole words (*word-blocks*). Successively, a further aggregation of *word-blocks* could be performed to identify text lines (*line-blocks*). As to the grouping of blocks into lines, since techniques based on the mean distance between blocks proved unable to correctly handle cases of multi-column documents, Machine Learning approaches were applied in order to automatically infer rewriting rules that could suggest how to set some parameters in order to group together rectangles (words) to obtain lines. To do this, a kernel-based method was exploited to learn rewriting rules able to perform the bottom-up construction of the whole document starting from the basic/word blocks up to the lines. Specifically, such a learning task was cast to a Multiple Instance Problem and solved by exploiting the kernel-based algorithm proposed in [32].

The next step towards the discovery of the high-level layout structure of a document page consists in applying an improvement of the algorithm reported in [33]. To this aim, DOC analyzes the whitespace and background structure of each page in the document in terms of rectangular covers and identifies the white rectangles that are present in the page by decreasing area, thus reducing to the 'Maximal White Rectangle problem' as follows:

**Given**    a set of rectangular content blocks ('obstacles') $C = \{r_0, \ldots, r_n\}$, all placed inside the page rectangular contour $r_b$,

**Find**    a rectangle $r$ contained in $r_b$ whose area is maximal and that does not overlap any $r_i \in C$.

The algorithm exploits a representation based on the following data structures:

**Rectangle**    represented by the coordinates of its *Left-Top* and *Right-Bottom* corners in the plane, i.e. $(x_L, y_T, x_R, y_B)$ respectively;

**Bound**    a pair $(r, O)$ made up of a Rectangle $r$ and a Set of obstacles $O$, each of which is in turn a Rectangle overlapping $r$;

**Priority Queue** of Bounds    $Q$, whose elements are organized according to the area of $r$ and providing direct access to the element having the maximum value for the area of $r$;

**Set** of Rectangles    $B$, that collects the pieces of background as long as they are identified.

Elements are iteratively extracted from the queue: if the set of obstacles corresponding to the extracted element is empty, then it represents the maximum white rectangle still to be discovered, so it is saved into $B$ and added as an additional obstacle to all bounds in the current queue to which it overlaps; otherwise one of its obstacles is chosen as a *pivot* and the contour is consequently split into four regions (those resting above, below, to the right and to the left of the pivot), some of which partially overlap. Each such region, along with the obstacles that fall in it, represents a new structure to be inserted in the priority queue. In the end, computing the complement of the areas in $B$ would yield the document content blocks. The described procedure is formally specified in Algorithm 1.

---

**Algorithm 1.**    Maximal White Rectangle

---

1: $B \leftarrow \emptyset$
2: $Q \leftarrow \{(r_b, C)\}$
3: **while** $Q \neq \emptyset$ **do**
4:     $(r, O) \leftarrow$ dequeue(Q)
5:     **if** $O = \emptyset$ **then**
6:         $B \leftarrow B \cup \{r\}$
7:         Add $r$ as a new obstacle to all elements in $Q$ overlapping it
8:     **else**
9:         $p \leftarrow$ pivot(O)
10:        $Q \leftarrow Q \cup \{( (r.x_L, r.y_T, r.x_R, p.y_T), \{b \in O | b \cap r_A \neq \emptyset\} )\}$
11:        $Q \leftarrow Q \cup \{( (r.x_L, p.y_B, r.x_R, r.y_B), \{b \in O | b \cap r_B \neq \emptyset\} )\}$
12:        $Q \leftarrow Q \cup \{( (r.x_L, r.y_T, p.x_L, r.y_B), \{b \in O | b \cap r_L \neq \emptyset\} )\}$
13:        $Q \leftarrow Q \cup \{( (p.x_R, r.y_T, r.x_R, r.y_B), \{b \in O | b \cap r_R \neq \emptyset\} )\}$
14: **return** $B$

---

However, taking the algorithm to its natural end and then computing the complement would result again in the original basic blocks, while the layout analysis process aims at returning higher-level layout aggregates, such as single blocks filled with the same kind of content and rectangular *frames* (meaningful collections of objects completely surrounded by white space that may be made up of many blocks of the former type). This raised the problem of identifying a stop criterion to end this process. An empirical study carried out on a set of 100 documents of three different categories revealed that the best moment to stop the algorithm is when the ratio of the last white area retrieved with respect to the total white area in the current page of the document decreases up to 0, since before it the layout is not sufficiently detailed, while after it useless white spaces are found. Indeed, this is the point in which all the useful white spaces in the document, e.g. those between columns and sections, have been identified. Such a consideration is generally valid for all the documents except for those having a scattered appearance. It is worth noting that this value is reached very early (around 1/4 of the steps needed by the algorithm to reach its natural end), and before the size of the structure containing the blocks waiting to be processed starts growing dramatically, thus saving lots of time and space resources.

Additional improvements are:

- choosing as a pivot a 'side' block (i.e., the top or bottom or leftmost or rightmost block), or even better a 'corner' one (one which is at the same time both top or bottom and leftmost or rightmost), which results in a quicker retrieval of the background areas with respect to choosing it at random or in the middle of the bound;
- considering horizontal/vertical lines in the layout as a natural separators, and hence adding their surrounding white space to the background before the algorithm starts;
- discarding any white block whose height or width is below a given threshold as insignificant (this should avoid returning inter-word or inter-line spaces).

In the end, each black block should correspond to a section or column in the page, depending on the layout detail level. The lower the threshold, the more white spaces should be identified. Starting from these blocks, the frames can be found.

## 4  Learning Layout Correction Theories

The strategy for automatically assessing a threshold to stop the background retrieval loop allows to immediately reach a layout. Often the identified layout is already good for many further tasks of document image understanding. Nevertheless, such a threshold is clearly a tradeoff between several document types and shapes, and hence in some cases the layout needs to be slightly improved through a fine-tuning step that must be specific for each single document. A first, straightforward way to perform this adjustment is allowing the user to take the loop some additional steps forward with respect to the stop condition, adding more background, or some steps backward, removing the most recently found background. However, there are cases in which the greedy technique it implements does not guarantee an optimal outcome. For instance, retrieving useful (e.g., those that concur to separate different target frames in the document layout) but very small pieces of background (possibly due to excessive fragmentation operated by the algorithm) might require, using the normal priority-based behavior, to preliminary retrieve several meaningless ones as a side-effect. Even worse, some pieces of background might not be retrieved at all, being smaller than the minimum size threshold. To handle these cases, a tool was provided that allows the user to directly point out useful background fragments that were not yet retrieved from the queue and add them explicitly, or, conversely, to select useless ones that were erroneously retrieved and remove them from the final layout. Let us call these types of manual intervention '*white forcing*' and '*black forcing*', respectively. In the rest of the chapter, the following groups of terms will be considered as synonyms, referred to the document layout:

- white, background;
- black, foreground, content;
- contour, area, rectangle.

## 4.1 From Manual to Automatic Improvement of the Layout Correction

The forcing functionality allows the user to interact with the layout analysis algorithm and suggest which specific blocks are to be considered as background or content. To see how it can be obtained, let us recall that the algorithm, at each step, extracts from the queue a new area to be examined and can take three actions correspondingly:

1. if the contour is not empty, it is split and the resulting fragments are enqueued;
2. if the contour is empty and fulfils the constraints, it is added to the list of white areas;
3. if the contour is empty but does not fulfil the constraints, it is discarded.

Allowing the user to interact with the algorithm means modifying the algorithm behavior as a consequence of his choices. First of all, the white areas that were discarded because too small must be stored in a list. This would allow the user to retrieve any white space in the document page. Thus, since in general, due to the stop criterion, the algorithm terminates much before the queue is completely emptied, the following structures will be available as its outcome:

- a queue $Q$ of the contours yet to be examined, some of which actually white, others to be further split;
- a list $B$ of the white areas (background blocks) retrieved thus far;
- a list $D$ of the discarded white areas (background blocks) deemed as insignificant because not satisfying the constraints.

Now, the interactive extension allows the user to make the algorithm perform some more steps forward or backward, causing a corresponding movement of blocks among these three structures. For instance, when a step forward is requested, $Q$

is processed until the first white contour that satisfies the constraints is found and put in $B$. As already noticed, this functionality is still not sufficient, because significant but very small white rectangles might require several steps forward to be retrieved, causing the previous retrieval of several insignificant ones in the meantime. In the worst case, such significant rectangles might not be retrieved at all (they would be discarded if not satisfying the constraints), preventing a correct layout to be returned. In these cases, the user would directly 'force' the retrieval or removal of specific blocks from the document layout, by pointing with the mouse the corresponding area. If he wants to force the retrieval of a background piece (white forcing), all the white blocks in $Q$ and $D$ are scanned to find one underlying the pointed spot, and in case it is found it is added to the background $B$ (even if it does not satisfy the requirements). A technical trick is needed to properly manage these blocks in case of backward/forward steps by the user. Indeed, if put back into the queue by taking a step backward, the blocks would be placed at the bottom, and hence a subsequent step forward would not retrieve them immediately, as expected (being considered as the inverse of the backward step). To avoid this, they are assigned as priority the area of the rectangle placed in that moment at the top of the queue, instead of their actual area. This also solves the problem of losing meaningful whites. Conversely, the 'black forcing' functionality allows the user to retrieve and remove a white block from $B$ in order to include it in a frame, placing it in $D$. Both these operations can be undone, this way restoring the original algorithm behavior.

From the above discussion, it turns out that the relevance of a (white or black) block to the overall layout can be assessed based on its position inside the document page and its relationships with the other layout components. According to this assumption, each time the user applies a manual correction the information on his actions and on their effect can be stored in a logfile for subsequent analysis. In particular, each manual correction (user intervention) can be exploited as an example from which learning a model on how to classify blocks as meaningful or meaningless for the overall layout. For instance, non-forced white blocks can be considered as negative examples for '*black forcing*', while the discarded blocks and the non-forced blocks in the queue can be assumed to be negative examples for '*white forcing*'. Applying the learned models in subsequent incoming documents, it would be possible to automatically decide whether or not any white (resp., black) block is to be included as background (resp., content) in the final layout, this way reducing the need for user intervention.

## 4.2   Tool Architecture

The whole procedure of preliminary layout analysis and automatic layout correction, endowed with learning functionality to improve future performance, was implemented in the Java language. A Document is considered as an aggregate of Pages, each of which specifies six lists of components, one for each of the following kinds:

Box         a fragment of text as extracted from the source file (possibly a single letter or just a fragment of word);

Word      a set of overlapping or adjacent, and co-linear, Boxes;

Line        a set of co-linear Words belonging to the same text column;

Stroke    a graphical line in the document;

Fill         a solid rectangle in the document;

Image    a raster image extracted from the source document, of an aggregation of overlapping graphical items (such as Fills and Strokes);

and separately undergoes the layout analysis process as described before. This process results in two kinds of aggregate structures, namely blocks and frames, represented as their minimum bounding box. Boxes, words, lines, images, strokes, fills, blocks, frames and pages are described as generic rectangles, according to the coordinates of their top-left and bottom-right corners, plus a unique identifier. This way, the computation of areas, distances, overlappings etc. can be easily carried out as geometrical operations. More specifically, the following main functionality is provided for:

stepForward       applies a single step of the layout analysis algorithm, by processing the queued boundaries until the first white is found;

findLayout         repeatedly applies 'stepForward's until the pre-specified threshold is reached, and returns a contour representing the page, containing a set of background blocks ('whites') retrieved;

findInverse        returns a contour representing the page, containing a set of content blocks ('blacks') obtained as the complement of the background;

stepBackward     removes the last white found from the current layout reconstruction;

forceWhite        forces the retrieval of a white block, overriding the normal behavior of the findLayout algorithm;

forceBlack        forces the removal of a specific white block previously found, in order to consider it as content ('black');

undoForceWhite  withdraws the forcing of a white block;

undoForceBlack  withdraws the forcing of a black block.

Additional service functionality allows, given a coordinate of the page, to scan the queue in order to retrieve an underlying queued boundary, or to search for an underlying white block from the current background.

DOC also maintains the log of manual corrections performed by the expert user. After he acknowledges the final layout, a translation functionality is activated to transform all manual applied corrections for black and white blocks into positive examples, and to produce a corresponding set of negative examples, for the two classes 'force_white' and 'force_black'. Hence, on such a log file of manual corrections, a typical problem of supervised inductive learning is set up and, since the relationships between different objects are significant, a first-order logic description language is used to represent these logs. At this stage, a first-order logic learner is exploited. Specifically, here we adopt INTHELEX that was already successfully exploited for document image understanding tasks [34] and hence turns out to be a natural candidate for application. Indeed, the incremental characteristic of the system allows to exploit each manual correction to refine the learned theory, and hence contributes to progressively optimize the system behavior and avoid further user intervention.

## 4.3 The Learning System

INTHELEX is an Inductive Logic Programming [35] system that learns hierarchical first-order logic theories from positive and negative examples. It is fully incremental (in addition to the possibility of refining previously generated hypotheses/definitions, learning can also start from an empty theory), and adopts a representation language that ensures effectiveness of the descriptions and efficiency of their handling, while preserving the expressive power of the unrestricted case [36]. It can learn simultaneously multiple concepts/classes, possibly related to each other, and it guarantees validity of the learned theories on all the processed examples.

The system is able to exploit feedback on performance to activate the theory revision phase, as described in the following. An initial theory is set up for the target concepts, learned by the system from a previous set of examples selected from the environment and classified by an expert, or provided directly by the expert, or even empty. Subsequently, such a theory can be applied to new available observations, producing a decision. If an oracle is available, that compares such a decision to the correct one, whenever the prediction is incorrect, the cause of such a wrong decision can be automatically pointed out by the system and the proper kind of correction applied by a theory revision process. In this way, it is able to incrementally modify incorrect theories according to a data-driven strategy.

Specifically, when a positive observation is not covered, a revision of the theory to restore its completeness is performed as follows: replacing a rule in the theory with one of its least general generalizations against the problematic observation; adding a new rule to the theory, obtained by properly turning constants into variables in the problematic example; adding the problematic observation as a positive exception. On the other hand, when a negative observation is covered, the system revises the theory to restore consistency by performing one of the following actions: adding *positive* information able to characterize all the past positive observations (and exclude the problematic one) to the rule that covers the example; adding *negative* information to discriminate the problematic observation from all the past positive ones to the rule that covers the problematic observation; adding the problematic observation as a negative exception.

Another peculiarity in the learning system is the embedding of multistrategy operators that may help in solving the theory revision problem. Induction, Deduction, Abduction and Abstraction were integrated according to the Inferential Theory of Learning theoretical framework [37]. For the present study, abstraction, in particular, is a precious support to tackle the complexity of the domain and of the related descriptions. It is a pervasive activity in human perception and reasoning, and aims at removing superfluous details from the description of both the examples and the theory. Thus, the exploitation of abstraction results in the shift from the language in which the theory is described to a higher level one. According to the framework proposed in [38], in INTHELEX abstraction takes place by means of a set of operators that replace a number of components by a compound object, or decrease the granularity of a set of values, or ignore whole objects or just part of their features, or neglect the number of occurrences of some kind of object.

## 5  Description Language

Now let us turn to the way in which the manual corrections are to be described. The assumption is that the user changes the document layout when he considers that the proposed layout is wrong (by observing it *before* applying the correction), then he forces a specific block because he knows the resulting effect on the document (he foresees the situation *after* the correction) and considers it as satisfactory. Thus, to properly learn rules that can help in automatically fixing and improving the document layout analysis outcome, one must consider what is available before the correction takes place, and what will be obtained after it is carried out. For this reason, each example, representing a correction, will include a description of the blocks' layout both before and after the correction. However, the modification is typically *local*, i.e. it does not affect the whole document layout, but involves just a limited area surrounding the forced block. This allows to limit the description to just such an area. To sum up, the log file of the manual corrections, applied by the user after the execution of the layout analysis algorithm, will include both the white and the black blocks he forced, and will record, for each correction, information about the blocks and frames surrounding the forced block, both before and after the correction.

Each learning example is represented as a first-order logic clause $H : - B$, whose head reports whether the description in the body represents a positive or negative example for the forcing of a black or white block. Positive examples are denoted by the following predicates:

- `force_white(forced_block, document)`
- `force_black(forced_block, document)`

while negative examples are expressed as negations thereof. The clause body is built on the set of predicates reported in Table 1. It contains information about the page in which the correction took place, i.e. horizontal/vertical size and position in the overall document (whether it is at the beginning, in the middle or at the end of the document, and specifically whether it is the first or last one), furthermore it describes the forced block and the layout situation both before and after the correction. Specifically, for a given correction *id_correction*, to denote these two moments, corresponding identifiers, *id_correction_*before and *id_correction_*after, are generated and introduced by two specific literals, respectively:

- `before(idCorrection, idCorrection_before)`
- `after(idCorrection, idCorrection_after)`

**Table 1** First-order logic descriptors for layout correction observations: Attributes (bold) and Relations (italic) to be applied to entities (Document, Page, Block, Frame)

| Correction identification | |
|---|---|
| *before*(idCorrection,idCorrection_before) | identifies the layout description before correction |
| *after*(idCorrection,idCorrection_after) | identifies the layout description after correction |
| **Page General Information** | |
| **page**(idDocument,idPage) | correction applied to page idPage in document idDocument |
| **page_height**(idPage,*h*) | *h* is the height of page idPage |
| **page_width**(idPage,*w*) | *w* is the width of page idPage |
| **first_page**(idPage) | idPage is the first page of the document |
| **last_page**(idPage) | idPage is the last page of the document |
| **first_pages**(idPage) | idPage belongs to the first 1/3 pages of the doc |
| **middle_pages**(idPage) | idPage belongs to the middle 1/3 pages of the doc |
| **last_pages**(idPage) | idPage belongs to the last 1/3 pages of the doc |
| **Rectangle (Block or Frame) General Information (% wrt the page dimensions)** | |
| **pos_x**(idRectangle,*x*) | horizontal position *x* of the centroid of rectangle |
| **pos_y**(idRectangle,*y*) | vertical position *y* of the centroid of rectangle |
| **width**(idRectangle,*w*) | width *w* of rectangle idRectangle |
| **height**(idRectangle,*h*) | height *h* of rectangle idRectangle |
| **Rectangle Typology** | |
| *frame*(idPage,idFrame) | for each frame idFrame that touches or overlaps the forced block |
| *block*(idPage,idBlock) | for each block idBlock that touches or overlaps the forced block |
| **type**(idRectangle,Type) | Type $\in$ {text, line, image, mixed} |
| **Relationships between Rectangles (Blocks or Frames)** | |
| *belongs*(idBlock,idFrame) | block idBlock belongs to frame idFrame |
| *overlaps*(idForcedBlock,idBlock, *p*) | *p* is the percentage of overlapping between the two blocks idForcedBlock and idBlock |
| *touches*(idForcedBlock, idRectangle) | rectangle idRectangle touches (but does not overlap) the forced block idForcedBlock |
| *overlap_part_N*(idRectangle1, idRectangle2) | spatial relation ($N \in \{1,\ldots,25\}$) between idRectangle1 and idRectangle2 according to the 25-plane model; specifically, between a block or frame and the forced block, or between two frames involved in the correction being described (each of which touches or overlaps the forced block), or between two blocks in the same frame (each of which touches the forced block) |

The description of each of the two situations (before and after the correction) is based on literals expressing the page layout. While the predicates, on which such literals are built, are mostly the same that can be used for document image understanding purposes, for the specific task aimed at learning correction rules it is

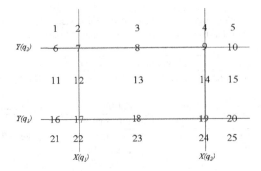

**Fig. 2** Partition of the plane with respect to a rectangle according to [39]

sufficient to express only a subset of the whole layout, and specifically the neighborhood of the forced block that is considered as the most significant and useful to understand why that block is being forced. In our case, the focus was put on the blocks and frames surrounding the forced block, and, among them, only on those touching or overlapping the forced block. The relationships between these components are described by means of a set of predicates representing the spatial relationships existing among all considered frames and among all blocks belonging to the same frame, that touch or overlap the forced block. Furthermore, for each frame or block that touches the forced block, a literal specifying that they touch is introduced. Finally, for each block of a frame that overlaps the forced block, the percentage of overlapping is reported.

It is fundamental to completely describe the mutual spatial relationships among all involved elements. All, and only, the relationships between each block/frame and the forced blocks are expressed, but not their inverses (i.e., the relationships between the forced block and the block/frame in question). To this aim, the model proposed in [39] for representing the spatial relationships among the blocks/frames was considered. Specifically, according to such a model, once a rectangle is fixed, its plane is partitioned in 25 parts (as shown in Figure 2) and its spatial relationships to any other rectangle in the plane can be specified by simply listing the parts with which the other rectangle overlaps. From this basic representation, higher-level topological relations [40, 39], such as closeness, intersection and overlapping between rectangles can be automatically derived using deductive capabilities.

For instance, the following fragment of background knowledge could be provided to the learning system:

$$overlap\_part\_14(B1,B2) \land \neg overlap\_part\_13(B1,B2) \Rightarrow touch(B1,B2)$$
$$overlap\_part\_17(B1,B2) \land \neg overlap\_part\_13(B1,B2) \Rightarrow touch(B1,B2)$$
$$overlap\_part\_18(B1,B2) \land \neg overlap\_part\_13(B1,B2) \Rightarrow touch(B1,B2)$$
$$overlap\_part\_19(B1,B2) \land \neg overlap\_part\_13(B1,B2) \Rightarrow touch(B1,B2)$$

and, given a description involving two blocks $b1$ and $b2$, and including a literal $overlap\_part\_14(b2,b1)$, but not a literal $overlap\_part\_13(b2,b1)$, it would be able to automatically recognize that $touch(b2,b1)$.

Finally, each involved frame or block is considered as a rectangular area of the page, and described according to the following parameters:

- horizontal and vertical position of the rectangle centroid with respect to the top-left corner of the page (coordinates expressed as percentages of the page dimensions),
- height and width of the rectangle (expressed as percentages of the page dimensions), and
- content type (text, graphic, line).

Note that, as regards the relationships between frames and blocks, since only those frames that touch or overlap the forced block are taken into account, if the forced block touches or overlaps just one frame no instance of such relationships will be included; the same holds for the blocks in a frame that touch or overlap the forced block. This often happens when the correction significantly changes the layout structure, modifying the number of blocks or frames. To clearly distinguish the various corrections, they are identified by the document identifier, followed by the page number and lastly by the progressive block number in that page.

The example depicted in Figure 3 shows a case in which the user must force a white block (b2 in figure) to become black because it is interpreted as a spacing. The reference frame (f1 in figure) contains two other blocks, one of which (b1) overlaps the rectangle to be forced (b2), while the other (b3) touches it. After the correction, the situation is as follows: there are two frames (f2 and f3) that touch the forced rectangle, both containing a block (b4 and b5 respectively) that touches it. The corresponding clause is:

```
[not(force_black(b2,d1),force_white(b2,d1)]:-
      ... general information of the block to force b2
      before(d1,d1_before), page_1(d1_before,d1_before_p1),
      ... general information of the page
      frame(d1_before_p1,f1), overlaps(b2,f1,100%),
      .... general information of the frame f1
      // spatial relationships between blocks f1 and b2
      overlap_part_2(f1,b2), overlap_part_3(f1,b2),
      overlap_part_4(f1,b2), overlap_part_5(f1,b2),
      overlap_part_7(f1,b2), overlap_part_8(f1,b2),
      overlap_part_9(f1,b2), overlap_part_10(f1,b2),
      overlap_part_12(f1,b2), overlap_part_13(f1,b2),
      overlap_part_14(f1,b2), overlap_part_15(f1,b2),
      overlap_part_17(f1,b2), overlap_part_18(f1,b2),
      overlap_part_19(f1,b2), overlap_part_20(f1,b2),
      overlap_part_22(f1,b2), overlap_part_23(f1,b2),
      overlap_part_24(f1,b2), overlap_part_25(f1,b2),
      block(d1_before_p1,b3), belongs(b3,f1), touches(b2,b3),
      .... general information of the block b3
      block(d1_before_p1,b1), belongs(b1,f1),overlaps(b2,b1,10%),
      .... general infomation on the block b1
      ... spatial relationships between blocks b3 and b2
      ... spatial relationships between blocks b1 and b2
```

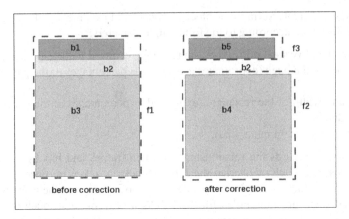

**Fig. 3** Example of a situation before and after the correction

```
... spatial relationships between blocks b3 and b1
after(d1,d1_after), page_1(d1_after,d1_after_p1),
frame(d1_after_p1,f2),  touches(b2,f2),
.... general infomation on the frame f2
block(d1_after_p1,b4), belongs(b4,f2),  touches(b2,b4) ,
.... general infomation on the block b4
frame(d1_after_p1,f3),  touches(b2,f3),
.... general infomation on the frame f3
block(d1_after_p1,b5), belongs(b5,f3),  touches(b2,b5) ,
.... general information on the block b5
... spatial relationships between blocks f2 and b2
... spatial relationships between blocks b4 and b2
... spatial relationships between blocks f3 and b2
... spatial relationships between blocks b5 and b2
... spatial relationships between blocks f2 and f3
```

## 6 Experiments

The proposed description language was used to run two experiments aimed at checking whether it is possible to learn a theory that can profitably automatize, at least partially, the layout correction process. Two target concepts were considered: 'force white' (corresponding to the fact that a block discarded or not yet retrieved by the layout analysis algorithm must be forced to belong to the background) and 'force black' (corresponding to the fact that a white rectangle found by the lauyout analysis algorithm must, instead, be discarded). In both cases, a 10-fold cross-validation technique was exploited to obtain the training and test sets. The experiments were run on a PC endowed with WindowsXP running on an Intel 2.4 GHz Core 2 Duo Processor and 2 GB RAM. INTHELEX was set so as to force each generalization to preserve not less than 40% of the original clause description and no more that 70%

thereof (computed as the number of literals in the body), in order to limit the search space and, as a consequence, the computational requirements.

The experimental dataset concerned the corrections applied to obtain the correct layout on about one hundred documents (specifically, papers published in scientific journals and conference proceedings), evenly distributed in four categories (ECAI, Elsevier, ICML, Springer-Verlag Lecture Notes). According to the strategy described above, the examples concern significant background blocks that were not retrieved ('white forcing') or useless white blocks erroneously considered as background ('black forcing') by the basic layout analysis algorithm. Since the layout analysis algorithm already treats isolated graphic lines as separators, it was necessary to force white blocks around such lines. The experimental evaluation was summarized in tables organized as follows. For each fold (specified in column *Fold*), the following figures are reported: the number of corresponding positive (*PosEx*) and negative (*NegEx*) training examples, the number of clauses in the learned theory (*NewCl*), the number of generalizations performed (*Lgg*), the number of positive exceptions introduced in the theory (*Pexc*), the number of positive (*Plit*) and negative (*Nlit*) literals added to specialize the theory, the number of negative exceptions introduced (*Nexc*), runtime in seconds (*Time*), predictive accuracy of the learned theories on the corresponding test sets (% in percentage), the ratio between the positive examples covered by the theory and the total number of positive examples ($TP = Pcov/TotExPos$) and, lastly, the ratio of rejected negative examples over the total number of negative examples ($TN = NNCov/TotExNeg$).

For the first dataset the layout correction activity resulted in a set of 786 examples of block correction, of which 263 for 'force white' and 523 for 'force black'. Positive examples for 'force white' were considered as negative for 'force black' and *vice versa*, this way exploiting the whole dataset. Thus, each single correction was interpreted from two perspectives: obviously as a positive example for the kind of forcing actually carried out by the user, and additionally as a negative example for the other kind of forcing. The head for a block forced_block forced white is as follows:

```
[
not(force_black(forced_block, document)),
force_white(forced_block, document)
]
```
while for a forced black is:
```
[
not(force_white(forced_block, document)),
force_black(forced_block, document)
]
```

Table 2 reports the results for concept 'force white'. Given the amount of available examples in the dataset, the experimental outcome reveals good predictive accuracy of the learned theories, that is 98,47% on average, never falling below 94.87%, and reaching 100% in 4 cases out of 10. Also runtimes are satisfactory: an average of 175.13 seconds (i.e., about 3 minutes), mostly due to a single problematic case that required 790.17 seconds (about 13 minutes) for accomplishment. INTHELEX added 1.4 positive literals and 1.5 negative ones on average during the specialization steps. The number of generalizations performed (16 on average) and the small number of clauses for each theory (5.4 on average) are good, because they indicate that the positive examples shared common features. A possible weakness is the number of negative exceptions (5 in the case of fold fb04), but a possible explanation is that the parameter that requires each generalization to preserve at most 70% of the description length tends to yield too general theories. Lastly, the ratios of correctly classified test examples (True Positives, TP, and True Negatives, TN, respectively) reveals that the various theories never failed on more than two negative examples, indicating a cautious behavior, as desired. In general, it might happen that too cautious theories cover too few positive examples, and hence might be useless, but the number of covered test examples (Pcov) shows this is not the case (the theories miss at most 4 thereof). In particular, the theory learned in fold fb05 might be selected as the best theory to be exploited, having predictive accuracy of 100%, just 4 clauses and no exception.

**Table 2** Experimental outcomes for concept 'force white'

| Fold | TP | TN | PosEx | NegEx | NewCl | Lgg | Pexc | Plit | Nlit | Nexc | Time | % |
|------|-----|-----|-------|-------|-------|-----|------|------|------|------|--------|-------|
| fb01 | 27/27 | 52/53 | 236 | 470 | 6 | 17 | 0 | 2 | 0 | 0 | 120.54 | 98.75 |
| fb02 | 26/27 | 53/53 | 236 | 470 | 7 | 16 | 0 | 2 | 4 | 0 | 271.89 | 98.75 |
| fb03 | 27/27 | 53/53 | 236 | 470 | 5 | 17 | 0 | 1 | 1 | 0 | 83.71 | 100 |
| fb04 | 26/26 | 50/52 | 237 | 471 | 6 | 21 | 0 | 2 | 6 | 5 | 790.17 | 97.44 |
| fb05 | 26/26 | 52/52 | 237 | 471 | 4 | 10 | 0 | 2 | 1 | 0 | 36.87 | 100 |
| fb06 | 22/26 | 52/52 | 237 | 471 | 6 | 19 | 0 | 0 | 0 | 0 | 171.74 | 94.87 |
| fb07 | 24/26 | 51/52 | 237 | 471 | 6 | 19 | 0 | 2 | 0 | 0 | 72.84 | 96.15 |
| fb08 | 25/26 | 52/52 | 237 | 471 | 4 | 11 | 0 | 1 | 1 | 0 | 46.55 | 98.72 |
| fb09 | 26/26 | 52/52 | 237 | 471 | 5 | 17 | 0 | 2 | 1 | 0 | 90.56 | 100 |
| fb10 | 26/26 | 52/52 | 237 | 471 | 5 | 13 | 0 | 0 | 1 | 0 | 68.12 | 100 |
| Average | | | | | 5.4 | 16 | 0 | 1.4 | 1.5 | 0.5 | 175.13 | 98.47 |
| Mean Dev. | | | | | 0.97 | 3.59 | 0 | 0.84 | 1.96 | 1.58 | 226.98 | 1.79 |
| Max | | | | | 7 | 21 | 0 | 2 | 6 | 5 | 790.17 | 100 |
| Min | | | | | 4 | 10 | 0 | 0 | 0 | 0 | 36.87 | 94.87 |

Table 3 reports the results for concept 'force black'. In this case, there were more positive examples than negative ones. The figures are again satisfactory, although slightly worse than those obtained for the previous concept. Indeed, the predictive accuracy of the various theories amounts to 97.82% on average, never falls below 94.87% and only in two cases reaches 100% (fn02 and fn03). Runtimes are worse than those of the previous experiment (394.41 seconds, i.e. about 6.5 minutes, on average). The number of generalizations carried out (17.6 on average) is quite similar for the previous concept. The number of negative exceptions is high (up to 13 in fold fn09), which means that there are negative examples (i.e., positive examples for 'force white') very similar to the positive ones, so that INTHELEX is not able to distinguish between them. However, it is encouraging that the best theories for predictive accuracy (fn02 and fn03) has very few (2 and 1 respectively). Lastly, by observing the ratio between covered examples and total examples during the test phase (Pcov and NNCov), one can note that the number of uncovered examples is quite stable. The best theory, among those produced by the ten folds, to be chosen for use seems to be that of fold fn03, that has 100% accuracy, just 2 clauses and one negative exception.

**Table 3** Experimental outcomes for concept 'force black'

| Fold | TP | TN | PosEx | NegEx | NewCl | Lgg | Pexc | Plit | Nlit | Nexc | Time | % |
|------|------|------|-------|-------|-------|------|------|------|------|------|--------|-------|
| fn01 | 52/53 | 27/27 | 470 | 236 | 4 | 12 | 0 | 2 | 0 | 0 | 85.62 | 98.75 |
| fn02 | 53/53 | 27/27 | 470 | 236 | 3 | 13 | 0 | 6 | 2 | 2 | 235.04 | 100 |
| fn03 | 53/53 | 27/27 | 470 | 236 | 2 | 8 | 0 | 2 | 2 | 1 | 105.34 | 100 |
| fn04 | 51/52 | 25/26 | 471 | 237 | 5 | 20 | 0 | 4 | 6 | 7 | 618.49 | 97.44 |
| fn05 | 51/52 | 25/26 | 471 | 237 | 8 | 30 | 0 | 7 | 1 | 0 | 309.46 | 97.44 |
| fn06 | 49/52 | 26/26 | 471 | 237 | 6 | 15 | 0 | 5 | 7 | 4 | 638.58 | 96.15 |
| fn07 | 50/52 | 24/26 | 471 | 237 | 7 | 26 | 0 | 4 | 0 | 1 | 257.05 | 94.87 |
| fn08 | 51/52 | 26/26 | 471 | 237 | 5 | 21 | 0 | 8 | 8 | 2 | 820.33 | 98.72 |
| fn09 | 51/52 | 25/26 | 471 | 237 | 2 | 13 | 0 | 3 | 4 | 13 | 462.44 | 97.44 |
| fn10 | 51/52 | 25/26 | 471 | 237 | 4 | 18 | 0 | 1 | 0 | 0 | 411.78 | 97.44 |
| Average | | | | | 4.6 | 17.6 | 0 | 4.2 | 3 | 3 | 394.41 | 97.82 |
| Mean Dev. | | | | | 2.01 | 6.79 | 0 | 2.3 | 3.06 | 4.14 | 241.88 | 1.61 |
| Max | | | | | 8 | 30 | 0 | 8 | 8 | 13 | 820.33 | 100 |
| Min | | | | | 2 | 8 | 0 | 1 | 0 | 0 | 85.62 | 94.87 |

The outcomes of the first experiment suggest that the description language proposed and the way in which the forcings are described are effective to let the system learn rules that can be successfully used for automatic layout correction. This suggested to try another experiment to simulate the actual behavior of such an automatic system, working on the basic layout analysis algorithm. Recall that, after finishing the execution of the layout analysis algorithm according to the required stop threshold, three queues are produced (the queued areas still to be processed, the white areas discarded because not satisfying the constraints and the white blocks selected as useful background). Among these, the last one contains white blocks that can be forced to black, while the other two contain rectangles that might be forced to white.

Since the rules needed by DOC to automatize the layout correction process must be able to evaluate each block in order to decide whether forcing it or not, it is not sufficient anymore to consider each white block forcing as a counterexample for black forcing and *vice versa*, but to ensure that all learned rules are correct, also all blocks in the document that have not been forced must be exploited as negative examples for the corresponding concepts. The adopted solution was to still express forcings as discussed above, including additional negative examples obtained from the layout configuration finally accepted by the user. Indeed, when the layout is considered correct, all actual white blocks that were not forced become negative examples for concept 'force black' (because they could be forced as black, but weren't), while all white blocks, discarded or still to be processed become negative examples for the concept 'force white' (because they weren't forced). The dataset for this experiment was obtained by running the layout analysis algorithm until the predefined threshold was reached, and then applying the necessary corrections to fix the final layout. The 36 documents considered were a subset of the former dataset, evenly distributed among categories (ICML, ECAI, Elsevier, Springer-Verlag Lecture Notes). Specifically, the new dataset included 113 positive and 840 negative examples for 'force black', and resulted in the performance reported in Table 4. The predictive accuracy was improved with respect to the previous experiment, reaching 99.16% on average, with several folds obtaining 100%, which is a very good result. The presence of very few negative exceptions (especially considering the number of examples), the presence of no more than 5 clauses in each theory along with the number of generalizations performed, suggest that the concept was properly learned and that the application of any of such theories to real cases may provide satisfactory results.

**Table 4** Experimental outcomes for concept 'force black' (no neg ex)

| Fold | TP | TN | PosEx | NegEx | NewCl | Lgg | Pexc | Plit | Nlit | Nexc | Time | % |
|---|---|---|---|---|---|---|---|---|---|---|---|---|
| fn01 | 12/12 | 84/84 | 101 | 756 | 4 | 12 | 0 | 0 | 1 | 0 | 77.81 | 100 |
| fn02 | 12/12 | 83/84 | 101 | 756 | 3 | 8 | 0 | 1 | 1 | 0 | 62.25 | 98.96 |
| fn03 | 12/12 | 84/84 | 101 | 756 | 4 | 11 | 0 | 1 | 0 | 0 | 57.73 | 100 |
| fn04 | 9/11 | 84/84 | 102 | 756 | 3 | 11 | 0 | 0 | 2 | 0 | 241.60 | 97.89 |
| fn05 | 11/11 | 84/84 | 102 | 756 | 3 | 9 | 0 | 0 | 0 | 1 | 88.52 | 100 |
| fn06 | 9/11 | 83/84 | 102 | 756 | 3 | 11 | 0 | 1 | 0 | 2 | 153.15 | 96.84 |
| fn07 | 11/11 | 84/84 | 102 | 756 | 4 | 12 | 0 | 0 | 1 | 0 | 94.16 | 100 |
| fn08 | 10/11 | 84/84 | 102 | 756 | 2 | 7 | 0 | 0 | 0 | 4 | 69.38 | 98.95 |
| fn09 | 11/11 | 84/84 | 102 | 756 | 5 | 14 | 0 | 4 | 1 | 0 | 264.99 | 100 |
| fn10 | 11/11 | 83/84 | 102 | 756 | 3 | 8 | 0 | 1 | 0 | 1 | 164.81 | 98.95 |
| Average | | | | | 3.4 | 10.3 | 0 | 0.8 | 0.6 | 0.8 | 127.44 | 99.16 |
| Mean Dev. | | | | | 0.84 | 2.21 | 0 | 1.23 | 0.7 | 1.32 | 75.70 | 1.09 |
| Max | | | | | 5 | 14 | 0 | 4 | 2 | 4 | 264.99 | 100 |
| Min | | | | | 2 | 7 | 0 | 0 | 0 | 0 | 57.73 | 96.84 |

As to the concept 'force white', the dataset was made up of 101 positive and 10046 negative examples. The large number of negative examples is due to the number of white blocks discarded or still to be processed being typically much greater than that of white blocks found. Since exploiting such a large number of negative examples might have significantly unbalanced the learning process, only a random subset of 843 such examples was selected, in order to keep the same ratio between positive and negative examples as for the 'force black' concept. The experiment run on such a subset provided the results shown in Table 5. The number of clauses in each theory is significantly larger than that resulting from the 'force black' experiment suggesting that the white forcing examples have less common features. This is most likely due to the intrinsic complexity of the current concept, that is more complex than the previous one because many different situations may lead to forcing a white, differently from blacks that are typically forced to remove indentations. Another explanation might be that the lower bound of 40% set for generalizations during the learning phase causes the production of theories that are not very general. The predictive accuracy of the various theories is very encouraging (98.10% on average, with peaks of 98.95%). This is due to the reduction of the set of negative examples, indeed the various theories fail more on positive examples than on negative ones.

**Table 5** Experimental outcomes for concept 'force white' (no neg ex)

| Fold | TP | TN | PosEx | NegEx | NewCl | Lgg | Pexc | Plit | Nlit | Nexc | Time | % |
|------|------|-------|-------|-------|-------|------|------|------|------|------|--------|-------|
| fb01 | 10/11 | 84/84 | 90 | 759 | 6 | 13 | 0 | 0 | 4 | 3 | 166.63 | 98.95 |
| fb02 | 9/10 | 85/85 | 91 | 758 | 9 | 20 | 0 | 2 | 2 | 1 | 249.31 | 98.95 |
| fb03 | 7/10 | 84/85 | 91 | 758 | 9 | 22 | 0 | 0 | 5 | 0 | 343.20 | 95.79 |
| fb04 | 8/10 | 84/85 | 91 | 758 | 7 | 14 | 0 | 1 | 6 | 2 | 375.85 | 96.84 |
| fb05 | 10/10 | 82/84 | 91 | 759 | 7 | 10 | 0 | 4 | 8 | 3 | 284.60 | 97.87 |
| fb06 | 9/10 | 84/84 | 91 | 759 | 10 | 19 | 0 | 1 | 3 | 1 | 335.25 | 98.94 |
| fb07 | 9/10 | 84/84 | 91 | 759 | 13 | 17 | 0 | 4 | 2 | 1 | 519.40 | 98.94 |
| fb08 | 10/10 | 83/84 | 91 | 759 | 8 | 14 | 0 | 0 | 3 | 2 | 163.16 | 98.94 |
| fb09 | 8/10 | 83/84 | 91 | 759 | 7 | 18 | 0 | 2 | 5 | 1 | 213.40 | 96.81 |
| fb10 | 9/10 | 84/84 | 91 | 759 | 9 | 13 | 0 | 2 | 2 | 5 | 525.50 | 98.94 |
| Average | | | | | 8.5 | 16 | 0 | 1.6 | 4 | 1.9 | 317.63 | 98.10 |
| Mean Dev. | | | | | 2.01 | 3.77 | 0 | 1.51 | 2 | 1.45 | 129.77 | 1.20 |
| Max | | | | | 13 | 22 | 0 | 4 | 8 | 5 | 525.50 | 98.95 |
| Min | | | | | 6 | 10 | 0 | 0 | 2 | 0 | 163.16 | 95.79 |

To confirm or reject this hypothesis, they were applied to the remaining negative examples previously discarded (9203), and the results (reported in Table 6) provide an ultimate confirmation of this: a predictive accuracy of 98.91% that, being obtained on so large a test set, is likely to correctly approximate the true behavior on real-world cases.

**Table 6** Experimental outcomes for the test of concept 'force white' on the discarded 9203 negative examples

| Fold | fb01 | fb02 | fb03 | fb04 | fb05 | fb06 | fb07 | fb08 | fb09 | fb10 | Av. | Max | Min |
|------|------|------|------|------|------|------|------|------|------|------|------|------|------|
| TN | 9089 | 9122 | 9153 | 9124 | 9077 | 9111 | 9059 | 9135 | 9084 | 9069 | | | |
| % | 98.76 | 99.12 | 99.46 | 99.14 | 98.63 | 99 | 98.44 | 99.26 | 98.71 | 98.54 | 98.91 | 99.46 | 98.44 |

# 7 Conclusions

The huge amount of documents available in digital form and the flourishing of digital repositories raise the need of effective retrieval techniques based on their contents. Automatic techniques able to properly identify and understand the structure of documents in order to focus on the most significant components only seem to be the most suitable solution as manual indexing is infeasible due to the amount of documents to be handled. Hence, the quality of the layout analysis outcome is crucial because it determines and biases the quality of the next understanding steps.

However, due to the variety of document styles and formats to be processed, the layout analysis task is a non-trivial one and often the automatically found structure needs to be manually fixed by domain experts. In this work we proposed a tool, embedded in a prototypical version of the document processing system DOMINUS, able to use the steps carried out by the domain expert with the aim of correcting the outcome of the layout analysis phase. Specifically, the tool is able to infer rules for the layout correction to be applied to future incoming documents. It makes use of a first-order logic representation of the document structure because corrections often depend on the relationships of the wrong components with the surrounding ones. Moreover, the tool exploits the incremental abilities of the system INTHELEX as the continuous flow of new and different documents requires the learned models to be updated and refined. Experiments in a real-world domain made up of scientific documents have been presented and discussed, showing the validity of the proposed approach.

# References

1. Wong, K.Y., Casey, R.G., Wahl, F.M.: Document analysis system. IBM Journal of Reserch and Development 26, 647–656 (1982)
2. Nagy, G., Seth, S.: Hierarchical representation of optically scanned documents. In: Proceedings of the 7th International Conference on Pattern Recognition, pp. 347–349. IEEE Computer Society Press, Los Alamitos (1984)
3. Wang, D., Srihari, S.N.: Classification of newspaper image blocks using texture analysis. Computer Vision, Graphics, and Image Processing 47, 327–352 (1989)
4. Nagy, G., Seth, S., Viswanathan, M.: A prototype document image analysis system for technical journals. Computer 25, 10–22 (1992)
5. Krishnamoorthy, M., Nagy, G., Seth, S., Viswanathan, M.: Syntactic segmentation and labeling of digitized pages from technical journals. IEEE Transactions on Pattern Analysis and Machine Intelligence 15, 737–747 (1993)
6. Sylwester, D., Seth, S.: A trainable, single-pass algorithm for column segmentation. In: Procedings of International Conference on Document Analysis and Recognition, vol. 2, pp. 615–618. IEEE Computer Society Press, Los Alamitos (1995)
7. Pavlidis, T., Zhou, J.: Page segmentation and classification. CVGIP: Graphical Models Image Process. 54, 484–496 (1992)
8. Jain, A.K., Bhattacharjee, S.: Text segmentation using gabor filters for automatic document processing. Machine Vision and Applications 5, 169–184 (1992)

9. Tang, Y.Y., Ma, H., Mao, X., Liu, D., Suen, C.Y.: A new approach to document analysis based on modified fractal signature. In: Procedings of International Conference on Document Analysis and Recognition, vol. 2, pp. 567–570. IEEE Computer Society Press, Los Alamitos (1995)

10. Normand, N., Viard-Gaudin, C.: A background based adaptive page segmentation algorithm. In: ICDAR 1995: Proceedings of the Third International Conference on Document Analysis and Recognition, vol. 1, pp. 138–141. IEEE Computer Society Press, Los Alamitos (1995)

11. Kise, K., Yanagida, O., Takamatsu, S.: Page segmentation based on thinning of background. In: ICPR 1996: Proceedings of the International Conference on Pattern Recognition (ICPR 1996), vol. III, 7276, pp. 788–792. IEEE Computer Society Press, Los Alamitos (1996)

12. Wang, S.-Y., Yagasaki, T.: Block selection: a method for segmenting a page image of various editing styles. In: ICDAR 1995: Proceedings of the Third International Conference on Document Analysis and Recognition, vol. 1, pp. 128–133. IEEE Computer Society Press, Los Alamitos (1995)

13. Simon, A., Pret, J.-C., Johnson, A.P.: A fast algorithm for bottom-up document layout analysis. IEEE Transactions on Pattern Analysis and Machine Intelligence 19, 273–277 (1997)

14. Sauvola, J., Pietikainen, M.: Page segmentation and classification using fast feature extraction and connectivity analysis. In: ICDAR 1995: Proceedings of the Third International Conference on Document Analysis and Recognition, vol. 2, pp. 1127–1131. IEEE Computer Society Press, Los Alamitos (1995)

15. Jain, A.K., Zhong, Y.: Page segmentation using texture analysis. Pattern Recognition 29, 743–770 (1996)

16. Shih, F.Y., Chen, S.S.: Adaptive document block segmentation and classification. IEEE Transactions on Systems, Man, and Cybernetics 26, 797–802 (1996)

17. Ittner, D., Baird, H.: Language-free layout analysis. In: ICDAR 1993: Proceedings of the Second International Conference on Document Analysis and Recognition, vol. 1, pp. 336–340. IEEE Computer Society Press, Los Alamitos (1993)

18. Lee, S.W., Ryu, D.S.: Parameter-free geometric document layout analysis. IEEE Transactions on Pattern Analysis and Machine Intelligence 23, 1240–1256 (2001)

19. O'Gorman, L.: The document spectrum for page layout analysis. IEEE Transactions on Pattern Analysis and Machine Intelligence 15, 1162–1173 (1993)

20. Liu, F.: A new component based algorithm for newspaper layout analysis. In: ICDAR 2001: Proceedings of the Sixth International Conference on Document Analysis and Recognition, pp. 1176–1180. IEEE Computer Society Press, Washington, DC, USA (2001)

21. Xi, J., Hu, J., Wu, L.: Page segmentation of chinese newspapers. Pattern Recognition 35, 2695–2704 (2002)

22. Chen, M., Ding, X., Liang, J.: Analysis, understanding and representation of chinese newspaper with complex layout. In: Proceedings of the 2000 International Conference on Image Processing (ICIP), pp. 90–93. IEEE Computer Society Press, Los Alamitos (2000)

23. Okamoto, M., Takahashi, M.: A hybrid page segmentation method. In: Proceedings of the Second International Conference on Document Analysis and Recognition, pp. 743–748. IEEE Computer Society Press, Los Alamitos (1993)

24. Liu, J., Tang, Y.Y., Suen, C.Y.: Chinese document layout analysis based on adaptive split-and-merge and qualitative spatial reasoning. Pattern Recognition 30, 1265–1278 (1997)

25. Chang, F., Chu, S.Y., Chen, C.Y.: Chinese document layout analysis using adaptive re-grouping strategy. Pattern Recognition 38, 261–271 (2005)
26. Etemad, K., Doermann, D., Chellappa, R.: Multiscale segmentation of unstructured document pages using soft decision integration. IEEE Transactions on Pattern Analysis and Machine Intelligence 19, 92–96 (1997)
27. Dengel, A., Dubiel, F.: Computer understanding of document structure. International Journal of Imaging Systems and Technology 7, 271–278 (1996)
28. Laven, K., Leishman, S., Roweis, S.: A statistical learning approach to document image analysis. In: ICDAR 2005: Proceedings of the Eighth International Conference on Document Analysis and Recognition, pp. 357–361. IEEE Computer Society Press, Los Alamitos (2005)
29. Malerba, D., Esposito, F., Altamura, O., Ceci, M., Berardi, M.: Correcting the document layout: A machine learning approach. In: ICDAR 2003: Proceedings of the Seventh International Conference on Document Analysis and Recognition, pp. 97–103. IEEE Computer Society Press, Los Alamitos (2003)
30. Wu, C.C., Chou, C.H., Chang, F.: A machine-learning approach for analyzing document layout structures with two reading orders. Pattern Recogn. 41, 3200–3213 (2008)
31. Esposito, F., Ferilli, S., Basile, T.M.A., Di Mauro, N.: Machine Learning for digital document processing: from layout analysis to metadata extraction. In: Marinai, S., Fujisawa, H. (eds.) Machine Learning in Document Analysis and Recognition. SCI, vol. 90, pp. 105–138. Springer, Heidelberg (2008)
32. Dietterich, T.G., Lathrop, R.H., Lozano-Perez, T.: Solving the Multiple Instance Problem with axis-parallel rectangles. Artificial Intelligence 89, 31–71 (1997)
33. Breuel, T.M.: Two geometric algorithms for layout analysis. In: Lopresti, D.P., Hu, J., Kashi, R.S. (eds.) DAS 2002. LNCS, vol. 2423, pp. 188–199. Springer, Heidelberg (2002)
34. Esposito, F., Ferilli, S., Fanizzi, N., Basile, T.M.A., Di Mauro, N.: Incremental multistrategy learning for document processing. Applied Artificial Intelligence Journal 17, 859–883 (2003)
35. Muggleton, S., Raedt, L.D.: Inductive logic programming: Theory and methods. Journal of Logic Programming 19/20, 629–679 (1994)
36. Semeraro, G., Esposito, F., Malerba, D., Fanizzi, N., Ferilli, S.: A logic framework for the incremental inductive synthesis of datalog theories. In: Fuchs, N.E. (ed.) LOPSTR 1997. LNCS, vol. 1463, pp. 300–321. Springer, Heidelberg (1998)
37. Michalski, R.S.: Inferential Theory of Learning. Developing foundations for Multistrategy Learning. In: Michalski, R., Tecuci, G. (eds.) Machine Learning. A Multistrategy Approach, vol. IV, pp. 3–61. Morgan Kaufmann, San Francisco (1994)
38. Zucker, J.D.: Semantic abstraction for concept representation and learning. In: Proceedings of the 4th International Workshop on Multistrategy Learning (MSL), pp. 157–164 (1998)
39. Papadias, D., Theodoridis, Y.: Spatial relations, minimum bounding rectangles, and spatial data structures. International Journal of Geographical Information Science 11, 111–138 (1997)
40. Egenhofer, M.J.: Reasoning about binary topological relations. In: Günther, O., Schek, H.-J. (eds.) SSD 1991. LNCS, vol. 525, pp. 143–160. Springer, Heidelberg (1991)

# Dataspaces: Where Structure and Schema Meet

Maurizio Atzori* and Nicoletta Dessì

**Abstract.** In this chapter we investigate the crucial problem that poses the bases to the concept of dataspaces: the need for human interaction/intervention in the process of organizing (getting the structure of) unstructured data. We survey the existing techniques behind dataspaces to overcome that need, exploring the structure of a dataspace along three dimensions: *dataspace profiling, querying and searching* and *application domain*. We will further explore existing projects focusing on dataspaces, induction of data structure from documents, and data models where data schema and documents structure overlaps will be reviewed, such as Apache Hadoop, Cassandra on Amazon Dynamo, Google BigTable model and other DHT-based flexible data structures, Google Fusion Tables, iMeMex, U-DID, WebTables and Yahoo! SearchMonkey.

## 1 Introduction

Data integration has emerged over the last few years as a challenge to improving search in vast collections of structured data that yield heterogeneity at scale unseen before. Current information systems and IT infrastructures are mainly based on the exchange of strongly-structured data and on well-established standards (database, XML files and other known data formats). Nevertheless, enterprise and personal data handled everyday are mostly unstructured (estimates range from 80 to 95%), i.e., their contents do not follow any rigid schema or format (e.g., text files or bitmap images), therefore not

Maurizio Atzori · Nicoletta Dessì
University of Cagliari
e-mail: `atzori@unica.it,dessi@unica.it`

* The work of Dr. Atzori has been done within the project *Unstructured Data Integration for Dataspaces (U-DID)* founded by "RAS PO Sardegna FSE 2007-2013 L.R.7/2007".

M. Biba and F. Xhafa (Eds.): Learning Structure and Schemas from Documents, SCI 375, pp. 97–119.
springerlink.com                                    © Springer-Verlag Berlin Heidelberg 2011

allowing complex queries or strong integration within automatic enterprise business processes or workflows. Data in enterprise computer systems (email, report, web pages, customers and supplier records) [1], in social networks [2] and in personal computers (documents, images, videos, chats, short messages, favourite web pages, appointments) [3] are often unused, or even forgotten, because of the weak support this generation of IT infrastructures (mainly based on databases management systems) offer for non-structured data.

Improving data integration in such heterogeneous information spaces leads to a fundamental question: the presence of a plethora of structures limits or makes too difficult the definition of some kind of global structure? In other words, are traditional approaches for engineering data over selected resources (i.e., mediator architectures and XML-based solutions) the only possible way for data integration or can we extend the data structuring concepts to improve data integration in such contexts?

The research community has recently proposed the concept of *dataspace* [4, 5] as a new scenario for structuring information relevant to a particular organization, regardless of its format and location, and capturing a rich collection of relationships between them. The elements of a dataspace [5] are a set of participants (i.e., individual data sources) and a set of relations denoting the relation in which the participants are. In this sense, a dataspace is an abstraction of databases that does not require data to be structured (i.e., in tabular form), with a minimal "off-the-shelf" set of search functions based on keywords. The key idea is to enhance the quality of data integration and the semantic meaning of information without an *a priori* schema for the data sources [6, 7]. Advanced DBMS-like functions, queries and mappings are provided over time by different components, each defining relationships among data when required. Integrated views over a set of data sources are provided following the so-called *pay-as-you-go* principle (i.e., the more you give the more you get, in an incremental and continuous fashion) that is currently emerging on the Web [8, 9].

The dataspace concepts have been presented in a visionary way [10, 4, 5] and their implementation on global scale opens new research challenges. From the point of view of metadata, dataspaces may be seen as a generalization of both DBMS and SE (search engines), where the schema of the former meets the document structure of the latter. In fact, in the context of databases, a schema is the set of metadata, relations and constraints regarding data. In other words, it is everything that is stored but the data itself. In search engines, the focus is instead on the documents, that are usually semi-structured (XHML), with ranking algorithms exploiting metadata such as links, headers and title. Despite this, the user cannot perform search using the structure of the document; they are just a set of words. Dataspaces aim at having the benefits of both DBMS and SE with minimal efforts for end users.

In this chapter we investigate the crucial problem that poses the bases to the concept of dataspaces: the need for human interaction/intervention in the process of organizing (getting the structure of) unstructured data. We

survey the existing techniques behind dataspaces to overcome that need, by
using algorithms for structure induction and flexible data models that take
into account the diversity among data sources (e.g., text documents, pictures,
tables) while leveraging all the knowledge about each document structure.

Our survey will explore the structure of a dataspace along three
dimensions:

*Dataspace profiling.* Analogous to databases profiling [11], this dimension
will analyze how recent proposals differ in defining the internal structure
(components and relationship) of a dataspace [12, 13].

*Querying and searching.* Querying and Searching represent one of the main
services supported by a dataspace [6, 7, 14]. Our review will investigate
how differences in dataspace profiling will affect the formulation of queries
on top of all participants in dataspaces.

*Application domain.* A dataspace defines a global "virtual location" connecting data from diverse application domains [4, 15]. This dimension will
explore the characteristics of the new types of functionalities the dataspace enables and the new possibilities it opens within domain-specific
applications.

Existing projects focusing on dataspaces, induction of data structure from
documents, and data models where data schema and documents structure
overlaps will be reviewed, such as Apache Hadoop [16, 17], Cassandra [18] on
Amazon Dynamo [19], Google BigTable model and other DHT-based flexible
data structures [20, 21], Google Fusion Tables [22], iMeMex [3, 23], U-DID,
WebTables [24] and Yahoo! SearchMonkey.

**Chapter Organization.** The chapter is organized as follows: in Section
2 an introduction to data structuring is given; Section 3 gives a review of
the approaches in data integration; Section 4 presents dataspaces, a new
concept born from the data integration community, further analyzed over
different dimensions in Section 5; finally, Section 6 describes existing and
ongoing projects in the field of massive structured data management with no
predetermined schema given.

## 2   Data Structuring

The definition of a data structure allows for organizing, storing querying data
in a software system so that it can be managed efficiently. Besides technical
aspects related with data to be stored, a data structure expresses an organization of logical concepts of data and its careful choice often allows the most
efficient algorithms to be used. As well, being some data structures highly specialized to well-defined tasks, the choice of a specific data structure depends
on the kind of application. For instance, hash tables are particularly suited
for compilers, while queries in very large databases are usually made more
effective by using B-trees. The implementation of a data structure usually

requires a schema describing how the instances of that structure can be stored, accessed and manipulated by applications. Thus a schema addresses both technical features related with the management of data (i.e., field formats and data types) and some aspects concerning the contents and the meaning of the data, such as the cardinality integrity, the referential constraints, etc. In some sense we may argue that the schema itself is a structure of metadata expressing semantic properties and exhibiting varying capabilities and expressiveness in supporting declarative access to and manipulation of data. This observation motivates the assertion that the efficiency of a schema cannot be analyzed separately from the operations that may be performed on its instances and the properties of those operations, including their efficiency and their cost.

To investigate more comprehensively the effectiveness of a schema, let us present the possible ways of structuring data and discuss the effectiveness of the related managing procedure. Regarding their structure, data can be classified into the following four groups.

**Structured data** is organized according to a schema in order to allow retrieval of data via a structured query language. The schema is defined in terms of constructs of some data model, for example, the relational model or the object oriented model. Data is formatted according to the schema prior to populate the database. The very front end of structured data deals with modifying the schema to make it more suitable as requirements for new types of entities and relationships arise. For example, in business environment, changes to the existing schema are rare and avoided wherever possible with new linked tables being created rather than exiting table modified.

**Unstructured data** includes word processing documents, pictures, digital audio and video, emails and PDF attachments, files and folders, PDFs as blobs. The conventional hypertext-based Web provides the illusion that the relationship between two linked documents is defined by some schema when in fact it is implicitly expressed by typed links in HTML.

**Semi-structured data** is generally regarded as data that is self-describing, i.e., the schema may not be known in advance but schema information may accompany the data, e.g., in the form of XML tags or RDF statements. In the context of the Semantic Web, researchers have recently sought to provide facilities for semantic annotation which assigns to an annotation a reference to a concept within an ontology rather then an entity type [25]. As well, data stored in a text form is considered semi-structured data in that it may contain a few structured fields (such as title, sections, authors etc.) that, usually, are not filled in. Without knowing the document content, it is difficult to formulate effective queries for analyzing and extracting useful information from data. Queries that contain keywords are the simplest search method on the Web whose growth results in impressive amounts of information potentially relevant to the user, but very disorganized at the moment. Usually, query results contain a large amount of redundant information and it is difficult to estimate relevant values from whole content that grew exponentially with the growth of Internet. Precision and Recall [26] are two basic measures for assessing the quality of query answers.

**Partially structured data** where information consists partly of some unstructured data conforming to a schema and partly as free text [27]. It is generally regarded as data that is self-describing: in semi-structured data there is not a schema defined but the data itself contains some structural information, such as XML tags. In contrast, the *text* in partially structured data has no structure, while other parts of the document may be structured. Examples include accidents reports, a databases such as SWISS-PROT [28] that includes comment fields containing unstructured information related to structured data.

## 3  Data Integration: The Story so Far

The advent of the Web led to the participation of the users into the content creation and application development process lowering the barrier to publish and accessing information. Data and information management is becoming increasingly complex as more computational resources are made available and more data is produced as part of large scale collaborative activities. This results in a rapid increasing of loosely structured heterogeneous collections of data and documents coming from a variety of information sources, and leads to two fundamental and related questions:

1. Can we define and/or discover latent structural characteristics and regularities on data coming from a constantly growing number of heterogeneous information sources?
2. How traditional techniques for information retrieval and data mining can be adapted with uniform capabilities in order to be effective over this web-scale heterogeneity?

The above questions have received considerable attention within the research community. The resulting proposals provide various benefits and articulate data integration according to the following approaches.

### 3.1  Schema Mapping

Schema mapping combines data residing in different sources under a single integrated global view that provides a uniform query interface by transforming the original query into specialized queries over the respective data sources. This means to construct data elements or a mediated schema between different data models to meet the requirements of end users. A well-defined set of queries can be formulated, each one having a precise answer. Since the source schemas are independently developed, they often have different structure and terminology. Thus, a first step in integrating the schemas is to identify and characterize these inter-schema relationships. This is the schema matching process. It is performed to find relationships between concepts in each schema, i.e., finding the semantic correspondences between elements of two schemas. The input information provides element names, data types, description, constraints and so on. The instance data is exploited to characterize the content and semantics of schema elements. Then the matching elements can be unified, resulting in a set of mapping items. Comprehensive surveys of schema matching approaches are presented in [29, 30, 31].

Before any service or application can be provided, a mapping schema must be defined, aware of the precise relationships between the terms used in each schema or data source. Schema matching nowadays is performed manually by domain experts. Obviously, it is a tedious, time consuming, error prone and expensive process. In general, it is not possible to fully automatically determine all correspondences between two schemas due to their semantic heterogeneity. Moreover autonomous changes in the sources require to modify the matching schema. In recent years, significant progresses in srning, database and data mining allow for partially automating schema matching. However, this approach is poorly suited in application environments (such as scientific communities, personal computing, virtual organizations, etc.) where it is difficult to capture consensus on a single schema.

## 3.2  Keyword-Driven Queries

Data integration appears with increasing frequency in a variety of situations and is critical when the volume and the need to share existing data tremendously increase according to the Web's unconstrained growth. It has been pointed out [10] that "traditional data integration techniques are no longer valid in the face of such heterogeneity and scale" and new proposal are needed. Schema matching solution is less effective for Web search engines and for on-the-fly data integration whose application scenario makes difficult and often impossible to get the right mappings. In this case, it is rather preferable an approach that does not require a matching process over the data source and allows the user to pose keywords query or to define structural search constraints. The query model is either based on keywords over all available sources (bag-of-words query model) or on defining XPaths expressions over a XML schema (XML query model). It is worth noting that the bag-of-words query model does not perform any data integration while the choice of a specific XML Schema can seriously impact the effectiveness of a query.

## 3.3  The Web of Data

Recent research aims to overcome some of the problems encountered in data integration by addressing not the definition of the best integrating structure, but instead the solution of semantic conflicts between heterogeneous data sources. In semantic web settings, a popular strategy to solve this problem, the approach involves the use of ontologies whose role is analogous to the mapping schema. Ontologies express semantic relationships over data in order to resolve semantic conflicts. As of 2006 the trend in semantic integration has favored integrating independently-developed ontologies into a single ontology according to the vision of a Web of Data connecting data from diverse domains with database-like functionality. The aim is replacing a global information space of linked documents to one where both documents and data are linked [32]. The following rules, also denoted as *Linked Data principles* [13], provide directions for publishing and interlinking data on the Web in a way that all published data become part of a single Web of Data:

1. Use the URIs as names for things
2. Use HTTP URIs so that people can look up those names
3. When someone looks up a URI, provide useful information, using the standards (RDF, SPARQL)
4. Include links to other URIs, so that they can discover more things.

Started in 2007 by the W3C Linking Open Data Community, the Web of Linked Data is a public repository consisting of hundred of datasets published by heterogeneous organizations (Universities, Companies, Governments) and comprising data of diverse nature such as music, films, scientific publications, genes, statistical data, television programs etc. The Linked Data approach allows data to be easily discovered and used by various applications. According to a navigation-based query model, RDF links allow client applications to navigate between data sources and to discover additional data. However few mechanisms exist for discovering relevant sources and getting the data automatically.

## 4 Dataspaces

By expressing a query to a web search engine  users pursue a variety of concerns. They interact opportunistically with the web environment, wondering what the components they encounter can contain and then exploring the consequence of various possible browsing actions. They search for information they imagine might exist, no matter how it is structured. They seek explanation for events and relationships they may note. Much of the interaction with the web environment results in carrying on browsing activity on linked sites and searching by keywords. The problem with this kind of approach is that it generates unmanageably huge datasets with no internal structure, no principle for integrating information since users cannot query data using a structured query. A recent proposal within database community is called *dataspaces* [4, 5] that try to offer uniform access to heterogeneous data sources. The key idea of a dataspace is to offer best-effort queries that return possibly related data. In contrast with the data integration approaches presented previously, a dataspace does not consider the existence of a matching schema nor assumes that complete semantic mappings have been specified. It raises the abstraction level at which data is managed in order to manage all information relevant to a particular organization regardless of its format and location, and relationships between them.

A dataspace consists of components (also called participants) and a set of relationships among them. The participants are individual data sources such as Relational Databases, XML schemas, unstructured or partially structured information. Each participant knows the relationships to other participants and their source, i.e., if the participant is added by a user or automatically generated. As well, each participant contains information about the kind of data it contains, the data allocation, the storage format and which querying mechanisms are allowed. In detail, a dataspace considers two models of queries [6]: predicate queries and neighborhood queries. A *predicate query* is formulated according to a simple structure and allow users to specify keyword-based query. An example of predicate query is as follows: "a Professor named `Smith`, teaching `Database course`, at `University X`".

A *neighborhood query* is also a keyword-based query, but it explores all possible associations between data items. For example, a neighborhood query searching for "`Smith`" will return all courses Smith teaches, all Smith publications, people working with Smith etc. Broadly speaking, data resides in distributed systems, but there is not a single schema to which all the data conform to.

A dataspace management system provides keyword search over all data sources: when the user formulates a more sophisticated query, tighter integrations are created only if their benefits balance the effort for their definition. The core idea of a dataspace is to start just with a simple data structure to formulate predicate queries over all its data source and to produce additional effort to semantically integrate those resources only when it is absolutely needed. This means that a dataspace addresses the data integration problem in a *pay-as-you-go* fashion by postponing difficult and expensive aspects related to the schema matching process. *The schema is defined incrementally by discovering the latent structure of data.* It is only available when the query requires a structure to be expressed over heterogeneous datasets. Therefore we believe dataspaces may be intended as an abstraction where schema meets structure.

## 5   Dataspace Dimensions

As described in the previous section, the goal of a dataspace is to discover and verify latent structural properties among its datasets in order to provide users with a query interface. Indeed, a measure of how well structural properties are defined is the extent to which the resulting user interface is effective in querying the dataspace. A dataspace is neither a data integration system nor a user interface. It merely provides the best effort to call attention to the aspects of a query that may still require work for structuring data. There is no guarantee that there will be a structured answer to the query but discovering latent structural properties of data explicitly helps to get out the relevant information.

Database dimensions are explored in [12] where the proposed framework considers the life cycle of a dataspace as composed by a set of phases. Just like any traditional data integration software, a dataspace is initialized, deployed, maintained and updated. The paper details each phase with a view of eliciting dimensions over which existing dataspace proposals have varied.

Here we propose a completely different approach that considers dataspaces dimensions as defined by three fundamental concerns that can affect their effectiveness: dataspace profiling, querying and searching, application domain.

## 5.1  Dataspace Profiling

Similar to database profiling [11], dataspace profiling concerns with defining a set of components and relationships among them in order to offer a rich and dynamic context for user queries. However, it differs from data integration approaches in taking the broader-scope perspective on user query activity. There is not a mediated schema to which all the data conform to and data is hosted in a plethora of systems. Conversely, a dataspace is a concrete representation supporting the query processes a user might engage in while pursuing a particular concern. As such, this representation is necessarily incomplete and there is an infinity of possible representations that depend on the relationships among components. The best effort in profiling a dataspace aims at having a representation that provides good coverage of the possible query activity. An approach to building a dataspace would start with a definition of the kinds of components in the system. The nature of relationships between components is structural and sacrifices much of their semantics. Relationships could be used to organize components, but also to generate other components. Given a set of components $S = (X, Y, Z)$, possible relationships are as follows:

- $X$ was manually created from $Y$ and $Z$
- $X$ and $Z$ came from the same source at the same time
- $X$ is a view or a replica of $Y$
- $X$ and $Y$ are created independently, but reflect the same physical system

The dataspace contains a detailed description of basic and semantic information about the data in the components. Specific applications automatically create relationships between the participants, and improve and maintain already existing relationships. Location and data mining techniques are used to discover relationships, but human attention is required for creation of relationships. Often, additional data management features (backup, recovery, and replication) are provided for components that have no or only limited data management functions. As well, cooperation among participants is enhanced by special support such as translation tables for coded values, classification and ratings of document, etc. Of course, relationships might fail in organizing even the majority of available components, but still it is useful both for providing structural insight about components and for generating new components. In terms of traditional database properties, a dataspace exhibits often common form of inconstancy about data (where the values come from and how they are created) that increases uncertainty. As stronger guarantees are required, it is necessary to develop formalisms to represent and reason about external lineage and promote agreements among the various owners of data sources. Among a few papers featuring dataspace profiling we mention [7] that underlines the importance of analyzing the structures and the properties exposed by an information source and investigates techniques for property and path analysis over a variety of data sources. Specifically, the paper presents *Quarry*, a software environment to browse and refine harvested metadata in scientific domains.

Originally conceived to provide browsing and querying services over a set of triples, Quarry accepts resource-properties-value triples (initially provided through scripts created by domain experts such as scientists) and automatically recognizes signatures, i.e., common property patterns for resources. Being each signature materialized in a multi-column table, a user can inspect a set of property-value conditions asserted on a collection of resources. Inspection is supported by an API that returns a set of unique properties used to describe a restricted set of resources and/or a set of unique values for a given property over a restricted set of resources. This allows the set of resources to be described by a path expression where each node of is a conjunctive query of propertyvalue pairs. From a merely technical viewpoint, a path type is a simple sequence of property names such as `belongs_to.is_author_of.has_title`. Profiling a dataspace means to explore the characteristics of these paths. While in a relational setting *following a path* generally means to join relations, in a dataspace *examining path characteristics* may mean, for instance, to explore whether instances of a specific type exist for every entity in a collection of participants, to calculate how many instances of a fixed type originate from participants, etc. Quarry follows a *pay-as-you-go* approach in discovering paths and aims at supporting users who don't have any technical knowledge about databases and, in general, about how the data is stored or managed.

## 5.2   Querying and Searching

A dataspace aims at supporting incremental refinement of automatic mapping using information from different resources with minimal effort. From the users perspective, the distinction between search and query disappears and the user should be able to iteratively refine and modify the previous query as well to formulate queries about the source of data that can be itself a new answer other than traditional answers. The focus is on finding methods to interpret queries in various languages on participants that support multiple data models and/or unstructured and heterogeneous. Due to data inconstancy, metrics must be defined for comparing the quality of answers and the efficiency of query processing techniques. To correct errors, results should be inspected by humans, but it is impossible at web scale. Hence, it is necessary to provide feedback from user on the quality of results. Popular approaches to get implicit feedback are to record which of the answers the user clicks on, to supply sample answer that would meet user needs, to observe which sources are viewed in the same session, etc. Proposed methods aim at managing the transitions among keyword querying, browsing and structured querying.

Following an information retrieval-based approach, [6] considers indexing support for predicate and neighborhood queries on heterogeneous data that are not semantically integrated. The basic idea is to build an index whose leaves are references to data items in the individual sources. Data is modelled by a set of triples either of the form (*instance, attribute, value*) or of the form (*instance, association, instance*). An instance is described by a set of attributes and linked to another instance by associations. Both instance and associations are extracted out of the data sources using a variety of methods. A query is formulated as a predicate that contains an attribute and some keywords $[k_1, k_2, \ldots, k_n]$. It is supported by an inverted list stored in a matrix where the cell at the $i$-th row and the $j$-th column memorizes the occurrence of the keyword $k_i$ in the instance $j$. Synonyms and hierarchies are used to accommodate heterogeneity.

A pay-as-you-go method is proposed in [23]. Data is represented using a graph model and a technique is proposed to gradually model relationships in a dataspace through *trails* that include traditional semantic attribute mapping as well as traditional keyword expansions as special cases. A trail asserts a (bidirectional) correspondence between two keywords $k_1$ and $k_2$ meaning that the query $Q_1$ ($k_1$) induces the query $Q_2$ ($k_2$), and vice versa. By mean of trails, it is possible to define equivalences between any two sets of elements in the dataspace. A query is modeled by a graph and the query process consists of three major phases. First, matching detects whether a trail should be applied to a given query by checking the query with the left side of the itrail. Then, the right side of the trail is used to transform the original query graph. Finally, the transformed query is merged with the information provided by itrail definition to obtain a new query that extend the semantic of the original one.

Leveraging ideas from keywords search in a databases, work in [33] proposes a method for automatic adding new data sources and relating them to the existing ones. Given a set of databases that contain known cross-references, the user specifies a keyword-based query that is dynamically expanded into a query graph by a system, called Q. From each graph,

$Q$ generates a conjunctive SQL query and associates a cost expression. Finally, $Q$ provides a ranked view consisting of a union of conjunctive queries over different combination of the sources. This view is materialized and refined through user feedback. When the user registers a new source, i.e., a new database, the new source relevance is evaluated to the existing ranked views that are updated as appropriate if the relevance is found to be significant. The paper incorporates state-of-the-art methods from database [34] and machine learning literature [35] not only to automatically discover semantic links among data sources, but also to combine information resulting from multiple matching methods. Through experiments on actual bioinformatics schemas, authors demonstrate that their strategy is an effective step towards the ultimate goal of automating data integration. However, the consideration of heterogeneous and unstructured dataset, including Web sources, is an ongoing work.

## 5.3  Application Domain

One challenge of Web-based environments is to move from the idea of a local site towards networked collaborative environments that are requested to provide access to a variety of information and data sources. These environments are constantly in evolution and keep on increasing the aggregation and sharing of heterogeneous and geographically dispersed resources via temporary collaboration. In practice, this means that data sources are more or less invisible to the users whose search and query processes do not occur at a single location in a single context, but rather spans a multitude of situations and locations covering a significant number of heterogeneous data sources. Often, results are composed of parts that may themselves be data sources and no limit exists in this deep structure. For these reasons, dataspaces encompass more than just an integration data approach aiming to be an highly versatile and dynamic paradigm for integrating data on-demand. There are a number of domains where they can be defined to promote and mediate people's interactions with computers and other peoples. A basic step is to recognize that queries cluster around pervasive user search. Because users are allowed to influence the behavior of a dataspace by customizing their usage on his preferences, a dataspace implementation has to consider the context in which user operates. This context can spawn a staggering number of aspects, but the dataspace could restrict attention to those concerns that can be classified at similar level of abstraction. Often these concerns involve datasets whose heterogeneity is not perceived by the user who is, in turn, overwhelmed by the similarities of information. Structural differences akin to the various ways of performing a musical piece: people perceives the leitmotif but often would be enabled to appreciate technical differences in various executions. As such, we believe that the principles enabling a dataspace for a research environment [36] can substantially differ from those supporting access and manipulation of all of the information on a person's desktop with possible extension to personal information on the Web. To explain these differences next section presents a survey on existing projects in different domains.

# 6    A Roundup of Existing Projects on Managing Structured Data

In this section we analyze existing and ongoing projects and proposals focusing on dataspace-like approaches to data management or correlated topics such as induction of data structure from documents and data models mixing schema and document structure. When observing data management systems from a higher point of view than enterprise DBMSs, *structured storage* is the term generally used in literature [37, 18] referring to a large set of data management systems of which standard DBMSs are just a subset. Structured storage generalize DBMSs by not requiring fixed schemas and usually making distributed storage as a built-in feature in order to manage massive datasets transparently to the client. If we think of data handled as tables, structured storage data tend to grow horizontally by having weak constraints on the structure (columns) of the tuples. Apache Cassandra [18] (backed by Amazon Dynamo [38]), Google's BigTable [20, 21] and Apache HBase [17] are considered the most significant examples of structured storage, currently used in production environments from major web players. In the following we summarize the main aspects of each project, with a focus on the data model adopted to store or represent data and the query interface made available to users.

## *6.1    Google BigTable*

BigTable [20, 21] is a data model and a structured storage system proposed by Google. It is built on top of Google File System (GFS for short) and Google Chubby locking service [39]. GFS [40] is a proprietary file system developed and used internally by Google to save data in commodity, cheap, with high failure-rate expected commodity servers. Data is usually appended to existing files, overwrites are rare. GFS is therefore distributed and like DBMSs it operates in the userspace (not being part of the underlying operating system kernel).

As in the Google paper of 2006 [21], BigTable is currently used by over 60 Google's products, including MapReduce, Google Reader, Google Maps, Google Book Search, "My Search History", Google Earth, Blogger.com, Google Code hosting service, Orkut, YouTube[1], and Gmail. According to the authors of the paper, it has been developed to enhance scalability, and better control performance characteristics, but in the conclusions they also

---

[1] Excluding the video storage.

state they "have gotten a substantial amount of flexibility from designing our own data model for BigTable". From the data model point of view, the main characteristic of this very influential work is that, despite standard relational databases, tables have not a fixed number of columns. They are sparse, distributed multi-dimensional sorted maps, sharing characteristics of both row-oriented and column-oriented databases. Thus, the data model behind BigTable can be summarized by the following function:

$$(row : string, column : string, time : int64) \rightarrow string$$

Collisions are not allowed (the corresponding item would be overwritten), and avoided by the model since the *time* key always change over time. The Application programming interface (API) offered to client applications is simple, while allowing very flexibility and full configurability. Rows are indexed by the *row* key, and are kept sorted by this key. Columns are also indexed, and kept grouped in a few families configurable in terms of access control, enabling compression, disk or memory storage, and data retention delays.

The usual operations performed against a BigTable are *get* and *put* to get/write an item with a specific key and *scan* to iterate over the items. Often the concept of BigTable is associated with a Google-patented approach to data analysis, called *MapReduce* [41, 42, 43]. Instead of accessing BigTable data with the basic API (get, put, scan), users exploit the functional programming approach of MapReduce, that forces the development of Divide-and-Conquer algorithms, therefore natively writing parallelizable data access.

After the publication of the paper, there has been a large number of attempts to implement the proprietary Google BigTable model, such as Hyper-Table, Open Neptune[2] and most notably the open source Apache's Cassandra and HBase (on top the Hadoop Core [16]), reviewed later in this section. Most of the code base of the Apache projects at hand has been donated and frequently contributed by web players like Facebook and Yahoo!, as described later on this section.

## 6.2   Apache Cassandra

Apache Cassandra is something similar to BigTable, initially developed and used by the Facebook team, and currently considered by the Apache Foundation a sort of open source implementation of the BigTable data model on top of a storage system similar to the DHT-featuring proprietary solution Amazon Dynamo [38]. A table in Cassandra is a distributed multidimensional map indexed by a key. The value is an object which is highly structured. Transactions are per-row, no matter the number of columns involved. Columns are grouped together into sets called column families, allowing personalization of column groups containing similar columns, as it happens in the BigTable

---

[2] See http://openneptune.com/

system. Cassandra also supports the use of multiple tables, although deployments often do not require this feature, expoiting instead the schema-free model of just one big table.

## 6.3   Apache Hadoop

Contributed by Yahoo! and used within their products, Hadoop is a Java framework on top of a number of supported file systems (including ad-hoc *Hadoop File System*, HDFS). As of 2008, Yahoo! Search Webmap was considered the largest in-production application featured by the Hadoop framework, with around 10K cores (approx 4K machines, also called nodes) powered by Linux. Facebook recently stated a 20 Petabyte Hadoop deployment for data warehouse of logs, being the largest known structured storage systems. Hadoop common framework exposes to clients a way to compute in parallel analysis over distributed data, in a MapReduce fashion [41, 42, 43] allowing, for instance, 1 terabyte of data sorted in order of 1 minute, assuming a reasonable amount of available nodes. Hadoop is an umbrella project for a number of subprojects; usually with the basic term of *Hadoop* it is meant the use of a framework to handle HDFS together to a job launcher/scheduler for MapReduce scripts. It is often distributed with some important subprojects, such as HBase, Hive, Pig and others. Interestingly, Hadoop has large support from open source community and big web companies, with further external projects such as *Ganglia* to monitor distributed Hadoop instances, and *Mahout* as a scalable machine learning library backed by Hadoop.

### 6.3.1   HBase

It is the component of Hadoop that, on top of HDFS, allows the BigTable data model in the Apache framework, with random read/write of the multidimensional associative array, leading to a column-oriented database management system [44, 17]. Efficient random access is not inherited by HDFS, but instead developed with appended-only log files, only occasionally merged with the remaining datafiles.

### 6.3.2   Hive

Hive [45, 46] is an extension of the Hadoop framework initially developed a few years ago by a Facebook team to data warehousing in a very scalable fashion, where a bunch of Mysql servers were failing in handling responsively lots of terabyte, with an average of 5Tb of data added every day. Hive adds to Hadoop the power of $HQL$, an SQL-like query language with a few extension w.r.t. Standard SQL. From a data model point of view it is interesting to note it handles sequences of data natively, by using commands Expand and Collapse. The former transforms the content of a column (actually a cell) into a sequence (i.e., another row with multiple columns), while the latter is the reverse operation, able to transform a multi-column row into a single value. The system also provides the command `Transform` to run MapReduce scripts within HQL.

As in standard SQL, the Hive data model consists of standard tables (i.e., typed columns, plus sequences and maps), and partitions (e.g., to partition tables by date) which are backed by directories in the HDFS subsystem. Bucketing can be applied to partitions (hash partitions within ranges, for sampling, optimizing hash joins, etc.). Regarding the implementation, tables and all the data of Hive are files in the HDFS (where directories determine the partitioning of tables), and the format of such files are arbitrary (e.g., comma separated, XML, YAML, JSON or any other chosen by the user). Sort of java "drivers" (usually very simple scripts, java classes called SerDe classes) must be provided in order to parse such unknown format files transforming them into Hive tables, although such parsers are already provided for most used formats. Another important component of the architecture is the *Metastore*, a component backed by any SQL-based management system as long as it provides a JDBC wrapper, such as MySQL or most notably Java Derby. Metastore is in charge of managing metadata, meaning for instance what in Oracle is contained in tables like `ALL_TAB_COLUMNS`; in the same way, Metastore saves metadata such as type of columns and any other information about the data which is not the data itself.

## 6.4 Apache CouchDB

Apache CouchDB is a document-based storage system characterized by its schema-free design, with a flat address space [47]. Documents can be considered semi-structured objects, essentially associative arrays where values can be either strings, dates, or even structured data like ordered lists or maps. The low level infrastructure allows data to be distributed among a number of server, with incremental replication and automatic bi-directional conflict detection between servers and user clients (in case of network failures or offline nodes). A free and commercial zero-configuration versions are currently supported by Cloudant [48].

## 6.5 DHT-Based Data Management Systems

The term Distributed Hash Table (DHT) is usually referred to systems able to handle very large hash tables over a medium-to-large set of nodes (machines). It is usually the ground level to construct higher level of abstraction on top of. One of the critical aspects of a DHT-based system is the routing protocol, that is, how to distribute (keys) over the network of nodes and how to search/recover them. Figure 1 shows a possible architecture featuring multi-level routing. Depending on the applications, needs of security

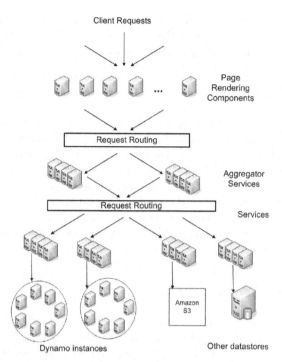

**Fig. 1** An example of distributed architecture with clients (e.g., web browsers) facing front-end nodes, multi-level routing nodes and cloud-based storage servers (e.g., Amazon Dynamo instances and others)

(untrusted nodes, no single point of failure) or high availability (redundancy, decentralization), DHT systems can be much more complicated than a basic distributed hash table structure. They are currently used in several projects, including the projects presented so far and most peer-to-peer projects such as BitTorrent, eMule and Freenet [49]. From the data model point of view, at the lowest level they basically represent and implement associative arrays. Partitioning, replication, versioning, failure management and low latency vs. high throughput are the key aspects to be considered for each DHT-Based Data Management System [50].

## 6.6  Google Fusion Tables

Fusion Tables [22] is a web-based application featured by Google that allows table data to be uploaded into Google servers and then shared, queried in a structured way (including joins with other table sources) or used within other web applications. For example, supposing each row of a table contains latitude-longitude pairs, they can be showed through the Google Map service. This kind of data visualization and integration of structured data are easy to obtain given the tight collaboration among different Google applications.

It is an interesting proposal on a possible UI for data, although being most data in form of tables, it reduces the range of applications to structured-only data and appears similar to spreadsheets (with no computed values).

## 6.7   WebTables

Webtables is a shortcut for Tables in the Web. According to [24], almost 15 billion tables have been crawled by Google, many of them used for web site layout, but still a subset of 154M tables found by the authors as containing high-quality relational data. One of the contribution of the paper is that web-scale analysis of tables allowed the definition of an *attribute correlation statistics database* (called AcsDB) containing statistics about correlated attributes in the tables, useful in a number of contexts, including "semantic analysis" of table contents such as finding attribute synonyms and automatic join traversal of the tables.

## 6.8   Yahoo! SearchMonkey

Yahoo! SearchMonkey (SM) is a framework in charge of transforming the way search results are displayed. It is an interesting way to show how external structured data can be integrated to personalize web search results. In fact users like web site owners should share structured data (they may be in microformats[3], XML feeds, semantic web RDF or links to other structured data), then create a SearchMonkey application that handles the provided structured data to show a personalized search results. For instance, when someone queries Yahoo! for "best pizza in NY", if the search results include the web page personalized with SM, then that single result will appear differently, depending on the SM application behaviour (e.g., with review stars, a picture, a menu of site contents, etc.). In order to personalize "snipets" results we will use Yahoo! and user provided data input sources. Interestingly, data are structured into trees, where each data input is the root of a tree containing different type of information. For instance, trees are labeled `yahoo:index` or `com.yahoo.uf.hcard`, while nodes like `yahoo:index/dc:language` or `com.yahoo.uf.hcard/rel:Card/vcard:fn` respectively contain the language of the url at hand (e.g., "en") and the author of the page at that url (e.g., "Sam"). Recently Yahoo! marked this service as deprecated [51], and going to shift from a model where developers build lightweight apps to install on Yahoo! to one where publishers enhance their own site markup to produce similar results.

---

[3] For further information on microformats, see `http://www.microformats.org/`

## 6.9  iMeMex

iMeMex is the first attempt to develop an opensource dataspace for personal information management (PIM), run at ETH Zurich by a team leaded by Dr. Jens-Peter Dittrich. It follows the dataspace ideas in [4, 5], enabling the loose integration of different data sources (primarily files of any kind, emails, documents) through the use of user-defined views that give structure interpretation of the data belonging to each source. The iMeMex data model (called iDM) [3] is essentially a tree-based structure (with each node being any kind of object, e.g., a directory, a PDF file, a subsection of a LaTeX file), with links among nodes, leading therefore to a graph.

Despite the fact that different sources can be distributed over a network, the core architecture of the iMeMex system is essentially focused on a main centralized index (customizable and enhanced by the Apache Lucene full-text search engine), containing the information to answer user queries or to locate the required resource over the different data sources. This fact makes iMeMex less scalable w.r.t. projects described so far, but also more flexible on the structure and storage of data. It also has a rule-based query processor that is able to operate in three different querying modes: warehousing (only local indexes and replicas are queried), mediation (local indexes are ignored, queries are shipped to the data sources), and hybrid (combination of the former methods). Pay-as-you-go query rewriting techniques such as *iTrails* [23] (we described them in Section 5.2) are currently under development.

## 7  Conclusions and Future Work

In this Chapter we discussed the problems arising when large data with diverse structures need to be stored, managed, queried and integrated to improve user utility. Dataspaces are a promising framework to manage all kind of data in a uniform way, based on a simple data structure to formulate predicate queries over all its data source and to produce additional effort to semantically integrate those resources only when needed. The schema in the data is defined incrementally by providing the latent structure of data, thus letting the schema meet the structure. We reviewed the state of the art of both research results and commercial/opensource projects in the area.

Although very advanced, we believe these systems are still lacking of automatic algorithms to exploit unstructured data in order to learn their hidden structure. In fact, as we noted in the previous sections, most of the existing proposals are based on a pay-as-you-go user intervention, allowing keyword-only search over the data unless structure is explicitly given by the user. In an attempt to further analyze these aspects, *Unstructured Data Integration for Dataspaces* (U-DID for short), a recent ongoing project carried out by the authors of this chapter at the University of Cagliari, aims to investigate the topic of dataspaces as a possible kernel for data management in next-generation infrastructures. The project activities are focusing on defining a common model for managed data, uniforming the heterogeneity of data but meanwhile being able to represent the characteristics of each data type. The use of *data wrappers*, i.e., middleware algorithms able to convert "sequence of bytes" to structured data, either centralized (within a data management system) or distributed, will be fundamental for investigating the various possible dataspace architectures.

This important step involves the study of algorithms that extract a set of structured and integrable data (through metadata) out of input sequences with no previously-established format. Such algorithms can be divided into two groups: *fully unsupervised*, i.e., able to automatically compute the output without human intervention, and *semi-supervised*, i.e., taking into account online information (interactively provided) or offline information (datasets the algorithm will use to learn how to structure data). As we said, this kind of *structure learning* algorithms have not been investigated enough in literature and the project is therefore committed to contribute to this lack in the dataspace community.

Another approach under consideration is about the use of data mining techniques, well-studied for model extraction in the context of structured data. Clustering and association rules for structure extraction are definitely innovative, and seem to be promising, though they will require heavy adaptation of current algorithms.

# References

1. Gounbark, L., Benhlima, L., Chiadmi, D.: Data integration system: toward a prototype. In: ACS/IEEE International Conference on Computer Systems and Applications, pp. 33–36 (2009)
2. Gatterbauer, W., Suciu, D.: Managing structured collections of community data. In: CIDR 2011, Fifth Biennial Conference on Innovative Data Systems Research, Online Proceedings, Asilomar (January 2011)
3. Dittrich, J.-P., Salles, M.A.V.: idm: A unified and versatile data model for personal dataspace management. In: Dayal, et al [52], pp. 367–378
4. Franklin, M.J., Halevy, A.Y., Maier, D.: From databases to dataspaces: a new abstraction for information management. SIGMOD Record 34(4), 27–33 (2005)
5. Halevy, A.Y., Franklin, M.J., Maier, D.: Principles of dataspace systems. In: Vansummeren, S. (ed.) PODS, pp. 1–9. ACM, New York (2006)
6. Dong, X., Halevy, A.Y.: Indexing dataspaces. In: Chan, C.Y., Ooi, B.C., Zhou, A. (eds.) SIGMOD Conference, pp. 43–54. ACM, New York (2007)
7. Howe, B., Maier, D., Rayner, N., Rucker, J.: Quarrying dataspaces: Schemaless profiling of unfamiliar information sources. In: ICDEW 2008: Proceedings of the 2008 IEEE 24th International Conference on Data Engineering Workshop, pp. 270–277. IEEE Computer Society Press, Washington, DC, USA (2008)
8. Jeffery, S.R., Franklin, M.J., Halevy, A.Y.: Pay-as-you-go user feedback for dataspace systems. In: Proceedings of the 2008 ACM SIGMOD International Conference on Management of Data, SIGMOD 2008, pp. 847–860. ACM, New York (2008)

9. Hedeler, C., et al.: Pay-as-you-go mapping selection in dataspaces. In: Proceedings of the 2011 ACM SIGMOD International Conference on Management of Data, SIGMOD 2011. ACM Press, New York (to appear 2011)
10. Madhavan, J., Halevy, A.Y., Cohen, S., Dong, X.L., Jeffery, S.R., Ko, D., Yu, C.: Structured data meets the web: A few observations. IEEE Data Eng. Bull. 29(4), 19–26 (2006)
11. Marshall, B.: Data quality and data profiling - a glossary (2007),
    http://www.w3.org/DesignIssues/LinkedData.html
12. Hedeler, C., Belhajjame, K., Fernandes, A.A.A., Embury, S.M., Paton, N.W.: Dimensions of dataspaces. In: Sexton, A.P. (ed.) BNCOD 26. LNCS, vol. 5588, pp. 55–66. Springer, Heidelberg (2009)
13. Lee, B.: Linked data - design issues (2006),
    http:/www.w3.org/DesignIssues/LinkedData.html
14. Liu, J., Dong, X., Halevy, A.Y.: Answering structured queries on unstructured data. In: WebDB (2006)
15. Halevy, A.Y., Rajaraman, A., Ordille, J.J.: Data integration: The teenage years. In: Dayal, et al [52], pp. 9–16
16. White, T.: Hadoop: The Definitive Guide, 1st edn. O'Reilly Media, Sebastopol (2009)
17. Apache Foundation Software. Apache hbase, subproject of hadoop (2006),
    http://hbase.apache.org/#Overview
18. Lakshman, A., Malik, P.: Cassandra: a structured storage system on a p2p network. In: auf der Heide, F.M., Bender, M.A. (eds.) SPAA, p. 47. ACM, New York (2009)
19. Decandia, G., Hastorun, D., Jampani, M., Kakulapati, G., Lakshman, A., Pilchin, A., Sivasubramanian, S., Vosshall, P., Vogels, W.: Dynamo: amazon's highly available key-value store. SIGOPS Oper. Syst. Rev. 41(6), 205–220 (2007)
20. Chang, F., Dean, J., Ghemawat, S., Hsieh, W.C., Wallach, D.A., Burrows, M., Chandra, T., Fikes, A., Gruber, R.E.: Bigtable: A distributed storage system for structured data. ACM Trans. Comput. Syst. 26(2) (2008)
21. Chang, F., Dean, J., Ghemawat, S., Hsieh, W.C., Wallach, D.A., Burrows, M., Chandra, T., Fikes, A., Gruber, R.: Bigtable: A distributed storage system for structured data (best paper award). In: OSDI [53], pp. 205–218
22. Gonzalez, H., Halevy, A.Y., Jensen, C.S., Langen, A., Madhavan, J., Shapley, R., Shen, W., Goldberg-Kidon, J.: Google fusion tables: web-centered data management and collaboration. In: Elmagarmid, Agrawal [54], pp. 1061–1066
23. Salles, M.A.V., Dittrich, J.-P., Karakashian, S.K., Girard, O.R., Blunschi, L.: itrails: Pay-as-you-go information integration in dataspaces. In: Koch, C., Gehrke, J., Garofalakis, M.N., Srivastava, D., Aberer, K., Deshpande, A., Florescu, D., Chan, C.Y., Ganti, V., Kanne, C.-C., Klas, W., Neuhold, E.J. (eds.) VLDB, pp. 663–674. ACM, New York (2007)
24. Cafarella, M.J., Halevy, A.Y., Wang, D.Z., Wu, E., Zhang, Y.: Webtables: exploring the power of tables on the web. PVLDB 1(1), 538–549 (2008)
25. Uren, V.S., Cimiano, P., Iria, J., Handschuh, S., Vargas-Vera, M., Motta, E., Ciravegna, F.: Semantic annotation for knowledge management: Requirements and a survey of the state of the art. J. Web Sem. 4(1), 14–28 (2006)
26. Tan, P.-N., Steinbach, M., Kumar, V.: Introduction to Data Mining. Addison-Wesley, Reading (2005)

27. King, P.J.H., Poulovassilis, A.: Enhancing database technology to better manage and exploit partially structured data. Technical report bbkcs-00-14, Birkbeck University of London (2000),
    http://www.dcs.bbk.ac.uk/research/techreps/2000/bbkcs-00-14.pdf
28. Bairoch, A., Boeckmann, B., Ferro, S., Gasteiger, E.: Swiss-prot: Juggling between evolution and stability. Briefings in Bioinformatics 5(1), 39–58 (2004)
29. Doan, A., Halevy, A.Y.: Semantic-integration research in the database community. AI Mag. 26, 83–94 (2005)
30. Kalfoglou, Y., Schorlemmer, M.: Ontology mapping: the state of the art. Knowl. Eng. Rev. 18, 1–31 (2003)
31. Choi, N., Song, I.-Y., Han, H.: A survey on ontology mapping. SIGMOD Rec. 35, 34–41 (2006)
32. Bizer, C., Heath, T., Berners-Lee, T.: Linked data - the story so far. Int. J. Semantic Web Inf. Syst. 5(3), 1–22 (2009)
33. Talukdar, P.P., Ives, Z.G., Pereira, F.: Automatically incorporating new sources in keyword search-based data integration. In: Elmagarmid, Agrawal [54], pp. 387–398
34. Do, H.H., Rahm, E.: Matching large schemas: Approaches and evaluation. Inf. Syst. 32(6), 857–885 (2007)
35. Talukdar, P.P., Reisinger, J., Pasca, M., Ravichandran, D., Bhagat, R., Pereira, F.: Weakly-supervised acquisition of labeled class instances using graph random walks. In: EMNLP, pp. 582–590. ACL (2008)
36. Dessì, N., Pes, B.: Towards scientific dataspaces. In: Web Intelligence, IAT Workshops, pp. 575–578. IEEE, Los Alamitos (2009)
37. Hamilton, J.: Perspectives: One size does not fit all (2009),
    http://perspectives.mvdirona.com/
    CommentViewguidafe46691-a293-4f9a-8900-5688a597726a.aspx
38. DeCandia, G., Hastorun, D., Jampani, M., Kakulapati, G., Lakshman, A., Pilchin, A., Sivasubramanian, S., Vosshall, P., Vogels, W.: Dynamo: amazon's highly available key-value store. In: Bressoud, T.C., Frans Kaashoek, M. (eds.) SOSP, pp. 205–220. ACM, New York (2007)
39. Burrows, M.: The chubby lock service for loosely-coupled distributed systems. In: OSDI [53], pp. 335–350
40. Ghemawat, S., Gobioff, H., Leung, S.-T.: The google file system. In: Scott, M.L., Peterson, L.L. (eds.) SOSP, pp. 29–43. ACM, New York (2003)
41. Dean, J., Ghemawat, S.: Mapreduce: Simplified data processing on large clusters. In: OSDI 2004, pp. 137–150 (2004)
42. Dean, J., Ghemawat, S.: Mapreduce: a flexible data processing tool. Commun. ACM 53(1), 72–77 (2010)
43. Dean, J.: Experiences with mapreduce, an abstraction for large-scale computation. In: PACT 2006: Proceedings of the 15th International Conference on Parallel Architectures and Compilation Techniques, p. 1. ACM Press, New York (2006)
44. George, L.: Hbase architecture (2009),
    http://www.larsgeorge.com/2009/10/
    hbase-architecture-101-storage.html
45. Apache Foundation Software. Apache hive, data warehouse infrastructure built on top of apache hadoop (2010), http://hive.apache.org/
46. Thusoo, A., Sarma, J.S., Jain, N., Shao, Z., Chakka, P., Anthony, S., Liu, H., Wyckoff, P., Murthy, R.: Hive: a warehousing solution over a map-reduce framework. Proc. VLDB Endow. 2(2), 1626–1629 (2009)

47. Apache Foundation Software. The couchdb project (2008),
    http://couchdb.apache.org/
48. Cloudant.com. Cloudant bigcouch (2008), https://cloudant.com/
49. Evans, N.S., GauthierDickey, C., Grothoff, C.: Routing in the dark: Pitch black.
    In: ACSAC, pp. 305–314. IEEE Computer Society, Los Alamitos (2007)
50. Balakrishnan, H., Frans Kaashoek, M., Karger, D., Morris, R., Stoica, I.: Look-
    ing up data in p2p systems. Commun. ACM 46, 43–48 (2003)
51. Yahoo! Searchmonkey (2011), http://developer.yahoo.com/searchmonkey/
52. Dayal, U., Whang, K.-Y., Lomet, D.B., Alonso, G., Lohman, G.M., Kersten,
    M.L., Cha, S.K., Kim, Y.-K. (eds.): Proceedings of the 32nd International Con-
    ference on Very Large Data Bases, Seoul, Korea, September 12-15. ACM, New
    York (2006)
53. Symposium on Operating Systems Design and Implementation (OSDI 2006),
    November 6-8. USENIX Association, Seattle (2006)
54. Elmagarmid, A.K., Agrawal, D. (eds.): Proceedings of the ACM SIGMOD In-
    ternational Conference on Management of Data, SIGMOD 2010, June 6-10.
    ACM, USA (2010)

# Transductive Learning of Logical Structures from Document Images

Michelangelo Ceci, Corrado Loglisci, and Donato Malerba

**Abstract.** A fundamental task of document image understanding is to recognize semantically relevant components in the layout extracted from a document image. This task can be automatized by learning classifiers to label such components. The application of inductive learning algorithms assumes the availability of a large set of documents, whose layout components have been previously labeled through manual annotation. This contrasts with the more common situation in which we have only few labeled documents and an abundance of unlabeled ones. A further degree of complexity of the learning task is represented by the importance of spatial relationships between layout components, which cannot be adequately represented by feature vectors. To face these problems, we investigate the application of a relational classifier that works in the transductive setting. Transduction is justified by the possibility of exploiting the large amount of information conveyed in the unlabeled documents and by the contiguity of the concept of positive autocorrelation with the smoothness assumption which characterizes the transductive setting. The classifier takes advantage of discovered emerging patterns that permit us to qualitatively characterize classes. Computational solutions have been tested on document images of scientific literature and the experimental results show the advantages and drawbacks of the approach.

## 1 Introduction

In document image understanding, one of the fundamental tasks is to recognize semantically relevant components in the layout extracted from a document image. This recognition process is based on domain-specific knowledge, which is represented in very different forms (e.g. formal grammars or production rules). Several prototypical document image understanding systems have been developed by

Michelangelo Ceci · Corrado Loglisci · Donato Malerba
Dipartimento di Informatica, Università degli Studi di Bari "Aldo Moro"
via Orabona 4 - 70125 Bari
e-mail: {ceci,loglisci,malerba@}di.uniba.it

M. Biba and F. Xhafa (Eds.): Learning Structure and Schemas from Documents, SCI 375, pp. 121–142.
springerlink.com

manually encoding the required knowledge in specific formalisms (e.g., DeLoS [34]). However, the layout of documents, even for the same publisher, may change considerably with time. Moreover, there might be a drift in the type of documents of interest in a specific context. To prevent obsolescence of the developed systems it is necessary to continuously update the required knowledge, which is unfeasible if based only on manual encoding. *Versatility*, that is, guaranteed competence over a broad and precisely specified class of document images, has been recently recognized as a key requirement for document image analysis systems [7]. In order to deal with this requirement, the application of machine learning methods has been advocated both to build models of image quality, layout and language and to infer the parameters of such models. Starting from a training set of document images, learning methods are able to extract a large number of features relevant for understanding the document structure.

The application of machine learning methods to document image understanding has been investigated for almost two decades [19]. From an operational viewpoint, a human operator provides a document image analysis system with image samples of documents and then detects and labels semantically relevant layout components, from which models of document structures are induced. This supervised learning approach, even though providing some flexibility, still does not ensure the key requirement of versatility. Indeed, to acquire the necessary knowledge on a really broad class of documents, supervised learning methods may require a large set of labeled documents. This contrasts with the more common situation in which only few labeled training documents are available, due to the significant cost of manual annotation. Therefore, it is important to exploit the large amount of information potentially conveyed by unlabeled documents to better estimate the data distribution and to build more accurate recognition models.

Two main settings have been proposed in the literature to exploit information contained in both labeled and unlabeled data: the *semi-supervised* setting and the *transductive* setting [40]. The former is a type of inductive learning, since the learned function is used to make predictions on any possible example. The latter requires less - it is only interested in making predictions for the given set of unlabeled data. When the set of documents to label is known a priori, the transductive setting is more suitable, since it appears to be an easier problem than (semi-supervised) induction.

In this chapter, we investigate the problem of transductive learning of document image understanding models. In the proposed method, unlabeled documents are used to reprioritize models learned from labeled documents alone. Indeed, while discriminative learning methods base their decisions on the posterior probability $p(y|x)$, the transductive learning method uses unlabeled documents to improve the estimate of the prior probability $p(x)$, and hence correct the posterior probability $p(y|x)$ by assuming some form of dependence with $p(x)$.

The proposed learning method follows a logic-based approach in which models are represented by a set of rules expressed in relational (or first-order) logic. In order to "understand" the layout structure of an unlabelled document, rules are matched against the relational description of the document layout. The *relational representation* of the document layout and rules is motivated by the fact that layout objects can be related by a number of spatial relationships, such as distance, directional and topological relationships. Standard feature vectors or equivalent propositional representations do not allow relationships in general, and spatial relationships in particular, to be straightforwardly represented.

The study of relational learning in a transductive setting has received little attention in the research community (notable exceptions are [11] for classification tasks, and [4] for regression tasks), although the transductive setting seems especially suitable for relational datasets which are characterized by positive autocorrelation [30]. The application of transductive relational learning to bootstrap the labelling process of document image collections remains an unexplored research direction.

The paper is organized as follows. In Section 2, the present work is motivated and the problem is defined. In Section 3, related works are discussed. Sections 4 and 5 are devoted to the presentation of the method. Finally, experimental results are reported in Section 6 and then some conclusions are drawn.

## 2 Motivation and Problem Definition

The recognition of semantically relevant layout components in document images is part of a complex transformation process of document images into a structured symbolic form which facilitate the modification, storage, retrieval, reuse and transmission of documents themselves [33]. This transformation is articulated in several steps. Initial processing steps include binarization, skew detection, and noise filtering. Then the document image is segmented into several layout components, such as text lines, half-tone images, line drawings or graphics (this step is called *layout analysis*). The understanding of document images follows layout analysis. It aims to associate a "logical label" (e.g. title, abstract of a scientific paper, picture in a newspaper) to semantically relevant layout components (called *logical components*), as well as to extract relevant relationships between logical components (e.g., reading order).

Document image understanding is typically based on layout information, such as the relative positioning of layout components or the size of layout components, as well as on content information (e.g., textual, graphical). This is the case of the work reported in this chapter, where the association of logical labels to layout components is based on both layout information and textual information. However, the novelty here is mainly in the strategy applied to build a classifier which can be used to recognize semantically relevant components.

In the literature, several methods have been proposed for building classifiers to be used in document understanding. They are briefly reviewed in the next Section. However, most of them assume that training data are represented in a single table of a relational database, such that each row (or tuple) represents an independent example (a layout component) and the columns correspond to properties of the example (e.g., height of the layout component). This single-table assumption, however, is too strong for at least three reasons. First, layout components cannot be realistically considered independent observations, because their spatial arrangement is mutually constrained by formatting rules typically used in document editing. Second, spatial relationships between a layout component and a variable number of other components in its neighborhood cannot be properly represented by a fixed number of attributes in a table. This is even more the case when layout components are heterogeneous (e.g., half-tone images, text lines) and have different properties (e.g., brightness, font size). Third, logical components may be mutually related (e.g., the title of a paper "is followed" by the author list). Since the single-table assumption limits the representation of relationships (spatial or non) between examples, it also prevents the discovery of these mutual dependencies which can be useful in document image understanding [31].

The above considerations motivate us to consider a relational representation of document layouts. In fact, document layout structures are a kind of spatial data, and relational approaches have been recently advocated as useful in spatial domains [30]. Document layout structures are also subject to spatial autocorrelation, i.e., a property value of a layout component depends on the property values observed for other layout components in the neighborhood. Spatial autocorrelation clearly indicates a violation of the independence assumption of observations usually made in statistics. By considering this in the definition of the learning method, it is possible to improve the performance of the learned classifiers.

A second motivation for this work is to face the usual scarcity of labeled documents which prevent the application of inductive learning algorithms to generate accurate classifiers. Indeed, manual annotation of the many layout components in a document is very demanding. Therefore, it is important to exploit the large amount of information potentially conveyed by unlabeled documents to better estimate the data distribution and to build more accurate classification models [21]. This is possible in transductive learning, which is formalized as follows:

Let $D$ be a dataset labeled according to an unknown target function, whose domain is $\mathbf{X} = X_1, X_2, \ldots, X_m$ and whose range is a finite set $Y = \{C_1, C_2, \ldots, C_L\}$. Given:

- a training set $TS \subset D$, and
- the projection of the working set $WS = D - TS$ on $\mathbf{X}$,

the goal is to predict the class value of each example in the working set $WS$ as accurately as possible.

The learner receives full information (including labels) on the examples in $TS$ and partial information (without labels) on the examples in $WS$ and is required to predict the class values only of the examples in $WS$. The original formulation of the problem of function estimation in a transductive (*distribution-free*) setting requires $TS$ to be sampled from $D$ without replacement. This means that, unlike the standard inductive setting, the examples in the training (and working) set are supposed to be mutually dependent. Vapnik also introduced a second (*distributional*) transduction setting, in which the learner receives training and working sets, which are assumed to be drawn i.i.d. from some unknown distribution. As shown in ([44],

Theorem 8.1), error bounds for learning algorithms in the distribution-free setting also apply to the more popular distributional transductive setting. Therefore, in this work we focus our attention on the first setting.

There is an interesting convergence of opinions which motivates us to investigate relational learning in the transductive setting. On the one hand, it is claimed that transduction is most useful when the standard i.i.d. assumption is violated ([14]). On the other hand, it is observed that statistical independence of examples is contradicted by many relational datasets ([23]). Moreover, it has also been observed that the presence of (positive) spatial autocorrelation in a dataset entails the smoothness assumption of transductive learning [30]. Therefore, the application of transductive learning to sets of spatially related layout components seems appropriate and worth of being investigated.

In the case of relational data, the problem of transductive classification can be more precisely formulated as follows:
*Given:*

- a database schema $SC$ which consists of a set of $h$ relational tables $\{T_0, \ldots, T_{h-1}\}$, a set PK of primary keys on the tables in $SC$, and a set FK of foreign key constraints on the tables in $SC$,
- a target relation $T \in SC$ (that permits us to represent layout components) and a target discrete attribute $Y$ in $T$, different from the primary key of $T$, whose domain is the finite set $\{C_1, C_2, \ldots, C_L\}$ (Logical label),
- the projection $T'$ of $T$ on all attributes of $T$ except $Y$,
- a training (working) set that is an instance $TS$ ($WS$) of the database schema $SC$ with known (unknown) values for $Y$;

*Find:* the most accurate prediction of $Y$ for examples in $WS$.

In the proposed approach, the prediction of $Y$ is based on a classification framework that works in the relational data mining setting. It is mainly inspired by an associative classification framework proposed by Ceci and Appice [9] where association rules discovered from training datasets are used by a naïve Bayes classifier which operates on relational representations of spatial data.

More precisely, given an object $E$ to be classified, a classical naïve Bayes classifier assigns $E$ to the class $C_i$, that maximizes the *posterior probability* $P(C_i|E)$. By applying the Bayes theorem, $P(C_i|E)$ is expressed as follows:

$$P(C_i|E) = \frac{P(C_i) \cdot P(E|C_i)}{P(E)}. \tag{1}$$

In fact, the decision on the class that maximizes the posterior probability can be made only on the basis of the numerator, that is $P(C_i) \cdot P(E|C_i)$, since $P(E)$ is independent of the class $C_i$.

To work on relational representations, Ceci and Appice proposed considering a set $\Re$ of association rules, expressed as first order definite clauses, which are mined on the training set and can be used to define a suitable decomposition of the likelihood $P(E|C_i)$ à la naive Bayes, in order to simplify the probability estimation problem. In particular, if $\mathfrak{R}(E) \subseteq \mathfrak{R}$ is the set of first order definite clauses, whose antecedent covers $E$, the probability $P(E|C_i)$ is defined as follows:

$$P(E|C_i) = P( \bigwedge_{R_j \in \mathfrak{R}(E)} antecedent(R_j)|C_i). \tag{2}$$

The straightforward application of the naïve Bayes independence assumption to all literals in $\bigwedge_{R_j \in \mathfrak{R}(E)} antecedent(R_j)$ is not correct, since it may lead to underestimating $P(E|C_i)$ when several similar clauses in $\mathfrak{R}(E)$ are considered for the class $C_i$. To prevent this problem the authors resort to the logical notion of factorization [39]. Details are reported in [9].

This associative classification framework for relational classification has been subsequently extended in order to use Emerging Patterns (EPs) instead of association rules. EPs are introduced in [17] as a particular kind of pattern (or multi-variate features), whose support significantly changes from one data class to another: the larger the difference of pattern support, the more interesting the pattern. Change in pattern support is estimated in terms of support ratio (or *growth rate*). EPs with sharp change in support (high growth rate) are useful to discriminate a class from the other classes. This observation motivated Ceci et al. [10] to investigate both the discovery of relational EPs and their usage in the associative classification framework. Experimental results proved the effectiveness of this extension. Therefore, in this work, we consider emerging-pattern based associative classifiers to define a new transductive learning algorithm for document image understanding.

## 3   Related Work

In the literature there are already several works on automatic recognition of semantically relevant layout components. Akindele and Belaïd [2] proposed applying the R-XY-Cuts method on training data, in order to extract the layout structures and to match them against an initial model defined by an expert. The aim of the matching is to discard training documents whose layout structure is very different from the expected one. Then, a generic model of the logical structure is built by means of a tree-grammar inference method applied to validated layout structures with associated labels. Therefore, this approach is based on demanding human intervention, which is not only limited to layout labeling but also involves the specification of an initial model.

Walischewski [45] proposes representing each document layout by a complete attributed directed graph, with one vertex for each layout object. The vertex attributes are pairs $(l, c)$, where $l$ denotes the type of layout component (page, block, line, word, char), while $c$ denotes the logic label of the layout object (e.g. title, author). Edges have thirteen attributes corresponding to Allen's qualitative relations on intervals [3]. An attribute of the edge $(v_i, v_j)$ is a pair $(h, v)$ describing qualitatively the relative horizontal/vertical location between the two vertices $v_i$ and $v_j$. The learning algorithm returns triples $[(c_i, c_j), (h, v), (w_h, w_v)]$ stating that Allen's relation $h(v)$ holds between $c_i$ and $c_j$ along the horizontal (vertical) axis with strength $w_h$ ($w_v$). Altogether, the triples define an attributed directed graph representing the model. Recognition is based on an error tolerant subgraph isomorphism between the graphs representing the document and the model. This approach, although relational, only handles qualitative information and has been tested on simple layout structures extracted from envelopes.

In the work by Palmero et al. [35] a document can be considered as a sequence of objects, where the object labels depend both on the geometrical properties of the block (size, position, etc.) and on the decisions made for previous sequence items. As in the work by Walischewski, there is an implicit recognition of the importance of considering autocorrelation on logical labels, although, in this case, the original bidimensional spatial autocorrelation boils down to one-dimensional temporal autocorrelation, which is handled by a recursive neuro-fuzzy learning algorithm. The effect of sequence ordering on blocks is not examined.

Probabilistic relaxation [6] is a general approach to deal with autocorrelation on logical labels. Indeed, objects are initially classified on the basis of their properties and then their classification is iteratively adjusted by using compatibilities with other objects found in the neighborhood. Le Bourgeois et al. [41] tested this approach on blocks delimiting words and compared it to a naïve Bayesian classification, by taking into account both word features and features of neighboring (left/right) words.

Aiello et al. [1] applied the well-known decision tree learning system C4.5 [38] to learn classification rules for textual logical components (body, caption, title, page number). Only seven attributes are considered: two for the geometrical properties of the block (aspect ratio, area ratio), four for the textual content (font size ratio, font style, number of characters, number of lines) and one for spatial closeness to a figure. Experimental results show that these seven features are sufficient to learn a decision tree with a very high ($> 90\%$) recognition rate for body, title and page number. However, the experimentation is mostly based on ground truth data for layout structures and textual content, which is an ideal situation.

Interestingly, although there have been attempts to deal with a relational representation of data [12], studies reported in the literature for document image understanding, do not consider the transductive learning setting. Therefore, we intend this contribution to be a further step towards the investigation of methods which originate from the intersection of these three promising research areas: namely, transduction, relational data mining and document image understanding.

For transductive learning, several methods have been proposed in the literature. They are based on support vector machines ([8] [21] [24] [15]), on k-NN classifiers ([25]) and even on general classifiers ([27]). However, they do not work with relational data. For relational classification in the transductive setting, three methods have been reported in the literature.

Krogel and Scheffer ([26]) investigate a transformation (known as *propositionalization*) of a relational description of gene interaction data into a classical double-entry table and then study transduction with the well-known transductive support vector machines. Therefore, transduction is not explicitly investigated on relational representations and it is based on propositionalization, which is fraught with many difficulties in practice ([16, 22]).

Taskar et al. ([42]) build, on the framework of Probabilistic Relational Models, a *generative* probabilistic model which captures interactions between examples, either labeled or unlabeled. However, given sufficient data, a *discriminative* model generally provides significant improvements in classification accuracy over generative models ([43]). This motivates our interest in designing classifiers based on discriminative models.

Ceci et al. [11] propose a different probabilistic method that is based on an iterative approach that bootstraps classification labels on unlabeled examples. However, this approach does not permit us to exploit a preliminary descriptive phase (e.g. association rule discovery or emerging pattern discovery) that helps to obtain a twofold advantage. First, the user can decide to mine both a descriptive and a classification model in the same data mining process [29]. Second, we can solve the *understandability* problem [36] that may occur with some classification methods. Indeed, many rules produced by standard classification systems are difficult to understand because these systems often use only domain independent biases and heuristics, which may not fulfill users' expectations. With the descriptive classification approach, the problem of finding understandable rules is reduced to a postprocessing task [29].

Data mining research has provided several solutions for the task of emerging pattern discovery. In the seminal work by Dong and Li [17], a border-based approach is adopted to discover the EPs discriminating between separate classes. Borders are used to represent both candidates and subsets of Emerging Patterns (EPs); the border differential operation is then used to discover the EPs. Zhang et al. [46] have described an efficient method, called ConsEPMiner, which adopts a level-wise generate-and-test approach to discover EPs which satisfies several constraints (e.g., growth-rate improvement). Recently, Fan and Ramamohanarao [20] have proposed a method which improves the efficiency of EPs discovery by adopting a CP-tree data structure to register the counts of both the positive and negative classes. All these methods assume that data to be mined are stored in a single data table. An attempt to upgrade the emerging pattern discovery to deal with relational data has been reported in [5], where the authors propose adapting the *levelwise* method described in [32] to the case of relational emerging patterns.

## 4   Extracting Emerging Patterns with SPADA

In this work, we propose a modified version of the system SPADA [28], originally designed for *relational* frequent pattern discovery, that permits us to extract emerging patterns. SPADA represents relational data *à la* Datalog, a logic programming language with no function symbols specifically designed to implement deductive databases. Moreover, it takes into account a background knowledge (*BK*) expressed in Prolog and is able to mine relational patterns at multiple levels of granularity, in order to properly deal with hierarchies of objects. When these are available, it is important to take them into account, since patterns involving more abstract objects are better supported (although less precise), while patterns involving more specific objects have higher confidence values (although lower support values). Hence by efficiently exploring the pattern space at different levels of abstraction (or granularity) it is possible to find the right trade-off between these two conflicting criteria.

SPADA distinguishes between the set $S$ of *reference* (or target) *objects*, which are the main subject of analysis, and the sets $R_k$, $1 \le k \le m$, of *task-relevant* (or non-target) objects, which are related to the former and can contribute to account for the variation. Each unit of analysis includes a distinct reference object and many related task-relevant objects. Therefore, the description of a unit of analysis consists of both the properties of included reference and task-relevant objects as well as their relationships. From a database viewpoint, $S$ corresponds to the target table $T \in SC$ and each $R_k$ corresponds to a different relational table $T_i \in SC$. A unit of analysis corresponds to a tuple in $t \in T$ and to all the tuples in the databases related to $t$ according to foreign key constraints.

In the following sub-sections, the document description problem is presented and the learning strategy is described, as it has been modified in order to mine emerging patterns.

## 4.1 Document Description

In the logic framework adopted by SPADA, a relational database is boiled down into a deductive database. Properties of both reference and task-relevant objects are represented in the extensional part $D_E$, while the domain knowledge is expressed as a normal logic program which defines the intensional part $D_I$. For example, we report a fragment of the extensional part of a deductive database $D$ which describes spatial and textual information extracted from the document image reported in Figure 1:

*block(b1). block(b2). ...*
*height(b2,[11..54]). width(b1,[7..82]). ...*
*on_top(b2,b1). ... on_top(b2,b3). ...*
*part_of(b1,p1). part_of(b2,p1). page_first(p1). ...*
*abstract(b1). title(b2). ...*
*text_in_abstract(b1,'base'). text_in_title(b2,'model')....*

In this example, $b1$ and $b2$ are two constants which denote as many distinct layout components (reference objects) , while $p1$ denotes a document page (task-relevant object). Predicate *block* defines a layout component, *part_of* associates a block to a document page, *height* and *width* describe geometrical properties of layout components, *on_top* expresses a topological relationship between layout components, *page_first(p1)* refers to the position of the page in the document, *abstract* and *title* associate $b1$ and $b2$ with a logical label, *text_in_abstract* and *text_in_title* permit us to describe the textual content of the logical components.

**Fig. 1** Document Layout: logical components

The complete list of predicates is reported in Table 1. The aspatial feature *type_of* specifies the content type of a layout component (e.g. image, text, horizontal line). Logical features are used to associate a logical label to a layout object and depend on the specific domain. In the case of scientific papers (considered in this work), possible logical labels are: *affiliation*, *page_number*, *figure*, *caption*, *index_term*, *running_head*, *author*, *title*, *abstract*, *formulae*, *subsection_title*, *section_title*, *biography*, *references*, *paragraph*, *table*.

Textual content is represented by means of another class of predicates, which are true when the term reported as the second argument occurs in the layout component denoted by the first argument. Terms are automatically extracted by means of a text-processing module. All terms in the textual components are tokenized and the set of obtained tokens is filtered out in order to remove punctuation marks, numbers and tokens of less than three characters. Standard text preprocessing methods are used to:

**Table 1** Used predicates

| | | |
|---|---|---|
| Layout structure | Locational features | $x\_pos\_center/2$ |
| | | $y\_pos\_center/2$ |
| | Geometrical features | $height/2$ |
| | | $width/2$ |
| | Topological features | $on\_top/2$ |
| | | $to\_right/2$ |
| | Aspatial feature | $type\_of/2$ |
| Logical structure | Logical features | application dependent (e.g., *abstract/1*) |
| Text | Textual features | application dependent (e.g., *text\_in\_abstract/2*) |

1. remove stopwords, such as articles, adverbs, prepositions and other frequent words;
2. determine equivalent stems (stemming), such as "topolog" in the words "topology" and "topological", by means of Porter's algorithm for English texts [37].

Only relevant tokens are used in textual predicates. They are selected by maximizing the product $maxTF \times DF^2 \times ICF$ [13] that scores high terms appearing (possibly frequently) in a single logical component $c$ and penalizes terms common to other logical components. More formally, let $c$ be a logical label associated to a textual component. Let $d$ be the bag of tokens in a component labeled with $c$ (after the tokenizing, filtering and stemming steps), $w$ a term in $d$ and $TF_d(w)$ the relative frequency of $w$ in $d$. Then, the following statistics can be computed:

1. the maximum value $TF_c(w)$ of $TF_d(w)$ on all logical components $d$ labeled with $c$;
2. the document frequency $DF_c^2(w)$, i.e., the percentage of logical components labeled with $c$ in which the term $w$ occurs;
3. the category frequency $CF_c(w)$, i.e., the number of labels $c' \neq c$, such that $w$ occurs in logical components labeled with $c'$.

Then, the score $v_i$ associated to the term $w_i$ belonging to at least one of the logical components labeled with $c$ is:

$$v_i = TF_c(w_i) \times DF_c^2(w_i) \times 1/CF_c(w_i) \qquad (3)$$

According to this function, it is possible to identify a ranked list of "discriminative" terms for each of the possible labels. From this list, we select the best $n_{dict}$ terms in $Dict_c$, where $n_{dict}$ is a user-defined parameter. The textual dimension of each logical component $d$ labeled as $c$ is represented in the document description as a set of ground facts that express the presence of a term $w \in Dict_c$ in the specified logical component.

## 4.2   The Mining Step

In the original version of SPADA, the problem of mining frequent patterns can be formalized as follows:
*Given*

- a set $S$ of *reference objects*,
- some sets $R_k$, $1 \leq k \leq m$, of *task-relevant objects*,
- a background knowledge $BK$ including some hierarchies $H_k$ on objects in $R_k$,
- $M$ granularity levels in the descriptions (1 is the highest while $M$ is the lowest),
- a set of granularity assignments $\Psi_k$ which associate each object in $H_k$ with a granularity level
- a set of thresholds $minsup[l]$ for each granularity level
- a language bias $LB$ that constrains the search space;

*Find* frequent multi-level patterns, i.e., frequent patterns involving objects at different granularity levels.

Hierarchies $H_k$ define *is-a* (i.e., taxonomical) relations on task-relevant objects. The frequency depends on the granularity level $l$ at which patterns describe data. Therefore, a pattern $P$ with support $s$ at level $l$ is *frequent* if $s \geq minsup[l]$ and all ancestors of $P$ with respect to $H_k$ are frequent at their corresponding levels.

SPADA operates in two steps for each granularity level: i) pattern generation; *ii*) pattern evaluation. It exploits statistics computed at granularity level $l$ when computing the supports of patterns at granularity level $l + 1$. The expressive power of first-order logic is utilized to specify both the background knowledge $BK$, such as hierarchies and domain specific knowledge, and the language bias $LB$. The $LB$ is relevant to allow the user to specify his/her bias for interesting solutions, and then to exploit this bias to improve both the efficiency of the mining process and the quality of the discovered rules.

In our case, we modified SPADA in order to discover emerging patterns instead of frequent patterns. Accordingly, the mining problem is modified as follows:
*Given*

- a set $S$ of *reference objects*,
- a label value $y \in Y = \{C_1, C_2, \ldots, C_L\}$ associated to each reference object,
- some sets $R_k$, $1 \leq k \leq m$, of *task-relevant objects*,
- a background knowledge $BK$ including some hierarchies $H_k$ on objects in $R_k$,
- $M$ granularity levels in the descriptions,
- a set of granularity assignments $\Psi_k$ which associate each object in $H_k$ with a granularity level
- a couple of sets of thresholds $minsup[l]$ and $minGR[l]$ for each granularity level
- a language bias $LB$ that constrains the search space;

*Find* A set of multilevel emerging patterns $\{F \,|\, supp_{C_i}(F) \geq minsup[l], GR_{C_i}(F) \geq minGR[l]\}$

In this formulation, $supp_{C_i}(F)$ represents the support of pattern $F$ in the subset of reference objects labeled with $C_i$, while the growth rate $GR_{C_i}(F)$ is defined as:

$$GR_{C_i}(F) = \frac{supp_{C_i}(F)}{supp_{\neg C_i}(F)}$$

where $supp_{\neg C_i}(F)$ is the support of pattern $F$ in the subset of reference objects labeled with $c \in \{C_1, \ldots, C_{i-1}, C_{i+1}, \ldots C_L\}$.

To efficiently mine frequent patterns, SPADA prunes the search space by exploiting the monotonicity of support. Let $F'$ be a refinement of a pattern $F$ (i.e. $F'$ is more specific than $F$). If $F$ is an infrequent pattern for the class $C_i$ (i.e. $supp_{C_i}(F) < minsup$), then also $supp_{C_i}(F') < minsup$. This means that $F'$ cannot be an emerging pattern that permits us to distinguish $C_i$ from $\neg C_i$. Hence, SPADA does not refine patterns which are infrequent on $C_i$.

Unfortunately, the monotonicity property does not hold for the growth rate: a refinement of an emerging pattern whose growth rate is lower than the threshold $minGR$ may or may not be an EP. However, also in this case it is possible to prune the search space. According to [46], we modified the mining algorithm originally developed in SPADA in order to avoid generating the refinements of a pattern $F$ in the case that $GR_{C_i}(F) = \infty$ (i.e., $supp_{C_i}(F) > 0$ and $supp_{\neg C_i}(F) = 0$). Indeed, due to the monotonicity of support, for each pattern $F'$ obtained as a refinement of $F$: $supp_{C_i}(F) \geq supp_{C_i}(F')$, then $supp_{C_i}(F') = 0$. Hence, $GR_{C_i}(F') = 0$ in the case that $supp_{C_i}(F') = 0$, while $GR_{C_i}(F') = \infty$ in the case that $supp_{C_i}(F') > 0$. In the former case, $F'$ is not worth considering. In the latter case, we prefer $F$ to $F'$, based on the Occam's razor principle, according to which all things being equal, the simplest solution tends to be the best one ($F$ has the same discriminating ability as $F'$).

In our application domain, reference objects are all the logical components for which a logical label is specified. Task relevant objects are all the logical components (including undefined components), as well as pages and documents. The *BK* is used to specify the hierarchy of logical components (Figure 2). The *BK* also permits us to automatically associate information on page order to layout components, since the presence of some logical components may depend on the page order (e.g. the author is on the first page). This concept is expressed by means of the following Datalog rules stored in the intensional part $D_I$ of the deductive database $D$:

*at_page_first*$(X)$ :- *part_of*$(Y,X)$, *page_first*$(Y)$.
*at_page_intermediate*$(X)$ :- *part_of*$(Y,X)$, *page_intermediate*$(Y)$.
*at_page_last_but_one*$(X)$ :- *part_of*$(Y,X)$, *page_last_but_one*$(Y)$.
*at_page_last*$(X)$ :- *part_of*$(Y,X)$, *page_last*$(Y)$.

Moreover, in the *BK* we can also define the predicate *text_in_component*:

*text_in_component*$(X,Y)$ :- *text_in_index_term*$(X,Y)$.
*text_in_component*$(X,Y)$ :- *text_in_references*$(X,Y)$.
*text_in_component*$(X,Y)$ :- *text_in_abstract*$(X,Y)$.
*text_in_component*$(X,Y)$ :- *text_in_title*$(X,Y)$.
*text_in_component*$(X,Y)$ :- *text_in_running_head*$(X,Y)$.

```
article
+ – – heading
|   + – – identification
|   |   + – – (title, author, affiliation)
|   + – – synopsis
|       + – – (abstract, index_term)
+ – – content
|   + – – final components
|   |   + – – (biography, references)
|   + – – body
|       + – – (section_title, subsect_title, paragraph, caption, figure, formulae, table)
+ – – page_component
|   + – – running_head
|   + – – page_number
+ – – undefined
```

**Fig. 2** Hierarchy of logical components

It is noteworthy that hierarchies are defined on task relevant objects. This means that, in theory, it is not possible to consider the same reference object at different levels of granularity. To overcome this limitation, we introduced in the *BK* the fact *specialize*$(X,X)$, which allows SPADA to consider a reference object as a task-relevant object, and we forced SPADA (by means of *LB* constraints) to include the predicate *specialize*$/2$ in the emerging patterns.

We also extended the language bias of SPADA in order to deal properly with predicates representing textual features. Indeed, the SPADA language bias requires the user to specify predicates that can be involved in a pattern. For instance, if we are interested in patterns that contain the predicate *text_in_abstract*$(A, paper)$, where $A$ is a variable representing a task-relevant object (*tro*) already introduced in the pattern and *paper* is a constant value representing the presence of the term "paper" in $A$, we have to specify the following bias rule:

$$lb\_atom(text\_in\_component(old\ tro, paper))$$

This means that it is necessary to specify a rule for each constant value that could be involved in the predicate. However, there are hundreds of constants representing selected terms (the number depends on the $n_{dict}$ constant and on the number of user-selected logical labels for which the textual dimension is considered). To avoid the manual or semiautomatic specification of different *lb_atom*'s, we extended the SPADA syntax for *LB* in order to support anonymous variables:

$$lb\_atom(text\_in\_component(old\ tro, \_))$$

This means that we intend to consider in the search phase those patterns involving the predicate *text_in_component*$/2$, whose second argument is an arbitrary term.

The SPADA search strategy has been consequently modified in order to support this additional feature.

An additional aspect worth to be considered in this work is related to the possibility of dealing with numerical data. Indeed, although the application requires the manipulation of such data in order to consider geometrical features of a layout component, SPADA, in its original version, is not able to automatically deal with them. To avoid this problem, a simple equal-frequency discretization algorithm has been integrated. This allows the mining algorithm implemented in SPADA to process discrete intervals instead of numerical data.

## 5 Transductive Classification

The transductive classification implemented in our proposal upgrades the EP-based classifier CAEP [18] to the relational setting. It computes a membership score of an object to each class. The score is computed by means of a growth rate based function of the relational EPs covered by the object to be classified. The largest score determines the object's class.

The score is computed on the basis of the subset of relational emerging patterns that cover the object to be classified. Formally, let $o$ be the description of the object to be classified (an object is represented by a tuple in the target table and all the tuples related to it, according to foreign key constraints), $\Re(o) = \{F \in \Re | \exists \theta \ F\theta \subseteq o\}$ is the set of relational emerging patterns that cover the object $o$.

The score of $o$ on the class $C_i$ is computed as follows:

$$score(o, C_i) = \sum_{F \in \Re(o)} \frac{GR_{C_i}(F)}{GR_{C_i}(F) + 1} sup_{C_i}(F) \qquad (4)$$

This measure may result in an inaccurate classifier in the case of unbalanced datasets, i.e. when training objects are not uniformly distributed over the classes. In order to mitigate this problem, in [18] the authors proposed normalizing this score on the basis of the median of the scores obtained from training examples belonging to $C_i$. This results in the following classification function:

$$class(o) = argmax_{C_i} \frac{score(o, C_i)}{median_{ro \in TS}(score(ro, C_i))} \qquad (5)$$

where $TS$ represents the training set.

Although this normalization solves problems due to unbalanced class distribution, in our case the main problem arises from the different number of emerging patterns that are extracted from different classes. This means that, in our case, a different normalization that weights the number of emerging patterns is necessary:

$$score(o, C_i) = \frac{1}{|\Re(o)|} \sum_{F \in \Re(o)} \frac{GR_{C_i}(F)}{GR_{C_i}(F) + 1} sup_{C_i}(F) \qquad (6)$$

Since $sup_{C_i}(F)$ represents the probability that a reference object belonging to class $C_i$ is covered by $F$, Equation (6) can be transformed as follows:

$$score(o,C_i) = \frac{1}{|\Re(o)|} \sum_{F \in \Re(o)} \frac{GR_{C_i}(F)}{GR_{C_i}(F)+1} P(F|C_i) \tag{7}$$

By applying the Bayes theorem:

$$score(o,C_i) = \frac{1}{|\Re(o)|} \sum_{F \in \Re(o)} \frac{GR_{C_i}(F)}{GR_{C_i}(F)+1} \frac{P(C_i|F)}{P(C_i)} \times P(F) \tag{8}$$

where $P(C_i|F)$ can be estimated as the percentage of examples covering $F$ in $TS$ that belong to $C_i$, $P(C_i)$ can be estimated as the percentage of examples in $TS$ that belong to $C_i$, and $P(F)$ is the percentage of examples covering $F$. According to the transductive learning setting, this factor is estimated by considering the whole set of examples ($TS \cup WS$). This would provide a more reliable estimation of $P(F)$ (since it is obtained from a larger population of examples potentially coming from the same distribution).

$$P(F) = \frac{\#\{ro|ro \in TS \cup WS, \exists \theta \ F\theta \subseteq ro\}}{\#\{ro|ro \in TS \cup WS\}} \tag{9}$$

## 6   Experiments

To evaluate the viability of the proposed transductive approach, it has been evaluated on a real-world dataset, consisting of multi-page articles published in an international journal. In particular, we considered twenty-four papers, published as either regular or short, in the IEEE Transactions on Pattern Analysis and Machine Intelligence (TPAMI), in the January and February issues of 1996. Each paper is a multi-page document, therefore, we processed 217 document images containing 3611 layout components (examples). Among them, the user manually labeled 2,693 layout components, that is, on average, 112.2 components per document and 12.41 per page. The components that have not been labeled are "irrelevant" for the task in hand or are associated to "noise" blocks: they are automatically considered *undefined*. Overall, there are 918 unlabeled layout components.

The dataset is analyzed by means of a 4-fold cross-validation on documents. Each fold contains 902.75 layout components on average and the algorithm is run in order to collect predictive average accuracy and, for each class, average precision and recall. Unlike the standard cross-validation approach, here one fold at a time is set aside to be used as the *training set* (and not as the *test set*). Small training set sizes allow us to validate the transductive approach, but may result in high error rates as well.

Table 2 reports classes involved in the evaluation (logical labels) and the number of examples belonging to them. As can be seen, classes are highly unbalanced in the number of examples.

**Table 2** Class and example distribution

| logical label | No. of examples |
|:---:|:---:|
| *abstract* | 39 |
| *affiliation* | 23 |
| *author* | 28 |
| *biography* | 21 |
| *caption* | 202 |
| *figure* | 357 |
| *formulae* | 333 |
| *index_term* | 25 |
| *page_number* | 191 |
| *paragraph* | 968 |
| *references* | 45 |
| *running_head* | 230 |
| *section_title* | 65 |
| *subsection_title* | 27 |
| *table* | 48 |
| *title* | 91 |
| undefined | 918 |
| TOTAL | 3611 |

In the extraction of emerging patterns, SPADA has been run with different parameters: $minGR = \{2, 8, 68\}$ and $minsup = \{10\%, 20\%, 30\%\}$. In Table 3 the average number of emerging patterns extracted with different parameter values is reported. As expected, by increasing *minsup* and *minGR* values, the number of extracted emerging patterns is drastically reduced.

**Table 3** Average number of extracted emerging patterns

| *minGR* | *minsup* | | |
|:---:|:---:|:---:|:---:|
| | 10 | 20 | 30 |
| 2 | 22959.25 | 11769.5 | 7943.75 |
| 8 | 15608.5 | 8315.25 | 5947.25 |
| 64 | 10266.75 | 5306.5 | 3939.5 |

By looking at the distribution of emerging patterns over the classes (see Table 4), we note that classes for which the discrimination is simpler are characterized by a higher number of emerging patterns. It is also noteworthy that the number of patterns is not related to the number of examples. This means that the extracted emerging patterns do not suffer from overfitting problems.

An example of an emerging pattern for the class *abstract* is reported in the following:

$$is\_a\_block(A), specialize(A, B), is\_a(B, abstract), text\_in\_component(B, paper).$$
$$supp_{abstract} = 50\%; GR_{abstract} = +\infty$$

**Table 4** Average number of extracted emerging patterns per class (*minsup* = 10%, *minGR* = 2)

| logical label | Average No. of emerging patterns |
|:---:|:---:|
| *abstract* | 3235.25 |
| *affiliation* | 738.5 |
| *author* | 1569.25 |
| *biography* | 645 |
| *caption* | 512.5 |
| *figure* | 744.75 |
| *formulae* | 595.75 |
| *index_term* | 2099.25 |
| *page_number* | 1443 |
| *paragraph* | 780.75 |
| *references* | 2782.25 |
| *running_head* | 2751.25 |
| *section_title* | 995.5 |
| *subsection_title* | 773.5 |
| *table* | 1108.75 |
| *title* | 2184 |
| TOTAL | 22959.25 |

This pattern states that 50% of layout components labeled as *abstract* contains the term "paper". Moreover, this pattern is not satisfied by layout components labeled with a different label. This is probably due to the fact that term "paper" is not selected for other logical labels.

We have the same pattern at a higher level of the hierarchy:

$is\_a\_block(A), specialize(A,B), is\_a(B, synopsis), text\_in\_component(B, paper).$
$$supp_{synopsis} = 23.5\%, GR_{synopsis} = +\infty.$$

As we can see, support decreases since the term "paper" is not selected for the class *index_term*.

Another example of an emerging pattern for the class *abstract* is:

$is\_a\_block(A), specialize(A,B), is\_a(B, abstract), only\_left\_col(B,C), C \neq B,$
$text\_in\_component(C, index).$
$$supp_{abstract} = 100\%, GR_{abstract} = +\infty.$$

This emerging pattern shows the advantage of exploiting the relational nature of data. Indeed, it shows that layout components labeled as abstract are always aligned with components that contain the term "index" (probably belonging to the class "index_term").

Finally, the emerging pattern:

$is\_a\_block(A), specialize(A,B), is\_a(B, section\_title), height(B, [6..12]),$
$at\_page\_first(B).$
$$supp_{section\_title} = 55.5\%, GR_{section\_title} = 42.67$$

considers other types of task relevant objects (i.e. pages) by exploiting information in *BK*. In this pattern, height is expressed in number of pixels.

In Table 5, average classification accuracy is reported. Results are collected for different values of *minGR* and *minsup*. As we can see, more interesting results are obtained with small values of *minGR* and with high values of *minsup* or, alternatively, with high values of *minGR* and with low values of *minsup*. This means that predictive accuracy significantly depends on the number of extracted patterns. With a high number of patterns, probabilities are flattened and the system loses its discriminative capabilities. On the other hand, when the number of patterns decreases, the system has not enough information to discriminate among classes.

**Table 5** Average classification accuracy

| minsup | minGR | | |
|---|---|---|---|
| | 2 | 8 | 64 |
| 10 | 43.47 | 56.67 | 60.39 |
| 20 | 63.84 | 58.32 | 54.8 |
| 30 | 64.63 | 57.69 | 44.82 |

A different perspective of results is reported in Table 6, where precision and recall are reported for each class. Results are collected for $minsup = 10\%$ and $minGR = 2$. As can be noted, results vary significantly from one class to another. Indeed, some classes are more difficult to identify since they do not show regularities (e.g. *subsection_title* vs. *section_title*). Other classes can be more easily identified because the use of text is effective and/or because they are subject to formatting regularities.

**Table 6** Average precision and recall per class ($minsup = 10\%$, $minGR = 2$)

| logical label | Average precision | Average recall |
|---|---|---|
| abstract | 0.28593 | 0.89157 |
| affiliation | 0.11167 | 0.94759 |
| author | 0.54731 | 0.75189 |
| biography | 0.16565 | 0.87101 |
| caption | 0.47886 | 0.07607 |
| figure | 0.88800 | 0.794598 |
| formulae | 0.55592 | 0.31498 |
| index_term | 0.45324 | 0.96687 |
| page_number | 0.81994 | 0.92180 |
| paragraph | 0.94674 | 0.44204 |
| references | 0.85855 | 0.850724 |
| running_head | 0.92105 | 0.80608 |
| section_title | 0.47162 | 0.29978 |
| subsection_title | 0.07779 | 0.74408 |
| table | 0.15290 | 0.23070 |
| title | 0.40062 | 0.68959 |
| Average | 0.5085 | 0.6625 |

## 7 Conclusions

In this work, the induction of a classifier for the automated recognition of semantically relevant layout components has been investigated. In particular, we have investigated the combination of transductive inference with principled relational classification, in order to face the challenges posed by the application domain, characterized by complex and heterogeneous data, which are naturally modeled as several tables of a relational database and characterized by the availability of a small (large) set of labeled (unlabeled) data.

The experiments provide interesting qualitative and quantitative results. For future work, we intend to compare our approach with other competitive approaches proposed in the literature and to employ a different classification strategy that permits us to exploit labels confidently associated to working examples in the classification of other working examples.

**Acknowledgements.** This work is partial fulfillment of the research objectives of the project "ATENEO 2009 - Estrazione, Rappresentazione e Analisi di Dati Complessi". The authors gratefully acknowledge Dr. Lynn Rudd for reading the final version.

## References

1. Aiello, M., Monz, C., Todoran, L.: Document understanding for a broad class of documents. IJDAR 5(1), 1–16 (2002)
2. Akindele, O.T., Belaïd, A.: Construction of generic models of document structures using inference of tree grammars. In: ICDAR 1995: Proceedings of the Third International Conference on Document Analysis and Recognition, vol. 1, p. 206. IEEE Computer Society, Washington, DC, USA (1995)
3. Allen, J.F.: Maintaining knowledge about temporal intervals. Commun. ACM 26(11), 832–843 (1983)
4. Appice, A., Ceci, M., Malerba, D.: Transductive learning for spatial regression with co-training. In: Shin, S.Y., Ossowski, S., Schumacher, M., Palakal, M.J., Hung, C.-C. (eds.) SAC, pp. 1065–1070. ACM Press, New York (2010)
5. Appice, A., Ceci, M., Malgieri, C., Malerba, D.: Discovering relational emerging patterns. In: Basili, R., Pazienza, M.T. (eds.) AI*IA 2007. LNCS (LNAI), vol. 4733, pp. 206–217. Springer, Heidelberg (2007)
6. Rosenfeld, A., Hummel, R., Zucker, S.: Scene labeling by relaxation operations. J IEEE Transactions SMC 6(6), 420–433 (1976)
7. Baird, H.S., Casey, M.R.: Towards versatile document analysis systems. In: Bunke, H., Spitz, A.L. (eds.) DAS 2006. LNCS, vol. 3872, pp. 280–290. Springer, Heidelberg (2006)
8. Bennett, K.P.: Combining support vector and mathematical programming methods for classification, pp. 307–326. MIT Press, Cambridge (1999)
9. Ceci, M., Appice, A.: Spatial associative classification: propositional vs. structural approach. Journal of Intelligent Information Systems 27(3), 191–213 (2006)
10. Ceci, M., Appice, A., Malerba, D.: Emerging pattern based classification in relational data mining. In: Bhowmick, S.S., Küng, J., Wagner, R. (eds.) DEXA 2008. LNCS, vol. 5181, pp. 283–296. Springer, Heidelberg (2008)

11. Ceci, M., Appice, A., Malerba, D.: Transductive learning for spatial data classification. In: Koronacki, J., Raś, Z.W., Wierzchoń, S.T., Kacprzyk, J. (eds.) Advances in Machine Learning I. SCI, vol. 262, pp. 189–207. Springer, Heidelberg (2010)

12. Ceci, M., Berardi, M., Malerba, D.: Relational data mining and ILP for document image understanding. Applied Artificial Intelligence 21(4&5), 317–342 (2007)

13. Ceci, M., Malerba, D.: Classifying web documents in a hierarchy of categories: a comprehensive study. J. Intell. Inf. Syst. 28(1), 37–78 (2007)

14. Chapelle, O., Schölkopf, B., Zien, A.: A discussion of semi-supervised learning and transduction. In: Chapelle, O., Schölkopf, B., Zien, A. (eds.) Semi-Supervised Learning, pp. 457–462. MIT Press, Cambridge (2006)

15. Chen, Y., Wang, G., Dong, S.: Learning with progressive transductive support vector machines. Pattern Recognition Letters 24, 1845–1855 (2003)

16. De Raedt, L.: Attribute-value learning versus inductive logic programming: the missing links. In: Page, D.L. (ed.) ILP 1998. LNCS (LNAI), vol. 1446, pp. 1–8. Springer, Heidelberg (1998)

17. Dong, G., Li, J.: Efficient mining of emerging patterns: Discovering trends and differences. In: International Conference on Knowledge Discovery and Data Mining, pp. 43–52. ACM Press, New York (1999)

18. Dong, G., Zhang, X., Wong, L., Li, J.: CAEP: Classification by aggregating emerging patterns. In: Arikawa, S., Nakata, I. (eds.) DS 1999. LNCS (LNAI), vol. 1721, pp. 30–42. Springer, Heidelberg (1999)

19. Esposito, F., Malerba, D., Semeraro, G.: Multistrategy learning for document recognition. Applied Artificial Intelligence 8(1), 33–84 (1994)

20. Fan, H., Ramamohanarao, K.: An efficient singlescan algorithm for mining essential jumping emerging patterns for classification. In: Pacific-Asia Conference on Knowledge Discovery and Data Mining, pp. 456–462 (2002)

21. Gammerman, A., Azoury, K., Vapnik, V.: Learning by transduction. In: Proc. of the 14th Annual Conference on Uncertainty in Artificial Intelligence, UAI 1998, pp. 148–155. Morgan Kaufmann, San Francisco (1998)

22. Getoor, L.: Multi-relational data mining using probabilistic relational models: research summary. In: Knobbe, A., Van der Wallen, D.M.G. (eds.) Proc.of the 1st Workshop in Multi-relational Data Mining, Freiburg, Germany (2001)

23. Jensen, D., Neville, J.: Linkage and autocorrelation cause feature selection bias in relational learning. In: Proc. of the Nineteenth International Conference on Machine Learning (2002)

24. Joachims, T.: Transductive inference for text classification using support vector machines. In: Proc. of the 16th International Conference on Machine Learning, ICML 1999, pp. 200–209. Morgan Kaufmann, San Francisco (1999)

25. Joachims, T.: Transductive learning via spectral graph partitioning. In: Proc. of the 20th International Conference on Machine Learning, ICML 2003, Morgan Kaufmann, San Francisco (2003)

26. Krogel, M.-A., Scheffer, T.: Multi-relational learning, text mining, and semi-supervised learning for functional genomics. Machine Learning 57(1-2), 61–81 (2004)

27. Kukar, M., Kononenko, I.: Reliable classifications with machine learning. In: Elomaa, T., Mannila, H., Toivonen, H. (eds.) ECML 2002. LNCS (LNAI), vol. 2430, pp. 219–231. Springer, Heidelberg (2002)

28. Lisi, F.A., Malerba, D.: Inducing multi-level association rules from multiple relations. Machine Learning 55, 175–210 (2004)

29. Liu, B., Hsu, W., Ma, Y.: Integrating classification and association rule mining. In: Knowledge Discovery and Data Mining KDD 1998, New York, pp. 80–86 (1998)

30. Malerba, D.: A relational perspective on spatial data mining. IJDMMM 1(1), 103–118 (2008)
31. Malerba, D., Ceci, M., Berardi, M.: Machine learning for reading order detection in document image understanding. In: Marinai, S., Fujisawa, H. (eds.) Machine Learning in Document Analysis and Recognition. SCI, vol. 90, pp. 45–69. Springer, Heidelberg (2008)
32. Mannila, H., Toivonen, H.: Levelwise search and borders of theories in knowledge discovery. Data Min. Knowl. Discov. 1(3), 241–258 (1997)
33. Nagy, G.: Twenty years of document image analysis in pami. IEEE Trans. Pattern Anal. Mach. Intell. 22(1), 38–62 (2000)
34. Niyogi, D., Srihari, S.N.: Knowledge-based derivation of document logical structure. In: ICDAR 1995: Proceedings of the Third International Conference on Document Analysis and Recognition, vol. 1, p. 472. IEEE Computer Society Press, Washington, DC, USA (1995)
35. Palmero, G.I.S., Dimitriadis, Y.A.: Structured document labeling and rule extraction using a new recurrent fuzzy-neural system. In: ICDAR 1999: Proceedings of the Fifth International Conference on Document Analysis and Recognition, p. 181. IEEE Computer Society Press, Washington, DC, USA (1999)
36. Pazzani, M.J., Mani, S., Shankle, W.R.: Beyond concise and colorful: Learning intelligible rules. In: KDD, pp. 235–238 (1997)
37. Porter, M.F.: An algorithm for suffix stripping. Readings in information retrieval, 313–316 (1997)
38. Quinlan, J.R.: C4.5: programs for machine learning. Morgan Kaufmann Publishers Inc., San Francisco (1993)
39. Robinson, J.A.: A machine oriented logic based on the resolution principle. Journal of the ACM 12, 23–41 (1965)
40. Seeger, M.: Learning with labeled and unlabeled data. Technical report, Institute for Adaptive and Neural Computation. University of Edinburgh (2001)
41. Souafi-Bensafi, S., Parizeau, M., Lebourgeois, F., Emptoz, H.: Bayesian networks classifiers applied to documents. In: ICPR (1), p. 483 (2002)
42. Taskar, B., Segal, E., Koller, D.: Probabilistic classification and clustering in relational data. In: Nebel, B. (ed.) IJCAI, pp. 870–878. Morgan Kaufmann, San Francisco (2001)
43. Vapnik, V.: The Nature of Statistical Learning Theory. Springer, New York (1995)
44. Vapnik, V.: Statistical Learning Theory. Wiley, New York (1998)
45. Walischewski, H.: Automatic knowledge acquisition for spatial document interpretation. In: ICDAR, pp. 243–247. IEEE Computer Society Press, Los Alamitos (1997)
46. Zhang, X., Dong, G., Ramamohanarao, K.: Exploring constraints to efficiently mine emerging patterns from large high-dimensional datasets. In: KDD, pp. 310–314 (2000)

# Progressive Filtering on the Web:
# The Press Reviews Case Study

Andrea Addis, Giuliano Armano, and Eloisa Vargiu

**Abstract.** Progressive Filtering is a hierarchical classification technique framed within the local classifier per node approach where each classifier is entrusted with deciding whether the input in hand can be forwarded or not to its children. In this chapter, we illustrate the effectiveness of Progressive Filtering on the Web, focusing on the task of automatically creating press reviews. To this end, we present NEWS.MAS, a multiagent system aimed at: (i) extracting information from online newspapers by using suitable wrapper agents, each associated with a specific information source, (ii) categorizing news articles according to a given taxonomy, and (iii) providing user feedback to improve the performance of the system depending on user needs and preferences.

## 1 Introduction and Motivation

Information Retrieval (IR) is the task of finding, from large collections, documents of unstructured nature that satisfy an information need [?]. A typical task consists of retrieving textual documents, such as newspaper and magazine articles or web documents.

An IR task involves three main activities: (i) extracting the required information, (ii) encoding and processing it according to the specific application, and – optionally– (iii) providing suitable feedback mechanisms to improve the overall performances.

Information extraction is aimed at extracting data from information sources through specialized wrappers. In general, given an information source, a specific wrapper can be devised and implemented, able to map the available data to a suitable description that contains relevant information in a structured form. In the literature,

Andrea Addis · Giuliano Armano · Eloisa Vargiu
Dept. of Electrical and Electronic Engineering, University of Cagliari, Italy
e-mail: {addis,armano,vargiu}@diee.unica.it

M. Biba and F. Xhafa (Eds.): Learning Structure and Schemas from Documents, SCI 375, pp. 143–163.
springerlink.com         © Springer-Verlag Berlin Heidelberg 2011

several tools have been proposed with the goal of generating wrappers for web data extraction [29].

Encoding is applied to the information obtained from the selected sources, whereas processing is aimed at progressively filtering it while retaining only relevant data. The actual encoding strictly depends on the specific application (preprocessing activities, such as feature selection and feature extraction, are typically performed to prepare data). If needed, personalization can be performed according to user needs and preferences. When the IR process focuses on text categorization, feature selection or feature extraction techniques are typically applied to perform encoding [49], whereas learned systems perform processing (in this case, categorization) [48]. A module for user feedback can be provided to deal with any feedback optionally provided by the end-user. Simple but effective solutions can be implemented using machine learning techniques.

Many information sources are organized as hierarchies, e.g., web repositories, digital libraries, patent libraries, email folders, product catalogs. In particular, web repositories, such as DMOZ[1], Wikipedia[2], and Medical Subject Headings (MeSH) in MEDLINE[3], encompass an underlying taxonomy. Taxonomies are also very useful in the field of news categorization [5], such as those provided by the International Press Telecommunications Council[4] and the RCV-taxonomy, proposed by Lewis [30] to perform hierarchical text categorization on the Reuters standard document collection.

## 2   Mission

The World Wide Web provides a growing amount of information and data coming from different and heterogeneous sources. As a consequence, it becomes more and more difficult for web users to select contents that match their interests, especially if contents are frequently updated (e.g., news aggregators, newspapers, scientific digital archives, RSS feeds, and blogs). Supporting users in handling the enormous and widespread amount of web information is becoming a primary issue. To this end, several online services have been proposed (e.g., Google News[5] and PRESSToday[6]). Unfortunately, they allow users to choose their interests among macro-areas, which are often inadequate to express what the user is really interested in. Moreover, existing systems typically do not provide a feedback mechanism that permits the user to specify irrelevant or unwanted items, with the goal of progressively adapting the system to her/his actual interests.

In this chapter, we illustrate the effectiveness of Progressive Filtering (PF), a Hierarchical Text Categorization (HTC) technique, in the task of automatically

---

[1] http://www.dmoz.org
[2] http://www.wikipedia.org
[3] http://medline.cos.com
[4] http://www.iptc.org
[5] http://news.google.com
[6] http://www.presstoday.com

creating press reviews. Section 3 reports a brief survey of relevant related work. Section 4 summarizes the PF technique, which, in its simplest setting, decomposes a given rooted taxonomy into pipelines, one for each path that exists between the root and each node of the taxonomy, so that each pipeline can be tuned in isolation. To this end, a Threshold Selection Algorithm (TSA) has been devised, aimed at finding a sub-optimal combination of thresholds for each pipeline. Section 5 illustrates a case study, i.e., NEWS.MAS, a multiagent system that automatically creates press reviews [1]. Built upon X.MAS [2] –a generic multiagent architecture, aimed at retrieving, filtering and reorganizing information according to user interests– NEWS.MAS is devoted to: (i) extract information from online newspapers by using suitable wrapper agents, each associated with a specific information source (i.e., the Reuters portal[7], The Times[8], The New York Times[9]), (ii) categorize with PF news articles according to a given taxonomy, and (iii) provide user feedback to improve the performance of the system depending on user needs and preferences.

## 3   Related Work

The most relevant issues that help to clarify the contextual setting of the chapter are: (i) the work done on HTC, (ii) the work done on the input imbalance problem, and (iii) the work done on multiagent systems (MAS) for information retrieval.

### 3.1   Hierarchical Text Categorization

In recent years several researchers have investigated the use of hierarchies for text categorization, which is also the main focus of this chapter.

Until the mid-1990s researchers mostly ignored the hierarchical structure of categories that occur in several domains. In 1997, Koller and Sahami [25] carried out the first proper study on HTC on the Reuters-22173 collection. Documents were classified according to the given hierarchy by filtering them through the single best-matching first-level class and then sending them to the appropriate second level. This approach showed that hierarchical models perform well when a small number of features per class is used, as no advantages were found using the hierarchical model for large numbers of features.

McCallum et al. [33] proposed a method based on naïve Bayes. The authors compare two techniques: (i) exploring all possible paths in the given hierarchy and (ii) greedily selecting at most two branches according to their probability, as done in [25]. Results show that the latter is more error prone while computationally more efficient.

Mladenić [34] used the hierarchical structure to decompose a problem into a set of subproblems, corresponding to categories (i.e., the nodes of the hierarchy).

---

[7] http://www.reuters.com
[8] http://www.the-times.co.uk/
[9] http://www.nytimes.com/

For each subproblem, a naïve Bayes classifier is generated, considering examples belonging to the given category, including all examples classified in its subtrees. The classification applies to all nodes in parallel; a document is passed down to a category only if the posterior probability for that category is higher than a user-defined threshold.

D'Alessio [17] proposed a system in which, for a given category, the classification is based on a weighted sum of feature occurrences that should be greater than the category threshold. Both single and multiple classifications are possible for each document to be tested. The classification of a document proceeds top-down possibly through multiple paths. An innovative contribution of this work is the possibility of restructuring a given hierarchy or building a new one from scratch.

Dumais and Chen [18] used the hierarchical structure for two purposes: (i) training several SVMs, one for each intermediate node, and (ii) classifying documents by combining scores from SVMs at different levels. The sets of positive and negative examples are built considering documents that belong to categories at the same level, and different feature sets are built, one for each category. Several combination rules have also been assessed.

In the work by Ruiz and Srinivasan [36], a variant of the Hierarchical Mixture of Experts model is used. A hierarchical classifier combining several neural networks is also proposed in [43].

Gaussier et al. [22] proposed a hierarchical generative model for textual data, i.e., a model for hierarchical clustering and categorization of co-occurrence data, focused on documents organization.

In [35], a kernel-based approach for hierarchical text classification in a multi-label context has been presented. The work demonstrates that the use of the dependency structure of microlabels (i.e. unions of partial paths in the tree) in a Markovian Network framework leads to improved prediction accuracy on deep hierarchies. Optimization is made feasible by utilizing decomposition of the original problem and making incremental conditional gradient search in the subproblems.

Ceci and Malerba [13] presented a comprehensive study on hierarchical classification of web documents. They extend a previous work [12] considering: (i) hierarchical feature selection mechanisms; (ii) a naïve Bayes algorithm aimed at avoiding problems related to different document lengths; (iii) the validation of their framework for a probabilistic SVM-based classifier; and (iv) an automated threshold selection algorithm.

More recently, in [19], the authors proposed a multi-label hierarchical text categorization algorithm consisting of a hierarchical variant of ADABOOST.MH, a well-known member of the family of "boosting" learning algorithms.

Bennett et al. [9] studied the problem of the error propagation under the assumption that a mistake is made at "high" nodes in the hierarchy, as well as the problem of dealing with increasingly complex decision surfaces.

Brank et al. [11] dealt with the problem of classifying textual documents into a topical hierarchy of categories. They construct a coding matrix gradually, one column at a time, each new column being defined in a way that the corresponding binary classifier attempts to correct the most common mistakes of the current ensemble of binary classifiers. The goal is to achieve good performance while keeping reasonably low the number of binary classifiers.

## 3.2 The Input Imbalance Problem

High imbalance occurs in real-world domains where the decision system is aimed at detecting rare but important cases [27]. Imbalanced datasets exist in many real-world domains, such as spotting unreliable telecommunication customers, detection of oil spills in satellite radar images, learning word pronunciations, text classification, detection of fraudulent telephone calls, information retrieval and filtering tasks, and so on [46] [47]. Japkowicz [23] contributed to study the class imbalance problem in the context of binary classification, the author studied the problem related to domains in which one class is represented by a large number of examples whereas the other is represented by only a few.

A number of solutions to the class imbalance problem have been proposed both at the data- and algorithmic-level [28] [14] [26]. Data-level solutions include many different forms of resampling such as random oversampling with replacement, random undersampling, directed oversampling, directed undersampling, oversampling with informed generation of new samples, and combinations of the above techniques. To counteract the class imbalance, algorithmic-level solutions include adjusting the costs of the various classes, adjusting the decision threshold, and adopting recognition-based, rather than discrimination-based, learning. Hybrid approaches have also been used to deal with the class imbalance problem.

## 3.3 Agents and Information Retrieval

Autonomous agents and MAS have been successfully applied to a number of problems and have been largely used in different application domains [44].

As for MAS in IR, in the literature, several centralized agent-based architectures aimed at performing IR tasks have been proposed. Among others, let us recall NewT [38], Letizia [32], WebWatcher [7], and SoftBots [20].

NewT [38] is composed by a society of information-filtering interface agents, which learn user preferences and act on her/his behalf. These information agents use a keyword-based filtering algorithm, whereas adaptive techniques are relevance feedback and genetic algorithms.

Letizia [32] is an intelligent user-interface agent able to assist a user while browsing the web. The search for information is performed through a cooperative venture between the user and the software agent: both browse the same search space of linked web documents, looking for interesting ones.

WebWatcher [7] is an information search agent that follows web hyperlinks according to user interests, returning a list of links deemed interesting.

In contrast to systems for assisted browsing or IR, SoftBots [20] accept high-level user goals and dynamically synthesize the appropriate sequence of Internet commands according to a suitable ad-hoc language.

Despite the fact that a centralized approach could have some advantages, in IR tasks it may encompass several problems, in particular how to scale up the architectures to large numbers of users, how to provide high availability in case of constant demand of the involved services, and how to provide high trustability in case of sensitive information, such as personal data. Suitable MAS devoted to perform IR tasks have been proposed. In particular, Sycara et al. [41] proposed Retsina, a MAS infrastructure applied in many domains. Retsina is an open MAS infrastructure that supports communities of heterogeneous agents. Three types of agents have been defined: *(i) interface agents*, able to display the information to the users; *(ii) task agents*, able to assist the user in the process of handling her/his information; and *(iii) information agents*, able to gather relevant information from selected sources.

Other than Retsina, several MAS have been proposed and implemented. Among others, let us recall IR-agents [24], CEMAS [10], and the cooperative multiagent system for web IR proposed in [37].

IR-agents [24] implement an XML-based multiagent model for IR. The corresponding framework is composed of three kinds of agents: *(i) managing agents*, aimed at extracting the semantics of information and at performing the actual tasks imposed by coordinator agents, *(ii) interface agents*, devised to interact with the users, and *(iii) search agents*, aimed at discovering relevant information on the web. IR-agents do not take into account personalization, while providing information in a structured form without the adoption of specific classification mechanisms.

In CEMAS (Concept Exchanging MultiAgent System) [10] the basic idea is to provide specialized agents for each main task, which are: (i) exchanging concepts and links, (ii) representing the user, (iii) searching for new relevant documents matching existing concepts, and (iv) supporting agent coordination. Although CEMAS provides personalization and classification mechanisms based on a semantic approach, the main drawback is that it is not generic, being mainly aimed at supporting scientists while looking for comprehensive information about their topic area.

Finally, in [37] the underlying idea is to adopt intelligent agents that mimic everyday-life activities of information seekers. To this end, agents are also able to profile the user in order to anticipate and achieve her/his preferred goals. Although interesting, the approach is mainly focused on cooperation among agents rather than on IR issues.

## 4    Progressive Filtering in Text Categorization

As we are considering a web scenario, we are mainly interested in studying how to cope with input imbalance in a HTC setting. The underlying motivation is that, in real world applications such as the web, the ratio between interesting and uninteresting documents is typically very low, so that directly using classifiers trained with a balanced training set may be uneffective. To deal with the imbalance problem, we perform the training activity in two phases. First, each classifier is trained by using a balanced dataset. Then, a threshold selection algorithm is applied and thresholds are calculated taking into account the input imbalance.

## 4.1  The Approach

A simple way to categorize the various proposals that have been made in HTC is to focus on the mapping between classifiers and the underlying taxonomy. According to [39], the approaches proposed in the literature can be framed as follows: (i) local classifier per node, (ii) local classifier per parent node, (iii) local classifier per level, and (iv) global classifier.

PF is a simple categorization technique framed within the local classifier per node approach, which admits only binary decisions, as each classifier is entrusted with deciding whether the input in hand can be forwarded or not to its children. The first proposals in which sequential boolean decisions are applied in combination with local classifiers per node can be found in [17], [18], and [40] . In [45], the idea of mirroring the taxonomy structure through binary classifiers is clearly highlighted; the authors call this technique "binarized structured label learning".

In PF, given a taxonomy, where each node represents a classifier entrusted with recognizing all corresponding positive inputs (i.e., interesting documents), each input traverses the taxonomy as a "token", starting from the root. If the current classifier recognizes the token as positive, it is passed on to all its children (if any), and so on. A typical result consists of activating one or more branches within the taxonomy, in which the corresponding classifiers have accepted the token. A theoretical study of the approach is beyond the scope of this chapter, the interested reader could refer to [6] for further details.

A simple way to implement PF consists of unfolding the given taxonomy into pipelines of classifiers, as depicted in Figure 1 and in Figure 2. Each node of the pipeline is a binary classifier able to recognize whether or not an input belongs to the corresponding class (i.e., to the corresponding node of the taxonomy).

Let us note that, partitioning the taxonomy in pipelines gives rise to a set of new classifiers, each represented by a pipeline (as sketched in Figure 2).

Finally, let us note that the implementation of PF described in this chapter performs a sort of "flattening" though *preserving* the information about the

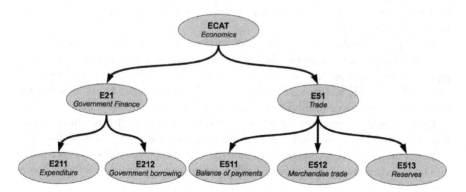

**Fig. 1** An example of taxonomy

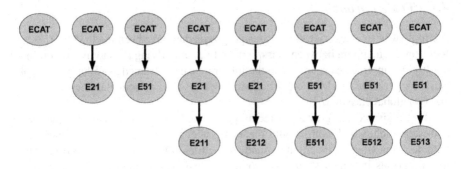

**Fig. 2** The pipelines corresponding to the taxonomy in Figure 1

hierarchical relationships embedded in a pipeline [3]. For instance, the pipeline $\langle ECAT, E21, E211 \rangle$ actually represents the classifier $E211$, although the information about the existing subsumption relationships (i.e., $E211 \leq E21 \leq ECAT$) is preserved.

## 4.2 The Threshold Selection Algorithm

As we know from classical text categorization, given a set of documents $D$ and a set of labels $C$, a function $CSV_i : D \rightarrow [0,1]$ exists for each $c_i \in C$. We assume that the behavior of $c_i$ is controlled by a threshold $\theta_i$, responsible for relaxing or restricting the acceptance rate of the corresponding classifier. Given $d \in D$, $CSV_i(d) \geq \theta_i$ permits to categorize $d$ under $c_i$, whereas $CSV_i(d) < \theta_i$ is interpreted as a decision not to categorize $d$ under $c_i$.

In PF, let us still assume that $CSV_i$ exists for each $c_i \in C$, with the same semantics adopted in the classical case. Considering a pipeline $\pi$, composed of $n$ classifiers, the acceptance policy strictly depends on the vector $\theta_\pi = \langle \theta_1, \theta_2, \cdots, \theta_n \rangle$ that embodies the thresholds of all classifiers in $\pi$. In order to categorize $d$ under $\pi$, the following constraint must be satisfied: $\forall k = 1 \ldots n$, $CSV_i(d) \geq \theta_k$; otherwise, $d$ is not categorized under $c_i$.

A further simplification of the problem consists of allowing a classifier to have different behaviors, depending on which pipeline it is embedded in. Each pipeline can be considered in isolation from the others. For instance, given $\pi_1 = \langle ECAT, E21, E211 \rangle$ and $\pi_2 = \langle ECAT, E21, E212 \rangle$, the classifier $ECAT$ is not compelled to have the same threshold in $\pi_1$ and in $\pi_2$ (the same holds for $E21$).

Given a utility function[10], we are interested in finding an effective and computationally "light" way to reach a sub-optimum in the task of determining the best vector of thresholds. Unfortunately, finding the best acceptance thresholds is a difficult task. Exhaustively trying each possible combination of thresholds (brute-force

---

[10] Different utility functions (e.g., precision, recall, $F_\beta$, user-defined) can be adopted, depending on the constraint imposed by the underlying scenario.

approach) is unfeasible, the number of thresholds being virtually infinite. However, the brute-force approach can be approximated by defining a granularity step that requires to check only a finite number of points in the range $[0,1]$, in which the thresholds are permitted to vary with step $\delta$. Although potentially useful, this "relaxed" brute force algorithm for calibrating thresholds (RBF for short) is still too heavy from a computational point of view. On the contrary, the threshold selection algorithm described in this chapter is characterized by low time complexity while maintaining the capability of finding near-optimum solutions.

Bearing in mind that the lower the threshold the less restrictive is the classifier, we propose a greedy bottom-up algorithm for selecting decision threshold that relies on two functions:

- *Repair* ($\mathscr{R}$), which operates on a classifier $C$ by increasing or decreasing its threshold –i.e., $\mathscr{R}(up,C)$ and $\mathscr{R}(down,C)$, respectively– until the selected utility function reaches and maintains a local maximum.

- *Calibrate* ($\mathscr{C}$), which operates going downwards from the given classifier to its offspring. It is intrinsically recursive and, at each step, calls $\mathscr{R}$ to calibrate the current classifier.

Given a pipeline $\pi = \langle C_1, C_2, \ldots, C_L \rangle$, TSA is defined as follows (all thresholds are initially set to 0):

$$TSA(\pi) := for \ k = L \ downto \ 1 \ do \ \mathscr{C}(up, C_k) \tag{1}$$

which asserts that $\mathscr{C}$ is applied to each node of the pipeline, starting from the leaf ($k = L$).

Under the assumption that $p$ is a structure that contains all information about a pipeline, including the corresponding vector of thresholds and the utility function to be optimized, the pseudo-code of TSA is:

```
function TSA(p:pipeline):
  for k := 1 to p.length do p.thresholds[i] = 0
  for k := p.length downto 1 do Calibrate(up,p,k)
  return p.thresholds
end TSA
```

The *Calibrate* function is defined as follows:

$$\begin{aligned}
\mathscr{C}(up, C_k) &:= \mathscr{R}(up, C_k), \quad k = L \\
\mathscr{C}(up, C_k) &:= \mathscr{R}(up, C_k); \mathscr{C}(down, C_{k+1}), \quad k < L
\end{aligned} \tag{2}$$

and

$$\begin{aligned}
\mathscr{C}(down, C_k) &:= \mathscr{R}(down, C_k), \quad k = L \\
\mathscr{C}(down, C_k) &:= \mathscr{R}(down, C_k); \mathscr{C}(up, C_{k+1}), k < L
\end{aligned} \tag{3}$$

where the ";" denotes a sequence operator, meaning that in "*a;b*" action $a$ is performed *before* action $b$.

The pseudo-code of *Calibrate* is:

```
function Calibrate(dir:{up,down}, p:pipeline, level:integer):
   Repair(dir,p,level)
   if level < p.length then Calibrate(toggle(dir),p,level+1)
end Calibrate
```

where *toggle* is a function that reverses the current direction (from *up* to *down* and vice versa). The reason why the direction of threshold optimization changes at each call of Calibrate (and hence of Repair) lies in the fact that increasing the threshold $\theta_{k-1}$ is expected to forward less FP to $C_k$, which allows to decrease $\theta_k$. Conversely, decreasing the threshold $\theta_{k-1}$ is expected to forward more FP to $C_k$, which must react by increasing $\theta_k$. The pseudo-code of *Repair* is:

```
function Repair(dir:{up,down}, p:pipeline, level:integer):
   delta := (dir = up) ? p.delta : -p.delta
   best_threshold := p.thresholds[level]
   max_uf := p.utility_function()
   uf := max_uf
   while uf >= max_uf * p.sf and p.thresholds[level] in [0,1]
      do p.thresholds[level] += delta
         uf := p.utility_function()
         if uf < max_uf then continue
         max_uf := uf
         best_threshold := p.thresholds[level]
   p.thresholds[level] := best_threshold
end Repair
```

The scale factor (*p.sf*) is used to limit the impact of local minima during the search, depending on the adopted utility function (e.g., a typical value of *p.sf* for $F_1$ is 0.8).

It is worth pointing out that, as also noted in [31], the sub-optimal combination of thresholds depends on the adopted dataset, hence it needs to be recalculated for each dataset.

To evaluate the expected running time and the computational complexity of TSA, let us define a granularity step that requires to visit only a finite number of points in a range $[\rho_{min}, \rho_{max}]$, $0 \le \rho_{min} < \rho_{max} \le 1$, in which the thresholds vary with step $\delta$. As a consequence, $p = \lfloor \delta^{-1} \cdot (\rho_{max} - \rho_{min}) \rfloor$ is the maximum number of points to be checked for each classifier in a pipeline. For a pipeline $\pi$ of length $L$, the expected running time for TSA, say $T_{TSA}(\pi)$, is proportional to $(L + L^2) \cdot p \cdot (\rho_{max} - \rho_{min})$. This implies that TSA has complexity $O(L^2)$, quadratic with the number of classifiers embedded by a pipeline. A comparison between TSA and the brute-force approach is unfeasible, as the generic element of the threshold vector is a real number. However, a comparison between TSA and RBF is feasible –although RBF is still computationally heavy. Assuming that $p$ points are checked for each classifier in a pipeline, the expected running time for *RBF*, $T_{RBF}(\pi)$, is proportional to $p^L$, so that its computational complexity is $O(p^L)$.

# 5   A Case Study: NEWS.MAS

To check whether the proposed approach can be adopted for the web, we devised a MAS to generate press reviews, which *(i)* extracts articles from Italian online newspapers, *(ii)* classifies them using text categorization according to user preferences (i.e., according to the classes selected by her/him), and *(iii)* provides suitable feedback mechanisms [1].

## 5.1   The Implemented System

To generate press reviews, we customized X.MAS [2], a generic multiagent architecture built upon JADE [8] devised to facilitate the implementation of information retrieval and information filtering applications. The motivation for adopting a MAS lies in the fact that a centralized classification system might be quickly overwhelmed by a large and dynamic document stream, such as daily-updated online news [21]. Furthermore, the Web is intrinsically a pervasive system and offers the opportunity to take advantage of distributed computing paradigms and spread knowledge resources.

Let us also note that a news retrieval system must take into account several issues, such as: *(i)* how to deal with different information sources and to integrate new information sources without re-writing significant parts of it, *(ii)* how to encode data while preserving the informative content useful to discriminate among categories, *(iii)* how to control the imbalance between relevant and irrelevant articles, *(iv)* how to allow the user to specify her/his preferences, and *(v)* how to exploit the user feedback to improve the overall performance of the system. These issues are typically strongly interdependent in state-of-the-art systems. To better concentrate on them separately, we adopted a layered multiagent architecture, able to promote the decoupling among all relevant concerns.

The system that has been implemented, namely NEWS.MAS, is organized in three "virtual" layers, each aimed at performing a specific IR step:

- *Information Extraction.* To perform information extraction, we use several wrapper agents, each associated with a specific information source: the Reuters portal[11], The Times[12], The New York Times[13], and the Reuters document collection. Once extracted, the information is suitably encoded to facilitate the categorization task. To this end, all non-informative words, e.g., prepositions, conjunctions, pronouns and very common verbs are removed using a stop-word list. After that, a standard stemming algorithm removes the most common morphological and inflectional suffixes. Then, feature selection based on the information-gain heuristics has been adopted for each category of the taxonomy, to reduce the dimensionality of the feature space.

---

[11] http://www.reuters.com

[12] http://www.the-times.co.uk/

[13] http://www.nytimes.com/

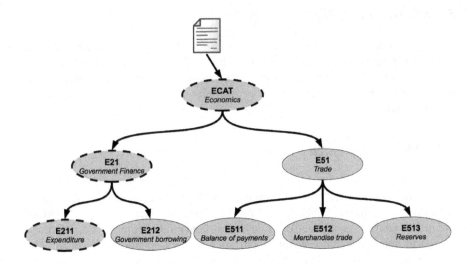

**Fig. 3** An example of pipeline (highlighted with bold-dashed lines)

- *Hierarchical Text Categorization.* To perform HTC, we adopted the PF approach recalled in Section 4. As an example of PF, let us assume that a document is to be categorized according to the taxonomy reported in Figure 1. Under the hypothesis that the document belongs to class E211 (expenditure/revenue), if correctly categorized, the document should be accepted by the classifiers ECAT (economics), E21 (government finance), and E211, i.e., it is expected to go along the pipeline $ECAT, E21, E211$ (Figure 3 highlights the pipeline).
- *User's Feedback.* When an irrelevant article is evidenced by the user, it is immediately embedded in the training set of a $k$-NN classifier that implements the user feedback. A suitable check performed on this training set after inserting the negative example allows to trigger a procedure entrusted with keeping the number of negative and positive examples balanced. In particular, when the ratio between negative and positive examples exceeds a given threshold (by default set to 1.1), some examples are randomly extracted from the set of true positive examples and embedded in the above training set.

The interface of the prototype is depicted in Figure 4. Through it the user can *(i)* set the source from which news will be extracted, and *(ii)* set the topics s/he is interested in. As for the newspaper headlines, the user can choose among the Reuters portal, The Times, and The New York Times. As for the topics of interest, the user can select one or more categories in accordance with the given RCV1 taxonomy [30]. First, information agents able to handle the selected newspaper headlines extract the news. Then, all agents that embody a classifier trained on the selected

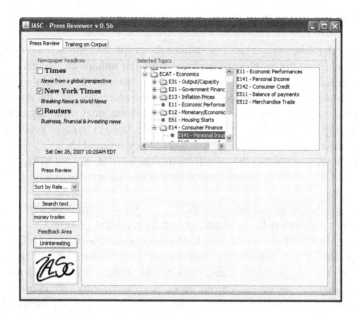

**Fig. 4** A snapshot of NEWS.MAS user interface

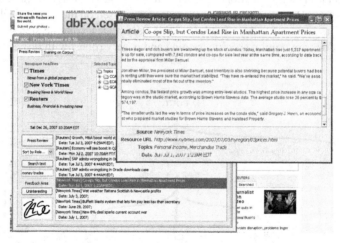

**Fig. 5** An example of results provided by the personalized press review system

topics are involved to perform text categorization. Finally, the system supplies the user with the selected news through suitable interface agents (see Figure 5). The user can provide a feedback to the system by selecting all non-relevant news (i.e., false positives). This feedback is important to let the system adapt to the actual interests of the corresponding user.

## 5.2 Experimental Results

Experiments have been performed to validate the proposed approach with respect to the impact of PF in the problem of dealing with input imbalance. To this end, three series of experiments have been performed: 1) performances calculated using PF have been compared with those calculated using the corresponding flat approach; 2) PF has been tested to assess the improvement of performances while augmenting the pipeline depth; and 3) performances have been calculated in terms of generalization- / specialization- / misclassification-error and unknown-ratio [13].

Furthermore, to show that the overall performances of PF are not worsened by the adoption of TSA vs. RBF, we performed further comparative experiments between TSA and RBF.

The system has been trained using RCV1-v2, the standard document collection proposed in [30], which is organized in four hierarchical groups: CCAT (Corporate/Industrial), ECAT (Economics), GCAT (Government/Social), and MCAT (Markets). A portion of the taxonomy is depicted in Figure 6. The complete list consists of 126 codes (i.e., the available categories). Not all of these were used in the coding phase. In particular, the codes "current news" and "temporary" are unused, as well as further 10 codes. Therefore, the total number of codes actually assigned to the data is 103.

The main issue being investigated is the effectiveness of the proposed approach with respect to flat classification. In order to make a fair comparison, the same classification system has been adopted, i.e., a classifier based on the $wk$-NN technology [16]. The motivation for the adoption of this particular technique stems from the fact that it does not require specific training and is very robust with respect to noisy

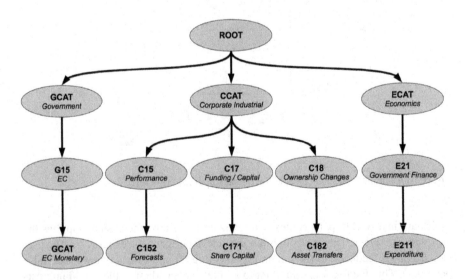

**Fig. 6** A portion of the RCV1-v2 taxonomy

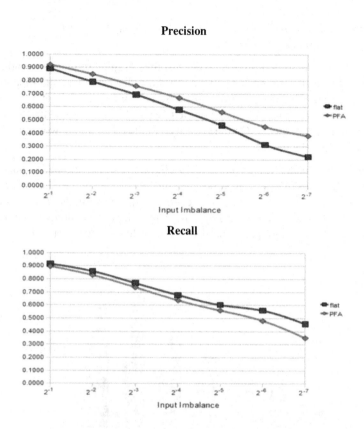

**Fig. 7** Comparison of precision and recall between PF and flat classification

data. In fact, as demonstrated in [42], *wk*-NN-based approaches can reduce the error rate due to robustness against outliers.

During the training activity, each classifier is first trained with a balanced data set of 1000 documents for Reuters, by using 200 (TFIDF) features selected in accordance with their information gain. For any given node, the training set contains documents taken from the corresponding subtree and documents of the sibling subtrees –as positive and negative examples, respectively. Then, the best thresholds are selected. Both the thresholds of the pipelines and of the flat classifiers have been found by adopting $F_1$ as utility function[14]. As for pipelines, we used a step $\delta$ of $10^{-4}$ for $TSA$.

Experiments have been performed by assessing the behavior of the proposed hierarchical approach in presence of different ratios of positive examples versus negative examples, i.e., from $2^{-1}$ to $2^{-7}$. We considered only pipelines that end with a

---

[14] The utility function is adopted depending on the constraint imposed by the given scenario. For instance, $F_1$ is suitable if one wants to give equal importance to precision and recall.

leaf node of the taxonomy. Accordingly, for the flat approach, we considered only classifiers that correspond to a leaf.

*PF vs Flat Classification.* Figure 7 shows macro-averaging of precision and recall. Precision and recall have been calculated for both flat classifiers and pipelines by varying the input imbalance. As pointed out by experimental results (for precision), the solution based on pipelines has reported results for precision better than those obtained with the flat model. On the contrary, results on recall are worse than those obtained with the flat model.

*Improving performance along the pipeline.* Figure 8 shows the performance improvements in terms of $F_1$ of the proposed approach with respect to the flat one. The improvement has been calculated in percentage with the formula $(F_1(pipeline) - F_1(flat)) \times 100$. Experimental results –having the adopted taxonomies a maximum depth of four show that PF performs always better than the flat approach.

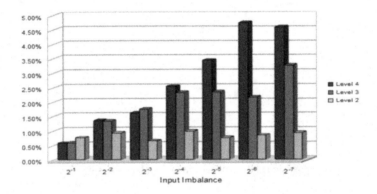

**Fig. 8** Performance improvement

*Hierarchical Metrics.* Figure 9 depicts the results obtained varying the imbalance on Reuters and DMOZ datasets. Analyzing the results, it is easy to note that the generalization-error and the misclassification-error grow with the imbalance, whereas the specialization-error and the unknown-ratio decrease. As for the generalization-error, it depends on the overall number of FNs, the greater the imbalance the greater the amount of FNs. Thus, the generalization-error increases with the imbalance. In presence of input imbalance, the trend of the generalization-error is similar to the trend of the recall measure. As for the specialization-error, it depends on the overall number of FPs, the greater the imbalance, the lower the amount of FPs. Hence, the specialization-error decreases with the imbalance. In presence of input imbalance, the trend of the specialization-error is similar to the trend of the precision measure. As for the unknown-ratio and misclassification-error, let us point out that an imbalance between positive and negative examples can be suitably dealt with by exploiting the filtering effect of classifiers in the pipeline.

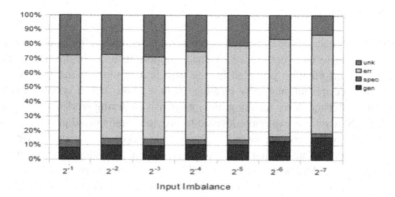

**Fig. 9** Hierarchical measures

**Table 1** Comparisons between TSA and RBF (in milliseconds), averaged on pipelines with $L = 4$

| Algorithm | $t_{lev4}$ | $t_{lev3}$ | $t_{lev2}$ | $t_{lev1}$ | $F_1$ |
|---|---|---|---|---|---|
| experiments with $\delta = 0.1, (p = 10)$ | | | | | |
| TSA | 33 | 81 | 131 | 194 | 0.8943 |
| RBF | 23 | 282 | 3,394 | 43,845 | 0.8952 |
| experiments with $\delta = 0.05, (p = 20)$ | | | | | |
| TSA | 50 | 120 | 179 | 266 | 0.8953 |
| RBF | 35 | 737 | 17,860 | 405,913 | 0.8958 |
| experiments with $\delta = 0.01, (p = 100)$ | | | | | |
| TSA | 261 | 656 | 1.018 | 1.625 | 0.8926 |
| RBF | 198 | 17,633 | 3.1E+6 | 1.96E+8 | 0.9077 |

### 5.2.1 TSA vs. RBF

Experiments aimed at comparing TSA and RBF have been carried out [4], and the time (in milliseconds) required to find the sub-optimal vector of thresholds for the selected utility function (i.e., $F_1$) is reported in Table 1. Different values of $\delta$ (i.e., 0.1, 0.05, 0.01) have been adopted as threshold increments during the search. Each pair of rows in Table 1 reports the comparison in terms of the time spent to complete each calibrate step ($t_{lev4} \ldots t_{lev1}$), together with the corresponding $F_1$. Results clearly show that the cumulative running time required for RBF tends to rapidly become intractable[15], while the values of $F_1$ are comparable.

---

[15] Note that 1.9E+8 millisecond $\approx$ 54.6 hours.

# 6 Conclusions

It becomes more and more difficult for web users to search for, find, and select contents according to their preferences. Hence, supporting users in the task of dealing with the information provided by the web is a primary issue. In this chapter, we focused on automatically creating press reviews with a system based on progressive filtering. After recalling how progressive filtering works, we presented NEWS.MAS, a system explicitly devoted to provide personalized press reviews to its users. The system encompasses three main tasks: (i) extracting articles from online newspapers, (ii) classifying them using hierarchical text categorization, and (iii) providing suitable feedback mechanisms. To validate the approach, we performed several experiments by adopting the standard document collection RCV1-v2. Results show that the approach is effective and can be adopted to create press reviews from the web.

# References

1. Addis, A., Armano, G., Mascia, F., Vargiu, E.: News retrieval through a multiagent system. In: WOA 2007 Dagli Oggetti agli Agenti: Agenti e Industria: Applicazioni tecnologiche degli agenti software, pp. 48–54 (2007)
2. Addis, A., Armano, G., Vargiu, E.: From a generic multiagent architecture to multiagent information retrieval systems. In: AT2AI-6, Sixth International Workshop, From Agent Theory to Agent Implementation, pp. 3–9 (2008)
3. Addis, A., Armano, G., Vargiu, E.: Assessing progressive filtering to perform hierarchical text categorization in presence of input imbalance. In: Proceedings of International Conference on Knowledge Discovery and Information Retrieval (KDIR 2010), pp. 14–23 (2010)
4. Addis, A., Armano, G., Vargiu, E.: A comparative experimental assessment of a threshold selection algorithm in hierarchical text categorization. In: Clough, P., Foley, C., Gurrin, C., Jones, G.J.F., Kraaij, W., Lee, H., Mudoch, V. (eds.) ECIR 2011. LNCS, vol. 6611, pp. 32–42. Springer, Heidelberg (2011)
5. Addis, A., Cherhi, G., Manconi, A., Vargiu, E.: A multiagent system for personalized press reviews. In: Soro, A., Armano, G., Paddeu, G. (eds.) Distributed Agent-Based Retrieval Tools, Polimetrica, pp. 67–86 (2006)
6. Armano, G.: On the progressive filtering approach to hierarchical text categorization. Tech. rep., DIEE - University of Cagliari (2009)
7. Armstrong, R., Freitag, D., Joachims, T., Mitchell, T.: Webwatcher: A learning apprentice for the world wide web. In: AAAI Spring Symposium on Information Gathering, pp. 6–12 (1995)
8. Bellifemine, F.L., Caire, G., Greenwood, D.: Developing Multi-Agent Systems with JADE. Wiley Series in Agent Technology. John Wiley and Sons, Chichester (2007)
9. Bennett, P.N., Nguyen, N.: Refined experts: improving classification in large taxonomies. In: SIGIR 2009: Proceedings of the 32nd International ACM SIGIR Conference on Research and Development in Information Retrieval, pp. 11–18. ACM, New York (2009)
10. Bleyer, M.: Multi-agent systems for information retrieval on the world wide web. Ph.D. thesis, University of Ulm, Germany (1998)
11. Brank, J., Mladenić, D., Grobelnik, M.: Large-scale hierarchical text classification using svm and coding matrices. In: Large-Scale Hierarchical Classification Workshop (2010)

12. Ceci, M., Malerba, D.: Hierarchical classification of HTML documents with webClassII. In: Sebastiani, F. (ed.) ECIR 2003. LNCS, vol. 2633, pp. 57–72. Springer, Heidelberg (2003)
13. Ceci, M., Malerba, D.: Classifying web documents in a hierarchy of categories: a comprehensive study. Journal of Intelligent Information Systems 28(1), 37–78 (2007)
14. Chawla, N.V., Bowyer, K.W., Hall, L.O., Kegelmeyer, W.P.: SMOTE: Synthetic minority over-sampling technique. Journal of Artificial Intelligence Research 16, 321–357 (2002)
15. Christopher, D., Manning, P.R., Schütze, H.: Introduction to Information Retrieval. Cambridge University Press, Cambridge (2008)
16. Cost, W., Salzberg, S.: A weighted nearest neighbor algorithm for learning with symbolic features. Machine Learning 10, 57–78 (1993)
17. D'Alessio, S., Murray, K., Schiaffino, R.: The effect of using hierarchical classifiers in text categorization. In: Proceedings of of the 6th International Conference on Recherche d'Information Assistée par Ordinateur (RIAO), pp. 302–313 (2000)
18. Dumais, S.T., Chen, H.: Hierarchical classification of Web content. In: Belkin, N.J., Ingwersen, P., Leong, M.-K. (eds.) Proceedings of SIGIR 2000, 23rd ACM International Conference on Research and Development in Information Retrieval, pp. 256–263. ACM Press, New York (2000)
19. Esuli, A., Fagni, T., Sebastiani, F.: Boosting multi-label hierarchical text categorization. Inf. Retr. 11(4), 287–313 (2008)
20. Etzioni, O., Weld, D.: Intelligent agents on the internet: fact, fiction and forecast. IEEE Expert 10(4), 44–49 (1995)
21. Fu, Y., Ke, W., Mostafa, J.: Automated text classification using a multi-agent framework. In: JCDL 2005: Proceedings of the 5th ACM, IEEE-CS Joint Conference on Digital Libraries, pp. 157–158. ACM Press, USA (2005),
http://doi.acm.org/10.1145/1065385.1065420
22. Gaussier, É., Goutte, C., Popat, K., Chen, F.: A hierarchical model for clustering and categorising documents. In: Crestani, F., Girolami, M., van Rijsbergen, C.J.K. (eds.) ECIR 2002. LNCS, vol. 2291, pp. 229–247. Springer, Heidelberg (2002),
http://link.springer.de/link/service/series/0558/
papers/2291/22910229.pdf
23. Japkowicz, N.: Learning from imbalanced data sets: a comparison of various strategies. In: AAAI Workshop on Learning from Imbalanced Data Sets (2000)
24. Jirapanthong, W., Sunetnanta, T.: An xml-based multi-agents model for information retrieval on www. In: Proceedings of the 4th National Computer Science and Engineering Conference, NCSEC 2000 (2000)
25. Koller, D., Sahami, M.: Hierarchically classifying documents using very few words. In: Fisher, D.H. (ed.) Proceedings of ICML 1997, 14th International Conference on Machine Learning, pp. 170–178. Morgan Kaufmann, San Francisco (1997)
26. Kotsiantis, S., Pintelas, P.: Mixture of expert agents for handling imbalanced data sets. Ann Math Comput Teleinformatics 1, 46–55 (2003)
27. Kotsiantis, S.B.: Local reweight wrapper for the problem of imbalance. Int. J. of Artificial Intelligence and Soft Computing 1, 25–38 (2008),
http://www.inderscience.com/link.php?id=21262
28. Kubat, M., Matwin, S.: Addressing the curse of imbalanced training sets: One-sided selection. In: Proceedings of the Fourteenth International Conference on Machine Learning, pp. 179–186. Morgan Kaufmann, San Francisco (1997)
29. Laender, A.H.F., Ribeiro-Neto, B.A., da Silva, A.S., Teixeira, J.S.: A brief survey of web data extraction tools. SIGMOD Rec. 31(2), 84–93 (2002),
http://doi.acm.org/10.1145/565117.565137

30. Lewis, D., Yang, Y., Rose, T., Li, F.: RCV1: A new benchmark collection for text categorization research. Journal of Machine Learning Research 5, 361–397 (2004)
31. Lewis, D.D.: Evaluating and optimizing autonomous text classification systems. In: SIGIR 1995: Proceedings of the 18th Annual International ACM SIGIR Conference on Research and Development in Information Retrieval, pp. 246–254. ACM, New York (1995), http://doi.acm.org/10.1145/215206.215366
32. Lieberman, H.: Letizia: An agent that assists web browsing. In: Mellish, C.S. (ed.) Proceedings of the Fourteenth International Joint Conference on Artificial Intelligence (IJCAI 1995), pp. 924–929. Morgan Kaufmann Publishers Inc., San Francisco (1995), citeseer.ist.psu.edu/lieberman95letizia.html
33. McCallum, A.K., Rosenfeld, R., Mitchell, T.M., Ng, A.Y.: Improving text classification by shrinkage in a hierarchy of classes. In: Shavlik, J.W. (ed.) Proceedings of ICML 1998 15th International Conference on Machine Learning, pp. 359–367. Morgan Kaufmann, San Francisco (1998)
34. Mladenić, D., Grobelnik, M.: Feature selection for classification based on text hierarchy. In: Text and the Web, Conference on Automated Learning and Discovery CONALD 1998 (1998)
35. Rousu, J., Saunders, C., Szedmak, S., Shawe-Taylor, J.: Learning hierarchical multicategory text classification models. In: ICML 2005: Proceedings of the 22nd international conference on Machine learning, pp. 744–751. ACM, New York (2005)
36. Ruiz, M.E., Srinivasan, P.: Hierarchical text categorization using neural networks. Information Retrieval 5(1), 87–118 (2002)
37. Shaban, K., Basir, O., Kamel, M.: Team consensus in web multi-agents information retrieval system. In: Team consensus in web multi-agents information retrieval system, pp. 68–73 (2004)
38. Sheth, B., Maes, P.: Evolving agents for personalized information filtering. In: Proceedings of the 9th Conference on Artificial Intelligence for Applications (CAIA 1993), pp. 345–352 (1993)
39. Silla, C., Freitas, A.: A survey of hierarchical classification across different application domains. Data Mining and Knowledge Discovery 22, 31–72 (2011); doi:10.1007/s10618-010-0175-9, http://dx.doi.org/10.1007/s10618-010-0175-9
40. Sun, A., Lim, E.: Hierarchical text classification and evaluation. In: ICDM 2001: Proceedings of the 2001 IEEE International Conference on Data Mining, pp. 521–528. IEEE Computer Society Press, Washington, DC, USA (2001)
41. Sycara, K., Paolucci, M., van Velsen, M., Giampapa, J.: The RETSINA MAS infrastructure. Tech. Rep. CMU-RI-TR-01-05, Robotics Institute Technical Report, Carnegie Mellon (2001), citeseer.ist.psu.edu/article/sycara01retsina.html
42. Takigawa, Y., Hotta, S., Kiyasu, S., Miyahara, S.: Pattern classification using weighted average patterns of categorical k-nearest neighbors. In: Proceedings of the 1th International Workshop on Camera-Based Document Analysis and Recognition, pp. 111–118 (2005)
43. Weigend, A.S., Wiener, E.D., Pedersen, J.O.: Exploiting hierarchy in text categorization. Information Retrieval 1(3), 193–216 (1999)
44. Wooldridge, M.J., Jennings, N.R.: Agent Theories, Architectures, and Languages: A Survey. In: Wooldridge, M.J., Jennings, N.R. (eds.) ECAI 1994 and ATAL 1994. LNCS, vol. 890, pp. 1–22. Springer, Heidelberg (1995), citeseer.ist.psu.edu/article/wooldridge94agent.html

45. Wu, F., Zhang, J., Honavar, V.G.: Learning classifiers using hierarchically structured class taxonomies. In: Zucker, J.-D., Saitta, L. (eds.) SARA 2005. LNCS (LNAI), vol. 3607, pp. 313–320. Springer, Heidelberg (2005)

46. Wu, G., Chang, E.Y.: Class-boundary alignment for imbalanced dataset learning. In: ICML 2003 Workshop on Learning from Imbalanced Data Sets, pp. 49–56 (2003)

47. Yan, A.R., Liu, Y., Jin, R., Hauptmann, A.: On predicting rare classes with svm ensembles in scene classification. In: Proceedings of IEEE International Conference on Acoustics, Speech, and Signal Processing (ICASSP 2003), vol. 3, pp. III-21–4 (2003)

48. Yang, Y.: An evaluation of statistical approaches to text categorization. Information Retrieval 1(1/2), 69–90 (1999),
`citeseer.ist.psu.edu/yang97evaluation.html`

49. Yang, Y., Pedersen, J.O.: A comparative study on feature selection in text categorization. In: Fisher, D.H. (ed.) Proceedings of ICML 1997, 14th International Conference on Machine Learning, pp. 412–420. Morgan Kaufmann, San Francisco (1997)

## Glossary of Terms

HTC: Hierarchical Text Categorization
IR: Information Retrieval
MAS: MultiAgent System
PF: Progressive Filtering
RBF: Relaxed Brute Force
RCV1-v2: Reuters Corpus Volume 1, version 2
SVM: Support Vector Machine
TFIDF: Term Frequency Inverse Document Frequency
TSA: Threshold Selection Algorithm
wk-NN: Weighted K-Nearest Neighbors

# A Hybrid Binarization Technique for Document Images

Vavilis Sokratis, Ergina Kavallieratou, Roberto Paredes,
and Kostas Sotiropoulos

**Abstract.** In this chapter, a binarization technique specifically designed for historical document images is presented. Existing binarization techniques focus either on finding an appropriate global threshold or adapting a local threshold for each area in order to remove smear, strains, uneven illumination etc. Here, a hybrid approach is presented that first applies a global thresholding technique and, then, identifies the image areas that are more likely to still contain noise. Each of these areas is re-processed separately to achieve better quality of binarization. Evaluation results are presented that compare our technique with existing ones and indicate that the proposed approach is effective, combining the advantages of global and local thresholding. Finally, future directions of our research are mentioned.

**Keywords:** Document Image Processing, Historical Document Images, Binarization Algorithm, Hybrid Algorithm.

## 1 Introduction

Documents can be a valuable source of information but often they suffer degradation problems, especially in the case of historical documents, such as strains, background of big variations and uneven illumination, ink seepage, etc. Binarization techniques

Vavilis Sokratis · Ergina Kavallieratou
Dept. of Information and Communication Systems Engineering
University of the Aegean
Greece
e-mail: sokratisvav@gmail.com, kavallieratou@aegean.gr

Roberto Paredes
PRHLT - Universidad Politecnica de Valencia
Spain
e-mail: rparedes@dsic.upv.es

Kostas Sotiropoulos
University of Patras
Greece
e-mail: kosotiro@upatras.gr

M. Biba and F. Xhafa (Eds.): Learning Structure and Schemas from Documents, SCI 375, pp. 165–179.
springerlink.com                                   © Springer-Verlag Berlin Heidelberg 2011

should be applied to remove the noise and improve the quality of the documents. A sample document image is shown in figure 1.

Document binarization is a useful and basic step of the image analysis systems. When applied, it converts gray-scale document images to black and white (binary) ones. In the transformed binary image the background is represented by white pixels and the text by black ones. By using binarization, the problems, mentioned before, are treated in order to provide a document form more suitable for further processing.

No matter how simple and straightforward, this procedure seems, it has been proved to be a complex task. The binary document image is essential to have good quality in order to proceed to the further stages of document analysis independent whether we are interested in performing OCR, or document segmentation, or just presentation of the document after some restoration stages [1]. Knowledge can be extracted from the documents and such systems are used in many applications, from electronic libraries or museums to search engines and other intelligent systems [2, 3]. Any remaining noise, due to bad binarization, could reduce the performance of the forthcoming processing stages and in many cases could even cause their failure.

**Fig. 1** Sample Document Image

In this chapter, a binarization technique that can be applied to both regular and historical document images is presented. It is a hybrid binarization approach that attempts to combine the advantages of both global and local thresholding. It is important to mention the basic information about the framework of this hybrid technique. First, a global thresholding technique is applied to the entire image.

Then, the image areas that still contain background noise are detected and the same technique is re-applied to each noisy area separately. The proposed hybrid framework is summarized as:

- Application of Global Thresholding Algorithm,
- Detection of "Noisy" Areas, and
- Application of Global Thresholding Algorithm to the detected areas.

By selecting only specific areas of the image for further thresholding, the cost of applying local thresholding is reduced. Moreover, the integral images [15] are used in order to further reduce the computational cost. Hence, by the hybrid approach, better adaptability of the algorithm is achieved in cases where various kinds of noise coexist on the same image, as in local binarization techniques, while keeping low the computational and time cost, as in global ones, since only a limited number of areas (instead of the entire image) need to be processed separately.

The rest of this chapter is organized as follows: Section 2 includes related work while the section 3 describes our approach in more detail. Section 4 presents comparative experimental results for the evaluation of the technique. Section 5 covers a discussion about the algorithm and the future work directions.

## 2 Related Work

Many different approaches and algorithms have been proposed for the document binarization task. Most of them can be categorized, according to the way they treat the problem, in two categories:

1. Global thresholding techniques: The pixels of the image are classified into text or background according to a global threshold. Usually, such techniques are fast. On the other hand, they are not effective in case the background noise is unevenly distributed in the entire image (e.g. presence of smear or strains) [4, 11].
2. Local thresholding techniques: The pixels of the image are classified into text or background according to a local threshold determined by their neighborhood area. Such techniques are more robust to the presence of different kinds of noise in the image. On the contrary, they are significantly more complex and time-consuming [5, 6].

Global thresholding binarization algorithms employ methods based on classification procedures, histogram, clustering, entropy and Gaussian distribution [7]. One of the oldest techniques is Otsu's [4] which calculates a global threshold based on a foreground and a background class. The threshold that minimizes the interclass variance of the thresholded black and white pixels, is selected by the algorithm. Clustering techniques are also used, e.g. methods based on K-means algorithm [8] to cluster the gray-level pixels in two clusters: background and foreground.

Such a global algorithm is the Iterative Global Thresholding (IGT) [9], an approach specifically designed for document images. This technique has the additional advantage of providing the option to maintain the image in grey-scale after the removal of background noise, a friendlier form for human readers.

Local thresholding binarization algorithms employ methods based on clustering procedures, local variation, entropy, neighborhood information, and Otsu's method [7]. Some techniques are Niblack's [6] that uses local mean and standard deviation, Bernsen's [5] which calculates local thresholds using neighbors and Sauvola's [10] which applies two different algorithms to determine a different threshold for each image and finally binarize the document image.

From a different point of view, binarization approaches can be divided as follows:

- General-purpose methods: Methods that may be applied to any image without taking into account specific characteristics of document images.
- Document image-specific methods: They attempt to take advantage of document image characteristics (e.g., background pixels is the majority, foreground pixels are in similar grey-scale tones, etc). In many cases, such methods are variations of general-purpose approaches [10].

Finally, another binarization category is the hybrid approach. Either being general or document specific binarization algorithm, these techniques combine element from both global and local thesholding techniques. They are based on the intuition that a hybrid algorithm can be fast like global thresholding ones while providing high quality binarization results like the local thresholding algorithms. Our proposal, as mentioned before, is based on this concept, and uses a global thresholding algorithm (IGT) and some noise area detection techniques.

## 3 Algorithm Description

In this section, the hybrid binarization approach that deals with document images is presented. As input, a grayscale document image is considered where document specific characteristics such as text or graphics are placed on the foreground and outrange over the background. The input images are described by the equation:

$$I(x, y) = r, r \in [0,1] \tag{1}$$

where x and y are the horizontal and vertical coordinates of the image, and r can take any value between 0 and 1 with 0 value representing a black pixel and 1 value a white pixel. The aim is the transformation of all the intermediate valued pixels to finally lie, on the background, pixel value 1 or, on the foreground, pixel value 0. The algorithm is based on the fact that a document image consists of few pixels of useful information placed on the foreground, such as characters, compared to the total size of the image [9]. Taking advantage of this fact, we assume that the average value of the pixel values of a document image is determined mainly by the background even if the document is quite clear of noise, which is quite helpful for a global threshold determination. Thus, a global technique may be used. In addition to this, a method for fixing specific document problems must be proposed which leads to the hybrid approach.

The algorithm first applies a global thresholding technique (the IGT), to the document image. Then, the areas that still contain noise are detected and reprocessed separately. In more detail, the proposed algorithm consists of the following steps:

- Application of IGT to the document image.
- Noisy area detection (areas with remaining noise).
- Application of IGT to each detected area separately.

Next, the analysis of the above steps follows.

## 3.1 Application of Iterative Global Thresholding

Iterative Global Thresholding (IGT) method is both simple and effective. It selects a global threshold for a document image based on an iterative procedure. In each iteration $i$, the following steps are performed:

Average pixel value calculation (Threshold $T_i$).

1. Subtraction of $T_i$ from each pixel.
2. The grayscale histogram is stretched so that the remaining pixels to be distributed in all the grey scale tones.
3. Repetition of steps 1-3 till the termination condition is fulfilled.
4. Binarization of the final image.

The calculation of the $T_i$, threshold used in i-th repetition, for an MxN document image, is given by the formula:

$$T_i = \frac{\sum_x \sum_y I_i(x, y)}{MxN} \qquad (2)$$

where $I_i(x,y)$ is the image after the (i-1)-th repetition. Keeping in mind that 1s stand for background and 0s for foreground the formula used for the subtraction that provides the after-subtraction and before-equalization image $I_s$ is:

$$I_s(x, y) = I_i(x, y) - T_i + 1 \qquad (3)$$

In each repetition, after the subtraction, a lot of pixels are moved to the side of the background and the rest of the pixels are fading. After the subtraction step, the intensity of the image is adjusted by using the histogram and extending the values to all the grey-scale range from 0 to 1. The background pixels retain their values while the rest of the pixel values should extend from 0 to 1. The relation used for the histogram stretching is:

$$I_{i+1}(x,y) = 1 - \frac{1 - I_s(x,y)}{1 - E_i} \qquad (4)$$

where $I_s$ is given by the equation (3) and $E_i$ is the minimum pixel value in the image $I_s$ during the i-th repetition, just before the histogram stretching. The whole procedure is repeated the necessary times till the document image is satisfactorily cleaned. Each repetition removes more stains from the image. The number of iterations depends on the image and the intensity of any existent stains, crumples and lighting effects on the image.

The necessary amount of repetitions depends very much on the document image, as well as on the required result, thus the termination condition of the algorithm or the specification of repetition boundary is the next step. In our implementation a maximum iteration upper bound of 20 rounds is posed, because in our experiments the algorithm never exceeded it. However, the process after the first repetitions is very slow. Finally, after many experiments, it was concluded that the amount of transformed pixels in each repetition compared to the previous one is an objective measure. Thus, the iterations stop based on the following criterion:

$$|T_i - T_{i-1}| < 0.05 \qquad (5)$$

Having already concluded to the appropriate final stage, the image is binarized by turning all the pixels that are not white (value 1) to black (value 0).

## 3.2  Noisy Area Detection

The detection of areas that need further processing is performed by using a simple method. The key idea is based on the fact that the areas that still contain background noise will include more black pixels on average in comparison with other areas, or the whole image. This is reasonable especially for document images that only include textual information.

The image is divided into segments, denoted by (S), of fixed size nxn. In each segment, the frequency of black pixels is calculated. The segments that satisfy the following criterion are, then, selected as:

$$f(S) > m + ks \qquad (6)$$

where $f(S)$ is the frequency of the black pixels in the segment S while m and s are the mean and the standard deviation of the black pixel frequency of the entire page, respectively. The selected segments form areas by connecting neighboring segments in respect to their original position in the image. The row-by-row labeling algorithm [12] is used for scanning the document by the nxn window.

The parameter k in the formula determines the sensitivity of the detection method. The higher the k, the less segments will be detected. This could mean that some of the areas that may still need further improvement will not be selected. On the other hand, a low k guarantees that all the areas that still need improvement will be selected however together with other areas in which the noise has already been removed. Moreover, the computational and time cost of the global thresholding step will increase. Therefore, an appropriate value of k should be selected to deal with this trade-off. The window segment size (nxn) is also an important factor both for the noisy areas selection and the successful re-application of IGT (as presented in the next subsection). For the area selection procedure a small window could select areas that not need further processing. If the window is really small, document information e.g. bold characters, may be marked as noisy areas because the amount of black pixel will be higher than the average of the image and this may lead to wrong clustering of more pixels to the background. In the proposed version of the hybrid algorithm k was assigned a value of 2 and the segment window size was set to 30x30 pixels.

## 3.3 Re-application of IGT (Local Thresholding)

The areas detected by the previously described procedure are separately reprocessed based on local thresholding. It is motivated by the belief that, re-application of the IGT method to a smaller and "noisy" segment will further clean the specified area. The algorithm may adjust better to the characteristics of the area and lead to a higher quality result. For a given image segment, the IGT global thresholding method is applied to the corresponding area of the original image. The process stops when either the termination condition (5) is satisfied or the number of iterations exceeds the corresponding number of iterations previously required for the global thresholding on the entire image.

Taking into account that the selected regions of the image have relatively high average density of black pixels, IGT removes a lot of black pixels during the first iterations. In comparison to the application of IGT to the entire page, the background noise in the selected areas is more likely to be removed since the area is likely to be more homogeneous than the entire image. In general, this procedure tends to move more pixels of the selected areas to the background in comparison with the previous application of IGT to the entire image.

As mentioned before the window size used for noisy area selection is a critical factor for the success of the iterative thresholding binarization of the selected areas. It is obvious that a small window size forms many but rather small segments. This strategy has the advantage of processing the areas in more detail and adapting parts that contain noise. On the other hand, the resulting areas are too small to provide the necessary information for successful application of the IGT. In case of large window size, fewer but bigger areas are detected. This provides enough information to the IGT algorithm in order to effectively remove background noise, but the areas cannot be easily adapted to a specific part of the image that still contains noise. As a consequence, the final image may contain neighboring areas that have dissimilar amount of background noise.

## 4  Experimental Results

In order to compare the proposed technique with other ones, the comparative results presented in detail in [7] are next shown.

The evaluation of the binarization methods was made on synthetic images. That is, starting from a clean document image (doc), which is considered as the ground truth image, noise of different types is added (noisy images). This way, during the evaluation, it is can be objectively decided for every single pixel if its final value is correct comparing it with the corresponding pixel in the original image. Two sets of images were combined by using image mosaicing techniques [13]. In the first case, the maximum intensity technique (max_int), the new image was constructed by picking up for each pixel in the new image, the darkest corresponding pixel of the two images. This means that in case of foreground, the doc would have a lead over the noisy, but in the background we would have the one from the noisy image since it is almost always darker than the document background that is absolutely white. This technique has a good optical result but it is not very natural as the foreground would be always the darkest, since it is not affected at all from the noise. This set permits us to check how much of the background can be detracted. However, in order to have a more natural result, we also used the image averaging technique (ave-int), where each pixel in the new image is the average of the two corresponding ones in the original images. In this case, the result presents a lighter background than that of the maximum intensity technique but the foreground is also affected by the level of noise in the image.

The intention is to check if every pixel was binarized in the right way. For the evaluation the following metrics were used: Pixel error, that is the total amount of pixels of the image that in the output image have wrong color: black if white in original document or white if black originally. Thus, the pixel error rate (PERR):

$$PERR = \frac{pixelerror}{MxN} \tag{7}$$

Also traditional measures of image quality description were used such as the square error (MSE), the signal to noise ratio (SNR) and the peak signal to noise ratio (PSNR) [14]. In tables 1 and 2 the evaluation result of several binarization algorithms using the metrics mentioned above are shown. For more details on the mentioned techniques please check [7].

Although there is a slightly better performance of the local binarization techniques vs. the global ones, the global ones based on histograms or classification techniques present almost as good results as the local ones. There is no obvious dependence of the algorithm performance on how recent the algorithm is.

During the evaluation procedure, Hybrid IGT exceeds the average performance. Although being very simple and fast, the algorithm in many cases performed better than very complex ones.

**Table 1** The evaluation metrics for ave-int technique

|              | MSE      | SNR      | PSNR     | PERR     | PERR variat. |
|--------------|----------|----------|----------|----------|--------------|
| Johansen     | 1030.09  | 18.29947 | 18.49771 | 1.584145 | 0.39702      |
| Li           | 1064.482 | 18.11257 | 18.31684 | 1.637035 | 0.39598      |
| Reddi        | 1067.702 | 18.1055  | 18.31057 | 1.641987 | 0.407469     |
| ALLT         | 1080.179 | 18.00511 | 18.20386 | 1.661175 | 0.374789     |
| Gatos        | 1082.475 | 18.13241 | 18.39107 | 1.664706 | 0.950808     |
| Vonikakis    | 1116.98  | 17.86367 | 18.08074 | 1.717771 | 0.416447     |
| Otsu         | 1136.256 | 17.82513 | 18.03767 | 1.747414 | 0.653088     |
| Fuz.C-means  | 1143.395 | 17.85714 | 18.04058 | 1.758393 | 0.521405     |
| Bernsen      | 1148.187 | 17.75559 | 17.96667 | 1.765763 | 0.443693     |
| Ramesh       | 1317.732 | 17.48979 | 17.65106 | 2.026501 | 1.453468     |
| Palumbo      | 1388.759 | 16.88965 | 17.10351 | 2.135731 | 0.461838     |
| Koh. SOM     | 1479.509 | 17.33808 | 17.57842 | 2.275293 | 11.58626     |
| Sauvola      | 1493.785 | 16.5649  | 16.80586 | 2.297247 | 0.77009      |
| Hybrid IGT   | 1592.008 | 16.22354 | 16.31476 | 2.448303 | 0.435930     |
| Black Perc.  | 1626.66  | 15.93992 | 16.15941 | 2.501591 | 0.354353     |
| Brink        | 1956.728 | 16.05018 | 16.15053 | 3.009194 | 3.234369     |
| Kapur        | 1958.988 | 15.41409 | 15.69104 | 3.012669 | 1.64126      |
| IIFA         | 2043.185 | 15.25285 | 15.58478 | 3.142154 | 1.827724     |
| Yen          | 2080.253 | 15.2717  | 15.55781 | 3.199158 | 4.127165     |
| Hist. peaks  | 2184.715 | 15.7486  | 15.91618 | 3.359808 | 15.59393     |
| Abutaleb     | 4079.849 | 11.6206  | 12.08744 | 6.274278 | 1.246441     |
| Parker       | 8455.937 | 8.187518 | 8.912703 | 13.00413 | 4.279259     |
| K-means      | 9069.963 | 14.99826 | 15.19859 | 13.94842 | 992.2235     |
| Kittler      | 14453.05 | 8.047554 | 9.957478 | 22.22692 | 639.8624     |
| Niblack      | 15780.57 | 4.806632 | 6.192451 | 24.26846 | 12.07129     |
| Riddler      | 15970.74 | 6.585206 | 8.091815 | 24.56092 | 213.2842     |
| Rosenf.Kak   | 18277.45 | 3.958347 | 5.613259 | 28.10834 | 36.88286     |
| Lloyd        | 19626.18 | 3.494714 | 5.314108 | 30.18251 | 46.62763     |
| Mardia       | 19973.03 | 3.291748 | 5.205405 | 30.71592 | 34.67375     |
| Pun          | 27847.65 | 0.950004 | 3.697339 | 42.82607 | 12.31683     |

**Table 2** The evaluation metrics for max-int technique

| | MSE | SNR | PSNR | PERR | PERR variat. |
|---|---|---|---|---|---|
| Johansen | 1105.647 | 17.9324 | 18.1326 | 1.700341 | 0.4167 |
| Li | 1176.348 | 17.66227 | 17.86199 | 1.80907 | 0.4898 |
| Reddi | 1712.938 | 16.28393 | 16.54482 | 2.634276 | 3.4973 |
| ALLT | 1772.267 | 15.51629 | 15.73644 | 2.725517 | 0.2751 |
| Gatos | 1843.791 | 15.66916 | 15.92546 | 2.83551 | 1.3631 |
| Vonikakis | 1875.928 | 15.88174 | 16.01977 | 2.884933 | 5.1598 |
| Otsu | 2350.078 | 14.64512 | 14.99464 | 3.614115 | 2.7403 |
| Fuz.C-means | 2587.894 | 16.40223 | 16.72078 | 3.979844 | 52.695 |
| Bernsen | 2595.835 | 14.18978 | 14.54335 | 3.992057 | 3.5038 |
| Ramesh | 2795.906 | 14.99625 | 15.36254 | 4.299741 | 15.996 |
| Palumbo | 2922.703 | 14.58387 | 14.9179 | 4.494738 | 14.289 |
| Koh. SOM | 4388.948 | 14.85988 | 15.36612 | 6.749631 | 161.52 |
| Sauvola | 4404.042 | 11.2722 | 11.73721 | 6.772845 | 0.9628 |
| Hybrid IGT | 5842.581 | 13.26888 | 13.87394 | 8.98513 | 150.38 |
| Black Perc. | 6242.384 | 12.80606 | 13.44569 | 9.599975 | 157.17 |
| Brink | 6356.625 | 12.45118 | 13.08459 | 9.775664 | 138.09 |
| Kapur | 8952.282 | 7.901008 | 8.661 | 13.76745 | 4.3483 |
| IIFA | 9014.171 | 3.062373 | 4.19233 | 13.86262 | 0.1431 |
| Yen | 9285.395 | 11.90665 | 12.82858 | 14.27973 | 314.35 |
| Hist. peaks | 11824.21 | 11.68115 | 12.30377 | 18.1841 | 683.19 |
| Abutaleb | 13901.2 | 9.544554 | 11.21195 | 21.37825 | 721.94 |
| Parker | 15288.62 | 5.023332 | 6.333367 | 23.51191 | 12.132 |
| K-means | 16567.28 | 5.726616 | 7.271567 | 25.47832 | 165.41 |
| Kittler | 18423.7 | 1.403792 | 8.922133 | 28.33326 | 1399.4 |
| Niblack | 18582.36 | 3.881317 | 5.582062 | 28.57725 | 49.992 |
| Riddler | 22771.88 | 2.429526 | 4.65036 | 35.0202 | 49.996 |
| Rosenf.Kak | 23270.71 | 2.789177 | 11.63249 | 35.78733 | 2026.4 |
| Lloyd | 23486.78 | -1.63902 | 7.439523 | 36.11962 | 1494.8 |
| Mardia | 27002.61 | 2.016883 | 5.234863 | 41.5265 | 549.26 |
| Pun | 29081.33 | 0.61298 | 3.506799 | 44.72331 | 11.337 |

## 5 Discussion – Future work

A binarization technique has been presented that combines the advantages of global and local binarization. As it has been shown the method consists of several modules, such as IGT (global thresholding), noisy area detection and others, that can be extended or changed. As future work, in order to make the algorithm more effective in terms of qualitative binarization, the following points will be further studied:

- Parameters tuning
- Changes in the global thresholding procedure
- Changes in the noisy area detection
- Post processing steps

It is expected that the proposed algorithm is able to provide better results if it is properly tuned. For example the algorithm parameters, such as k or the window size nxn, can be properly set according to the document image, in order to increase the binarization quality. This tuning can be the result of a series of experiments or a user feedback system.

Of particular interest is the dynamic determination of the parameters values by the algorithm on its own. The algorithm can collect information about the document image, like the image dimensions or the variance in the mean pixel value, and decide the proper parameter values. An interesting case could be that the dynamic selection of the threshold formula and the calculation of the scanning window used for the "noisy" area detection to be determined according to the image size, e.g. 10% of the image size on each dimension, and the average pixel value. This process can be also assisted by human feedback, in various ways, helping the algorithm by correcting inappropriate decisions.

The global thresholding procedure is one of the fundamental components of the hybrid algorithm. The overall result depends heavily on its effectiveness. Aspects that could be extended or changed concern the determination method of the threshold used for the "pixel" shifting to the background, after each iteration. An initial proposal could be the extension of the threshold by adding a variance factor this would result to this threshold formula:

$$T_i = m+ks \qquad (8)$$

where $T_i$ is the threshold, s is the standard derivation and k is a tuning parameter [6]. Another alternative could be the relation of the threshold to the noise of the image e.g. the Signal to Noise Ratio (SNR) could be used as a threshold.

Additionally, it has to be pointed out that the termination criterion of the iteration process is a critical factor of the proposed method. Currently, the criterion applied is the ratio of the pixels shifted between two successive iterations. A global boundary condition could be used instead, e.g. the amount of black pixels not to exceed the 10% of the total pixel count or by the use of SNR. In the former case, it is based on a ground truth hypothesis, that the document has a homogenized structure and the amount of black pixel ranges in a specific ratio. In the later case, the total remaining noise should be below a limit, or the reduction ratio of noise between two iterations should be satisfactory.

So far we mentioned that the detection of a problematic area in the image is based on the frequency of the black pixels inside the segment. Again the detection can be based on a ground truth belief of a global boundary e.g. the amount of black pixel doesn't exceed the 10% of the image. By setting a global boundary, the algorithm is forced to correct all the regions that break the condition. Here it is also possible to apply a detection condition based on the image noise, similarly to the termination criterion proposed before.

Although a post-processing step in the algorithm may reduce the performance by raising the total computation cost, it can improve the overall quality of the document. After the application of the hybrid binarization methodology, several filters could be applied to the image fixing specific problems. An illustrative filter could clean any black spots (salt-and-pepper noise) from the final image. This way, small black regions will be removed without destroying any letters or figures of the binarized image.

Summarizing, some alternatives that focus on key points of the method, are planned to be further studied:

*Proposal 1:* The threshold calculation and termination condition of the global thresholding iteration is changed. Initially, the threshold is calculated by the formula [6]:

$$T_i = m+ks \tag{9}$$

The threshold calculation includes not only the average pixel value m of the image but also the standard derivation s multiplied by the sensitivity parameter k. This way, the threshold represents a more representing barrier between background and foreground elements. In addition the termination condition is configured accordingly, to fit the new threshold, as mentioned before. All the other functionality of the algorithm remains the same.

*Proposal 2:* The noisy area detection criterion is changed. The detection of the problematic area is based on noise detection. As a criterion the Signal to Noise Ratio (SNR) is used. For each window, the SNR of the region is calculated and then compared to the SNR of the whole image. The condition is represented by the formula:

$$SNR(Region) <= SNR(Image) \tag{10}$$

All the other functionality of the algorithm remains the same.

*Proposal 3:* It is a combination of the proposals 1 and 2 described before. The threshold calculation is based on the formula (9) and the segment detection on the formula (10). All the remaining functionality remains the same. By combining the two different extensions, it is expected that the method will benefit from both, providing better binarization results.

In figures 2-4 the basic hybrid algorithm and the results of the three proposals are presented for the same document image.

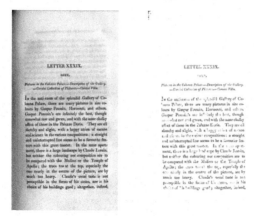

**Fig. 2** Document image (left) after applying the Hybrid Method (right)

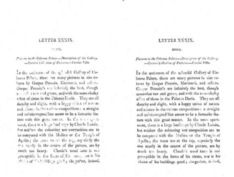

**Fig. 3** Results of proposal 1 (left) and 2 (right)

The original document image presents problems such as background of uneven illumination and seepage of ink. The hybrid algorithm fixes many of the problems and makes a good quality binarized images. By a closer look, we can see that a small region in the upper left corner remains, and that some of the characters in the sides of the image are a bit faded out. Although not a big problem for a human, this fading problem may be critical to an OCR tool. Proposal 1 solves this problem by providing more clear characters while keeping the background noise in the same level with the original hybrid approach. In the case of Proposal 2, it should be mentioned that even more solid characters are provided after the binarization, but the level of background noise is higher. Finally, the combination of both extensions leads to the best results. Background noise is kept low, while the quality of the characters is better than the other experiments.

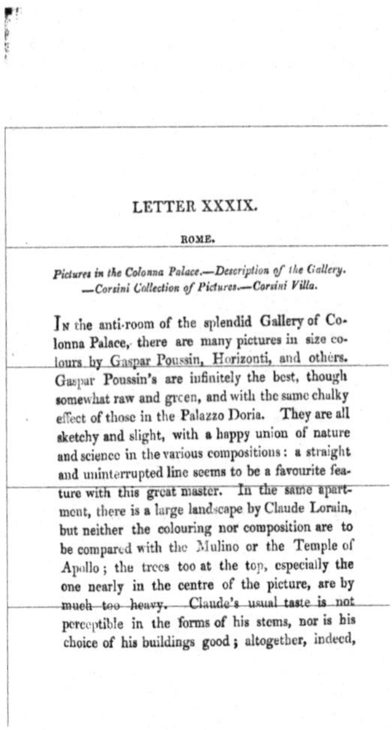

LETTER XXXIX.

ROME.

*Pictures in the Colonna Palace.—Description of the Gallery.—Corsini Collection of Pictures.—Corsini Villa.*

In the anti-room of the splendid Gallery of Colonna Palace, there are many pictures in size colours by Gaspar Poussin, Horizonti, and others. Gaspar Poussin's are infinitely the best, though somewhat raw and green, and with the same chalky effect of those in the Palazzo Doria. They are all sketchy and slight, with a happy union of nature and science in the various compositions: a straight and uninterrupted line seems to be a favourite feature with this great master. In the same apartment, there is a large landscape by Claude Lorain, but neither the colouring nor composition are to be compared with the Mulino or the Temple of Apollo; the trees too at the top, especially the one nearly in the centre of the picture, are by much too heavy. Claude's usual taste is not perceptible in the forms of his stems, nor is his choice of his buildings good; altogether, indeed,

**Fig. 4** Result of Proposal 3

It is obvious from the results, that the change in the calculation of the threshold makes the algorithm more objective. The distinction barrier between background and foreground elements is clearer, so the characters are not fading. Also the change in the selection of the noisy areas resulted in even more clear text on the binarized images but with the drawback of background noise. This happens because after the global thresholding step, fewer areas are selected by the algorithm (only those with relatively higher noise ratio). The combined Proposal 3 pointed out the best result among the other extensions. On the one hand, the change in the threshold calculation made the global thresholding procedure and every local iteration more efficient in the removal of background noise. On the other hand, the SNR noisy area detection technique chooses only areas with high noise ratio.

More experiments on the extension of the algorithm should be performed in order to expose the strength of the proposed hybrid binarization framework. Also the algorithm is planned to be extended to a more dynamic and human assisted procedure, in order to adapt to each image characteristics and provide better binarization.

## 6 Conclusion

In this Chapter we presented a hybrid binarization approach aiming at the removal of background noise from document images. This way, we attempt to combine the advantages of global and local thresholding, that is, better adaptability of various kinds of noise at different areas of the same image in low computational and time cost.

The evaluation results indicate that the proposed approach provides quality results and is comparable to other global and local thresholding techniques. Initial proposals for the extension of the hybrid approach were presented, in order to increase the efficiency and the quality of the binarized images. The next step is to study more extensions and combine them with dynamic and human assisted systems. Finally, we aim to provide the evaluation of such extensions, and a comparison with other binarization methods.

# References

1. Couasnon, B., Camillerapp, J., Leplumey, I.: Making handwritten archives documents accessible to public with a generic system of document image analysis. In: DIAL 2004, pp. 270–277 (2004)
2. Baird, H.S.: Difficult and Urgent Open Problems in Document Image Analysis for Libraries. In: DIAL 2004, pp. 25–32 (2004)
3. Marinai, S., Marino, E., Cesarini, F., Soda, G.: A general system for the retrieval of document images from digital libraries. In: DIAL 2004, pp. 150–173 (2004)
4. Otsu, N.: A threshold selection method from gray-level histograms. IEEE Trans. Systems Man Cybernet. 9(1), 62–66 (1979)
5. Bernsen, J.: Dynamic thresholding of grey-level images. In: 8th Int. Conf. on Pattern Recognition, pp. 1251–1255 (1986)
6. Niblack, W.: An Introduction to Digital image processing, pp. 115–116. Prentice-Hall, Englewood Cliffs (1986)
7. Stathis, P., Kavallieratou, E., Papamarkos, N.: An evaluation technique for binarization algorithms. Journal of Universal Computer Science 14(18), 3011–3030 (2008)
8. Jain, A.K., Dubes, R.C.: Algorithms for Clustering Data. Prentice Hall, Englewood Cliffs (1988)
9. Kavallieratou, E.: A Binarization Algorithm Specialized on Document Images and Photos. In: 8th Int. Conf. on Document Analysis and Recognition, pp. 463–467 (2005)
10. Sauvola, J., Pietikainen, M.: Adaptive document image binarization. Pattern Recognition 33, 225–236 (2000)
11. Leedham, G., Varma, S., Patankar, A., Govindaraju, V.: Separating Text and Background in Degraded Document Images. In: Proceedings Eighth InternationalWorkshop on Frontiers of Handwriting Recognition, pp. 244–249 (September 2002)
12. Shapiro, L., Stockman, G.: Computer Vision. Prentice-Hall, Englewood Cliffs (2001)
13. Gottesfeld Brown, L.: A survey of image registration techniques. ACM Computing Surveys 24(4), 325–396 (1992)
14. Kitte, T.D., Evans, B.L., Daamera-Venkata, N., Bovil, A.C.: Image Quality Assessment Based on Degradation Model. IEEE Trans. Image Processing 9, 909–922 (2000)
15. Veksler, O.: Fast Variable Window for Stereo Correspondence using Integral Images. In: Proceedings of IEEE Computer Society Conference on Computer Vision and Pattern Recognition 2003, vol. 1, pp. I-556 – I-561 (2003)

# Digital Libraries and Document Image Retrieval Techniques: A Survey

Simone Marinai, Beatrice Miotti, and Giovanni Soda

**Abstract.** Nowadays, Digital Libraries have become a widely used service to store and share both digital born documents and digital versions of works stored by traditional libraries. Document images are intrinsically non-structured and the structure and semantic of the digitized documents is in most part lost during the conversion. Several techniques related to the Document Image Analysis research area have been proposed in the past to deal with document image retrieval applications. In this chapter a survey about the more recent techniques applied in the field of recognition and retrieval of text and graphical documents is presented. In particular we describe techniques related to recognition-free approaches.

## 1 Introduction

Under a broad point of view a Digital Library (DL) can be seen like a more general document database. If all is known about the preservation, indexing, and retrieval of records belonging to structured and fielded data, maintenance and accessing to a full text document archive is a more challenging problem. Traditional relational databases store information in a structured way and each field of the records can be accessed, the queries can be formulated and records retrieved by indexing the involved fields [14]. In Digital Libraries, data are in most cases made by digitized documents and as such they are less structured [79][81]. Therefore, it is not easy to define suitable queries to this kind of archive because of the difficulties to understand the semantics of the stored data. If it is not easy to retrieve information from scanned documents processed by Optical Character Recognition (OCR) or from digital born documents, because of their limited structured nature, it is even more complicated when documents are stored as images and it is not possible to use recognition-based

Simone Marinai · Beatrice Miotti · Giovanni Soda
University of Florence
e-mail: simone.marinai@unifi.it

M. Biba and F. Xhafa (Eds.): Learning Structure and Schemas from Documents, SCI 375, pp. 181–204.
springerlink.com      © Springer-Verlag Berlin Heidelberg 2011

approaches. Storing the whole documents by their images preserves users from the lost of important and useful information which can not be suitably recognized by OCR such as graphics, pictures, and stylistic features like font and layout. In case of document images not processed by OCR the textual content is not explicitly available and this prevents users from performing queries on the basis of the text itself. Recent literature proves that this challenging problem has gained the interest of researchers and several approaches to automatic indexing and retrieval of document images have been proposed.

Nowadays, digital library technologies are well established and understood. This is proven by the large number of books and papers related to this topic and published in the last few years [40][67][73][81]. When DLs deal with scanned images of the works held in traditional libraries, Document Image Analysis and Recognition techniques (DIAR) can be applied to create, store, retrieve, and transmit electronic documents. When dealing with printed text the images can be processed by means of Optical Character Recognition systems to extract the textual content. These kinds of techniques are available, and perform well, on most document typologies, but a lot of information, e.g. related to layout or text style is likely lost. Furthermore, depending on the nature of the data to be accessed, OCR systems do not perform well unless some ad hoc training has been carried out. For instance in the case of handwritten documents and ancient manuscripts, as in case of tables or mathematical formulae. A different approach with respect to the document recognition advises in the analysis of document images a relevant alternative especially in the cases which bring to a failure of OCR systems. In this case each document is seen as a set of pixels without any known relationship among them. The retrieval of a semantic query on a document collection is in this case translated to a different domain because it must be disguised as a set of image features such as color, shape, texture, and spatial relations. This broad category of approaches, called Document Image Retrieval (DIR), is an important research line. DIR techniques identify relevant documents relying only on image features and are the main subject of this chapter.

## 2   Retrieval Paradigms

Several techniques which can be used to perform information retrieval from Digital Libraries have been proposed. The majority of these techniques follow a common paradigm: the documents are first stored and indexed in an offline phase; then the user formulates a query and the system evaluates its similarity with the stored documents and gives as output the ranked results. One important difference among the various techniques is the "level" at which the similarity computation occurs [44][78].

The simplest approach can be referred to as *free browsing*: a user browses through a document collection, looking for the desired information. In this case the similarity is evaluated by the user that visually identifies the most relevant documents.

The second approach is the *recognition-based retrieval* which relies on the complete recognition of the documents. According to it the similarity between documents is evaluated at the symbolic level and it is expected that a recognition engine can extract the full text from text-based documents or metadata from multimedia documents. The textual information is then indexed and the retrieval can be performed either by considering full text queries or by means of keywords provided by the user. The recognition-based approach has the advantage that the similarity computation and result ranking has a low computational cost. On the other hand it has some limitations when dealing with very noisy documents or containing multi-lingual texts printed with non-standard fonts and a variable layout, such as historical or damaged ones. Some of the earliest methods adopted for the recognition-based approach, and in particular for the OCR-based text retrieval, have been described in two comprehensive surveys [14][54]. Recently some works have proposed to use a mixed approach where document image analysis techniques are used together with OCR engines and metadata extraction. For instance, in [6] Belaid et al. propose an indexing and reformulation approach for ancient dictionaries, where OCR engines are trained to classify additional classes such as ligatures, gothic characters and specific shapes. In [9] and [29] the OCR engine is used to recognize words and perform layout analysis, while the vector model approach, derived from information retrieval, is used to index the dataset and retrieve the documents.

The last technique is the *recognition-free approach* which is related to content-based image retrieval. In this case the similarity is evaluated considering the actual content of the document images that is described by means of suitable features like colors, texture, or shape. An advantage of a content-based retrieval approach is the possibility of looking for information without the need of specific domain knowledge. For example, users may not be able to perform correct textual queries if they have no knowledge about the indexed language, but can perform layout queries in a language independent manner. On the other hand, even the Content-Based Image Retrieval (CBIR) approach has some problems, especially regarding the selection of appropriate features to represent the indexed objects. Most systems works with low level features such as color, texture, and shape, while only few systems attempt to extract high level or semantic features. Examples of these techniques are reported in [2][34][39] where keyword spotting techniques have been proposed considering a word-level representation on the basis of a set of low-level features.

Word spotting is one widely used approach to perform text retrieval in the recognition-free paradigm. Word-spotting was initially proposed by Jones in the field of speech processing [23], while later this definition was adopted by several researchers for printed [12] or handwritten [43] document indexing. This approach permits to localize a user selected word in a document without any syntactic constraint and without an explicit text recognition or training phase. The word-spotting technique draws all the word images belonging to indexed documents and returns a ranking of them according to a similarity measure with the query word image. This method is more closely related to CBIR than to word recognition because the matching is carried out considering image features only. Some word-spotting methods are based on the clustering of words belonging to the document collections, in order to create the index. The clustering divides the word images into equivalence classes. In each class we expect to have several instances of the same word. By assigning to one representative word for each cluster an ASCII interpretation, it is possible to index the word occurrences in the indexed documents. One disadvantage of this approach is that it can be sensitive to the style and font used for the template word especially when dealing with handwritten text. One important element of word-spotting methods is the segmentation technique used to extract the textual items from the documents. The segmentation can be carried out at local or global level. In the first case each word is segmented into characters and therefore one crucial step is the splitting of scanned images in elementary objects. On the contrary in the global approach the recognition takes place considering the whole word without attempting to perform the character segmentation.

The approaches proposed to represent words in keyword spotting methods can be roughly divided into two groups. The first class methods analyze a word image by means of global image-level features, such as intensity autocorrelation and moments that are used to represent each image. These approaches are suitable for low quality documents and are language independent but require a training phase to identify the best feature combination. The second group of methods is based on word shape coding where each word image is encoded as a sequence of symbols roughly corresponding to characters. In most cases the symbol set has a lower cardinality with respect to the character set in the original language and it is easier to recognize. Each word is in this case represented by a symbol string. Because of the reduced number of symbol classes, usually there is no guarantee of a one to one correspondence between a symbol and a character and therefore a symbol string can be mapped to several words. The main advantage of these approaches is the simple query formulation and the absence of a training phase. However, they are language dependent and are not as robust as the first group in case of poor quality images.

A common model used in Information Retrieval to represent documents is the Bag of Words (BOW) approach, early referred by Zellig H.[1] in [20]. According to this schema a document is represented by the occurrences of words in it regardless their position in the document. The Bag of Visual Word (BOVW) approach has been introduced by [70] in a paper on object and scene video retrieval and it is an extension of BOW to the case of images. The BOVW approach relies on three main steps: in the first a certain number of image keypoints or local interest points are automatically extracted from the image by means of an appropriate detector. Keypoints are salient image point rich of information content and for this reason, suitable to describe the whole image. In the second step keypoints, or in some cases shape descriptors evaluated on keypoints, are clustered and similar descriptors are assigned to the same cluster. Each cluster corresponds to a visual word that is a representation of the features shared by the descriptors belonging to that cluster. The cluster set can be interpreted as a visual word vocabulary.

---

[1] And this stock of combinations of elements becomes a factor in the way later choices are made ...for language is not merely a bag of words but a tool with particular properties which have been fashioned in the course of its use.

In the last step each image is described by a vector containing the occurrences of each visual word in that image. The most critical points of this approach are the detection of the local interest points (e.g. by means of Scale-Invariant feature transform (SIFT) or corner points) and the choice of the most suitable description of the regions of interest. Regarding the clustering method, K-means, K- Nearest Neighbor algorithm (K-NN), probabilistic Latent Semantic Analysis (pLSA) and Support Vector Machine (SVM) are the most popular techniques [58].

In the rest of this chapter the main steps of the recognition-free systems are presented. In Section 3 several types of features are presented according to the image level in which they are computed. In Section 4 the main representation models are described. In Section 5 and Section 6 the different kinds of distances and clustering techniques are presented. In Section 7 the matching approaches are described considering the different features and descriptors used in the previous steps. Some conclusions are then reported in Section 8.

## 3 Features

In a document image retrieval system the identification of features is a crucial task since it significantly affects its performance. Broadly speaking, the features can be divided into two main groups: the first is related to *local features*, according to which one feature is extracted for each point in the input domain, in the second group *global features* are evaluated on sets of pixels (e.g a word), on a region or even on the whole document.

### 3.1 Pixel Level

When features are computed at a local level some values are obtained for each pixel. In [33] Leydier et al. propose a word-spotting method to access the textual data of medieval manuscripts. This approach does not require image binarization and layout segmentation and it is tolerant to low resolution and image degradations. The informative parts of the images are represented through a set of features provided by gradient orientation. In [24], Journet et al. propose a method for the characterization of pictures of old printed documents based on a texture approach. For each pixel of the image, five features are extracted at different resolutions of the image for a total of 20 values. In particular, three features are related to the texture orientation and evaluated by means of an auto-correlation based approach while the other two are related to the properties of the pixels grey level transitions. The computation of some features in regions defined by a grid superimposed to the page is proposed in [22][76], while in [52] this zoning technique is applied to the connected components of symbols in the task of script recognition and writer identification. In the latter case for each cell the number of black pixels is computed. Something different is proposed by Delalandre et al. [13] regarding the retrieval of old printed graphics (initial letters) from a large database. A run length encoding algorithm is first used to compress the image and then a template matching between images is obtained evaluating the distance with a pixel to pixel comparison.

The above methods are in most cases characterized by a high computational cost for both the feature extraction (during the indexing) and marching (during the retrieval).

## 3.2 Column Level

Some approaches require a segmentation phase, such as the segmentation of words and characters. In this case a method based on the analysis of column pixels in segmented objects can be exploited. In [28], Khurshid et al. present a method for figure caption detection which is performed by word-spotting of figure labels. The segmentation of words and characters is done by finding the connected components and, for each pixel column of the character, a set of 6 features is calculated: vertical projection profile on the gray level image, upper character profile position, lower character profile position, vertical histogram, number of ink/non-ink transitions and middle row transition state. Similarly, in [11] words are analyzed column by column and the corresponding Hidden Markov Model is built.

## 3.3 Sliding Window

One technique related to column level representation adopts a sliding window . In this case a fixed size window is moved across the word image and some features are evaluated for each position. This strategy is frequently used to obtain the input descriptors for supervised classifiers such as the MultiLayer Perceptron (MLP) neural network [47]. The sliding window approach is also used in [75], where Teresawa et al. propose a method for word spotting in historical handwritten documents without performing word segmentation. In this case the text lines are scanned by a sliding window whose size depends on the character sizes and a low dimensional descriptor is generated for each slit image by applying the eigenspace method. A different task is addressed by Schomaker et al. in [69] where they propose a line strip retrieval on the basis of content similarity with respect to the query. Line strips are used as the basic objects for search and retrieval because they represent a good compromise between the reliability of segmentation and the recognition performance. The sliding window has a width defined on the basis of the query line and is moved along the indexed text lines. Some features related to the ink density, the connected component shapes and contours are computed at each step. In [36] Licata et al. develop a system to identify the provenance of ancient handwritten documents basing on the ink appearance similarity. The features considered in the sliding window are in this case first and second order statistical features such as histogram mean, skewness, contrast, entropy. Uttama et al. [77] exploit the sliding window approach to separate homogeneous and textured regions in images of historical documents. The features are based on the Gray-Level Co-occurrence Matrix.

## 3.4 Stroke and Primitive Level

When the objects in document images are complex and important spatial relationships among primitives are possible, such as in sketches [35] and in trademarks [80], one structural representation, which is able to represent the variety of connections, is essential. Liang et al. [35] and Wei et al. [80] propose the combination of structural and global features: the structural part describes the interconnections among primitives and the global features reflect the object as a whole. Liang et al. [35] exploit eight kinds of spatial relations between primitives: cross, half-cross, adjacency, parallelism, cut, tangency, embody and ellipse intersection. Global features are composed by seven types of descriptors: eccentricity, normalized distances of sketch centroid in major and minor axis orientations, average distance between centroids of sketch primitives and the number of primitives of each type as line, arc, and ellipse. Wei et al. [80] propose global features based on 15 Zernike moments of orders 0-4, the standard deviation of the curvature, the mean and the standard deviation of distance to centroids.

In the context of shape-based image retrieval, Wong et al. [82] propose a two step feature extraction process: the shape contour is first represented by the Freeman chain code as a connected sequence of straight line segments with specified lengths and directions. Then the relative spectrum is plotted as the normalized curve length with respect the normalized geometric moment, where the normalized curve length is the length of the segment between two keypoints of a shape divided by the total curve length and the normalized geometric moment is the distance between a dominant point and the geometric center. On this spectrum four features are evaluated: the total normalized moment variance, the total normalized area covered by the spectrum, the cross-sectional normalized area and the cross-sectional normalized moment variance.

In [16] Fonseca et al. present a shape classification technique based on topological and geometrical descriptors. In this case the spatial organization is described by the relationships of inclusion and adjacency, while the geometry of shapes is described by some geometric attributes like area and perimeter. In [64] Rusinol et al. deal with symbol recognition starting from a vectorial representation of the image. Information about constraints between segments such as parallelisms, straight angles and overlap-ratios are analyzed.

Dealing with symbol-spotting , Rusinol et al. [65] propose to detect graphical symbols in large and complex document images by techniques which do not need neither a segmentation step nor a priori knowledge. For each symbol its primitives are extracted by means of connected component analysis and contour detection then the Cassinian ovals parameters are evaluated. In a two-center bipolar coordinate system a Cassinian oval is described by all the points such that the product of their distances from the two centers is a constant. In this case, the parameters that characterized the minimum Cassinian oval which encompasses the normalized shape contour, are used as shape features. Similarly Zhang et al. [84] propose to extract a simple set of features from the vectorial representations of the symbols: in particular they propose to evaluate the angle between two line segments to represent the relationship between segments. The relationship between a line segment and an arc and between arcs is represented considering the angle information among arc starting points and arc centers. The relationship between a line segment and a circle is evaluated considering the tangency, intersection, and disjointness relations.

A similar approach can be even used in text document analysis. For instance in [10] Chellapilla et al. segment words in strokes and then the sequence of Chebyshev polynomial coefficients is evaluated for each segment. In [21] a word spotting for online handwritten documents is proposed. In this work strokes are sampled and for each sampled point three features are evaluated: the height of the sample point, the direction and the curvature of the stroke in that point.

## 3.5   Connected-Component Level

In the processing of handwritten or ancient printed documents, it is not always easy to segment a document and to identify text lines, words, and characters. Especially the character segmentation is a difficult task because of the variability of handwriting and the presence of touching characters. In [8] handwritten touching characters are addressed by identifying all the possible ligatures connecting two characters by heuristic analysis of the contour. In so doing a word image is divided in several pieces. Assuming that a character is made up by at most four consecutive pieces, a series of hypothesized character images are created. One way to segment text in objects, broadly corresponding to characters, is to find the connected components in the document image. Moghaddam et al. [55] present a line and word segmentation free method to perform word-spotting on old historical printed documents. After the detection of the connected components, a set of six features: aspect ratio, horizontal frequency, scaled vertical center of mass, number of branch points, height ratio to line height, and presence of holes, are extracted from them. Similarly, Marinai [45] propose to use the connected component clustering as a first step in indexing. Relevant words are identified considering a modified Dynamic Time Warping (DTW) algorithm that includes the word width in the distance computation. Barbu et al. [5] work on graphical document images considering graph-based representations: connected components in the image are represented by graph nodes and rotation and translation invariant features, based on Zernike moments, are extracted.

## 3.6   Word Level

In most word spotting applications it is possible to assume that the word segmentation in indexed documents is not problematic. In this case the retrieval is carried out considering the word as a whole. In [3] each word is described by profile-based and shape-based features, while in [60] and [61] Rath et al. present single value features based on Projection Profile, Word Profile, Background to Ink Transitions, and Grayscale Variance. In [83] Zhang et al. propose to use the Gradient-based binary features (GSC) evaluated under a $4 \times 8$ division of the word image. GSC features are based on the evaluation of the direction of the gradient, on structural information, and on concavity features. Structural and concavity features are evaluated by means of 12 rules applied to the image pixels and are based on the pixel density, on larger strokes in both horizontal and vertical directions, and on the direction of concavity each pixel belongs to. Similarly, in [31] and [27] the zoning technique is applied to word images and the density of the character pixels in each zone is evaluated. Subsequently a second group of features based on the area under the upper and lower word profiles, is considered. In [17] and [63] some sets of feature vectors are evaluated for not-overlapping windows on the query image: in [17] the features are based on pixel density, in [63] the Local Gradient Histogram (LGH) features, introduced by Rodriguez in [62], are used. Dealing with printed documents, Meshesha et al. [53] propose to describe words by means of word profiles, moments and transform domain representations. Similarly, Bai et al. [2] propose the extraction of seven features: character ascenders, descenders, deep eastward and westward concavity, holes, i-dot connectors and horizontal-line intersection. In [42] features are extracted for each word by means of the Left-to-Right Primitive String (LRPS) algorithm. This algorithm splits each word in primitives which can be described by means of two-tuples: the Line-or-Traversal Attribute (LTA) and the Ascender-and-Descender Attribute (ADA). In a more recent work [40], Lu et al. extend the previous approach to the case of printed document images captured by a digital camera. In this difficult task the set of extracted features includes three perspective invariants: holes, water reservoirs, and character ascenders and descenders. Nakai et al. [56] propose a mixed approach: they work at a lower level considering the centroids of connected components and some features related to the area of connected components while at a higher level analyze words and compute the word centroids.

## 3.7   Line and Page Level

Tan et al. [74] deal with the script identification among three different on-line handwritten scripts: Arabic, Roman and Tamil. After the detection of text lines, they extract a set of features at line-level such as the horizontal and vertical interstroke direction, horizontal and vertical stroke direction, average stroke length, stroke density and the reverse direction.

Some features can be extracted at page level by means of geometric transformations. In [25] Joutel et al. develop a system for paleographers and literary experts, to support their work on manuscripts dating and authentication through different historical periods. The approach is based on the Curvelet transform to compose an unique signature for each handwritten page.

**Table 1** Features

| Level | Data | Features | Ref. |
|---|---|---|---|
| Pixel | Medieval manuscripts | Gradient orientation. | [33] |
| | Old printed pictures | Auto-correlation based and pixels grey levels transitions. | [24] |
| Zoning | Printed | Pixel density. | [27][31][52] |
| | Layout | Row encoding. | [22][76] |
| Column | Historical printed | Projection profiles, vertical histogram, number of ink/non-ink transitions. | [33] |
| Sliding Window | Historical handwritten | A vector is generated for each slit image by applying the eigenspace method. | [75] |
| | Historical handwritten | Ink density, the connected component shapes and the contours. | [69] |
| | Historical handwritten | First and second order statistical features (e.g. histogram mean, skewness, entropy). | [36] |
| | Old printed pictures | Gray-Level Co-occurrence matrix and texture uniformity. | [77] |
| Stroke and primitives | Sketches/ on-line handwritten | Spatial relations between primitives (e.g. cross, adjacency, parallelism, tangency) and global features (e.g. eccentricity, number of primitives of each type). | [35][74] |
| | Trademarks | 15 Zernike moments of orders 0-4, standard deviation of the curvature, mean and standard deviation of distance to centroids. | [80] |
| | Shape | Normalized moment variance and the total normalized area. | [82] |
| | Graphics | Cassinian ovals. | [65] |
| | Handwritten | Sequence of Chebyshev polynomial coefficients. | [10] |
| Connected components | Printed | Aspect ratio, horizontal frequency, scaled vertical center of mass, number of branch points, and presence of holes. | [55] |
| | Graphics | Zernike moments. | [5] |
| Word | Handwritten | Single value features: projection profiles, background to ink transitions, grayscale variance, gaussian smoothing and gaussian derivatives. | [60][61] |
| | Printed | Area under the projection profiles. | [27] [31] |
| | Printed | Word profiles, moments and transform domain representations. | [53] |
| | Printed | Character ascenders, descenders, deep eastward and westward concavity, holes, i-dot connectors and horizontal-line intersection. | [2] [40] |
| Page | Historical manuscript | Curvelet transform. | [25] |

## 3.8    Shape Descriptor

Shape descriptors are frequently used in image analysis to compare 2D object silhouettes. Recently they have been adopted also in document image analysis to compare symbol images in a recognition-free approach. According to the object representation the shape descriptors are evaluated on, three main categories can be identified [32]. In the first category *contour based* descriptors are evaluated on the object contours; in the second, *image based* descriptors include the shape descriptors based on the overall image pixel values; in the last category, *skeleton based* descriptors are evaluated on the image skeletons. Since document image retrieval has to deal with images affected by scale and perspective changes, the shape descriptors must be invariant to similarity and affine transformations as well as rotation and scale. In most cases, descriptors are computed on keypoints, that are points of the image rich of information content. To reduce the complexity of the evaluation, images are usually preprocessed and the contour or the skeleton are detected. The interesting points are in his case extracted from the preprocessed image. An example of keypoints are corners (points with high curvature), which can be detected with the Harris-Laplace detector [66] or as in [82] on the basis of the curvature variance. In [57] Nguyen et al. propose the use of the Difference-of-Gaussian detector and the keypoints in an image are considered as the extrema in a scale-space pyramid built with DoG filters.

A first example of shape descriptor is the Scale Invariant Feature Transform (SIFT) proposed by Lowe [38]. SIFT are able both to localize keypoints and to evaluate a shape descriptor on these points according to the local gradient histogram. SIFT descriptors are proved to be invariant to rotation, scale changes and affine transformations. In [85] Zhang et al. propose the use of a modified version of SIFT descriptors applied to handwritten Chinese character recognition.

Another descriptor is the Shape Context, proposed in [7]. This descriptor is able to capture the spatial distribution of points in proximity of the point where it is

**Table 2**  Shape Descriptors

| Shape Descriptor | Data | Features | Ref. |
|---|---|---|---|
| Keypoints | Contour or Skeleton | Corners detected by means of the Harris-Laplace detector or considering the curvature variance. | [66][82] |
|  | Contour | Interest points are the extrema in a scale-space pyramid built with Difference-of-Gaussian filters. | [57] |
| SIFT | Character | Modified SIFT algorithm. | [85][66] |
| Shape Context | Contour or Skeleton | Descriptor captures the spatial distribution of points in proximity of the point where it is computed. | [37][51] [57][66] |

computed. The descriptor is represented as a logarithmic polar mask centered on the point of interest and divided in bins in the polar space. Each cell is populated according to the position of other image points with respect to the center of the mask. Shape Contexts are invariant to translation and scale. In [37] shape context descriptors are used to represent words and are computed on points belonging to the skeleton of word images, while in [51] they are computed on points on the contour of mathematical symbols. In [57] shape context are computed on points of interest detected by means of Difference-of-Gaussian detector. In [66]Rusinol et al. propose the SIFT and the shape context descriptors for the symbol spotting task and compare their performance with the results obtained with more simple descriptors such as geometric moments of steerable filters.

## 4    Representation

After a set of features has been extracted from the document image, it is essential to identify a suitable representation of their values. Some of the features presented in the previous section are naturally represented in the form of vectors. For instance in the case of zoning the features are extracted from each cell of a grid superimposed to the image and the resulting descriptor has the same dimensionality of the number of cells [17][22][52][76][85]. In case of a sliding window each slit is usually represented by a low dimensional descriptor [69][75], or by an high-dimensional descriptor such as in[36] where the intensity histogram statistics and co-occurrence statistics are concatenated. When the features are evaluated at pixel level, such as in[77], a feature vector of a certain dimension is created for any pixel of the input image. If the segmentation is performed at word level [3][27][31][83], some binary vectors are generated starting from the features related to the word profiles, while in [16] and [35] the same approach is proposed in case of shape classification where the features describe the spatial organization and structural characteristics of geometric shapes. Even in case of features computed at character or connected-component level, the vector representation is usually exploited. Some examples can be found in [8][21][49][55]. Regarding the automatic indexing and retrieval of graphical document images Barbu et al. [5] describe each connected component related to a graphical object as a feature vector. A similar approach is used in [37][51][57], but in this case Shape Context descriptors are used to describe the shape of a word and each word is represented as a collection of vectors. Rusinol et al [66] exploit a vector representation for each type of shape descriptor. In the field of shape matching, Super et al. [72] propose to describe the shape contour as a vector in a 2n- dimensional shape space, where $n$ is the number of contour sample points. An extension of the vectorial features representation is proposed in [84] where symbol signatures are represented in a matrix form and each bin of the matrix represents the relationships between strokes of the symbol, such as segments, arcs and circles.

When the number of documents in a database is large the retrieval process can be computational expensive. Some techniques use a compressed data structure to represent the images in order to decrease their handling times. In particular Delalandre

**Table 3** Representation

| Representation | Features | Method | Ref. |
|---|---|---|---|
| Vectors | Zoning | Vector corresponding to a grid overlapped to the image. | [17][22][52] [76][85] |
| | Sliding windows | Descriptors of low or high dimensionality. | [36][69][75] |
| | Pixel | One descriptor for each pixel. | [77] |
| | Word | Binary vectors generated from the profile features. | [3][27][31] [83] |
| | Word | Binary vectors describe the spatial organization and structural characteristics of geometric shapes. | [16][35] |
| | Character or connected-component | One dimensional vector for each character or connected component. | [8][21][49] [55] |
| | Shape descriptors | Describe the shape of a word. Each word is represented as a collection of vectors. | [37][51][57] [66] |
| Matrix | Strokes | Each bin of the matrix represents the relationships between strokes of the symbol such as segments, arcs and circles. | [84] |
| Probabilistic | Word | Non-Symmetric Half Plane Hidden Markov model and Semi-continuous HMM. | [11][63] |
| Character and Word shape coding | Strokes | A code is assigned to each primitive of the word image. | [34][39][40] [42][41] |
| Vectorial model | Symbol | Vector elements represent the frequency of visual terms in the documents and queries. | [51][52][57] [64][74] |
| Graph-based | Connected component or strokes | Each connected component is a graph node; arcs connect near primitives. | [5][18][26] [59][77] |

et al. [13] propose to use the run-lenght encoding algorithm to compress images. In [25], Joutel et al. propose an approach for retrieval of handwritten historical documents at page level based on the Curvelet transform to compose an unique signature for each page. In [10] words are segmented and each stroke is represented as the sequence of Chebyshev polynomial coefficients. Then each segment is processed by a Time Delay Neural Network (TDNN) to determine the probability the segment belongs to each possible character in the language. The TDNN outputs associate to each segment the most likely characters according to their membership probabilities.

When the vocabulary is limited, as in bank check recognition or in automatic handwritten mail sorting, probabilistic approaches can be used. In [11] Choisy et al. propose to model words by means of a Non-Symmetric Half Plane Hidden Markov Model (NSSP-HMM), while Rodriguez et al. [63] propose a system where typed text is used as support for handwritten recognition. They use robust LGH features to describe the word shapes and Semi-continuous HMM (SC-HMM) for modeling the link between typed and handwritten words.

Strictly connected to the vector representation of words or characters in case of word spotting, there are the character and word shape coding approaches. Generally character shape code encodes, in the form of a code string, the properties of each symbol such as e.g. whether or not the character in question fits between the baseline and the x-line, whether it has an ascender or descender and the number of spatial distribution of the connected components [2][71]. In [40] and [42] Lu et al. present a system based on a word image coding schema: the Left-to-Right Primitive String (LRPS), Line-or-Traversal Attribute (LTA) and the Ascender-and-Descender Attribute (ADA) features are used to assign a code to each primitive of the word images. Recently some extensions of the previous works have been presented by Lu et al. [39][41]. In [41] a word shape coding approach for documents in five different Latin languages has been presented. In this case each word is converted into a word shape code that is composed of two parts. In the first part, character extrema points, which are in the upward and downward text boundaries, are classified as belonging to one of three categories according to their position with respect to the base line. In the second part of the word code the number of horizontal word cuts is reported. A similar approach is proposed by Li et al. in [34] where they encode every word as a sequence of numbers and each number represents a character.

The vector space model was introduced by Salton et al. [68] in 1975. According to this model, documents and queries can be represented as vectors whose elements represent the frequency of each term in the document [41]. Some weighting schema can be applied to the vector model e.g. the tf-idf [15] and some modified versions of it [9][19][74]. The same approach can be extended to the case of vectors corresponding to occurrences of features [51][52][57][64][74].

Another kind of document representation is related to the graph-based approach. In [5] Barbu et al. describe a document image according to a graph-based representation at primitive level analyzing the relationships between connected components. Each connected component is a graph node and one arc between primitives exists only if they are spatially near. The same approach is used in [18] where Gordo et al. represent the document layout as a bipartite graph built considering the centroids of the regions on one side and the center of mass of all the regions on the other. A tree representation of the layout is considered in [48] and [50]. In symbol spotting applications the graph based approach is quite frequent. In [26], Karray et al.

develop a method for the analysis of initial letters which is based on the Attributed Relation Graph representation. In particular, after the segmentation step nodes represent regions and arcs express the relationships among regions. For the same task, in [77], Uttama et al. describe an initial letter by means of its signature according to two approaches: Minimum Spanning Tree and Pairwise Geometric Attributes. In [59] each symbol is described by means of a graph to capture the spatial and topological relationships between graphical primitives. Alajlan [1] propose to use the Curvature Tree hierarchical data structure which reflects the inclusion relationships among objects and holes.

## 5 Similarity Measure

In document image analysis and recognition, the retrieval of a query word can be performed considering the similarity or distance between two images: the reference image, given by the user, and the dataset images representing the indexed words. According to the different models used to represent keywords, a different similarity measure may be used. Moreover, in case a clustering phase is performed, the more appropriate distance to compare features has to be chosen such as Euclidean distance [51] or cosine distance [57].

When keywords are represented by feature vectors, the most common way to compare them is the Euclidean distance [16][35][80] or the L1 distance [27][31].

In case of template matching the distance among images is computed at pixel level [13] by a simple value comparison or considering a specific dissimilarity function [33]. When documents and keywords are represented by means of the vector space model in analogy with the vector model in Information Retrieval, to evaluate the similarity among elements it is frequently computed the cosine of the angle between two vectors. This approach is used in [15][39][57]. In [51] a modified version of the cosine similarity is proposed. Taking account of the properties of the clustering algorithm used to group features, Marinai et al. introduce an additional term to the formula which is necessary to deal with inexact match. In [74] the similarity is computed by means of the Chi-square distance which involves the distribution of tf-ifd vectors, while in [77] the Bhattacharyya distance between two histograms is

**Table 4** Similarity

| Similarity measure | Representation | Ref. |
| --- | --- | --- |
| Euclidean distance | Vectors/ Vector model | [80][35][16] |
| L1 distance | Vectors | [27][31] |
| Comparison | Pixels | [13][33] |
| Cosine similarity | Vector model | [39][15][57][51] |
| Chi-square distance | Vector model | [74] |
| Bhattacharyya distance | Vector model | [77] |
| Minimum Edit distance | Word coding | [55][34][15] |
| Hamming distance | Word coding | [40] |

used. To deal with a word shape coding representation, some distances may be used. In particular in [15][34] the use of the minimum Edit distance is proposed. The Edit distance between two strings is given by the minimum number of operations needed to transform one string into the other, where an operation is an insertion, deletion, or substitution of a single character. In [55] an enhanced version of the edit distance in proposed where stroke width is used as a priori information. In [40] Lu et. al propose the Hamming distance to compare word shape codes.

## 6  Clustering

In computer vision applications a family of methods based on the *bag of visual words* framework has been recently proposed [4][5][30] [51][57]. These methods extend the *bag of words model* used in textual information retrieval, that represents documents considering the number of occurrences of words, regardless of their position in the text. In the bag of visual words approach a visual vocabulary is constructed by clustering the feature vectors that represent symbols. Each cluster can be considered as a visual word and all the feature vectors belonging to that cluster can be represented by its centroid. The clustering can be performed by means of a semi-supervised learning, or by means of an unsupervised learning approach. To this second class of methods belongs the K-means algorithm used in [61][57] and its revisited version called K-medoids [5]. In [36] Licata et al. propose to use the K-Gaussian clustering where the number of clusters is determined using Minimum Description Length. In this case each feature vector is assigned to a cluster by selecting the component that maximizes the posterior probability. This algorithm may be more appropriate than K-means clustering when clusters have different sizes. Another approach to cluster is the Self Organizing Map (SOM) applied to the feature vector quantization [45][49][51][46][55]. The SOM map has the property that more

**Table 5** Clustering

| Clustering | Representation | Number of clusters | Ref. |
|---|---|---|---|
| K-mean | Feature vectors assigned to the nearest cluster. | 1-20, 20 | [57] [61] |
| K-medoids | A more robust version of the K-mean. | 16 | [5] |
| K-Gaussian | Feature vectors assigned to clusters by selecting the component that maximizes the posterior probability. | 200 | [36] |
| SOM | Feature vectors assigned to the nearest cluster according to the Euclidean distance. Closest neurons in the map are the most similar. | 10, > 100 | [45][49] [51][55] |

similar patterns are usually clustered in closer clusters. In [55] the SOM has been used to cluster the feature vectors belonging to connected components while in [51], the shape context descriptors, evaluated for each symbol, are clustered by means of a SOM and then each symbol is represented by the occurrences of shape contexts assigned to a particular centroid according to the Euclidean distance.

## 7 Matching

Feature matching deals with measuring the similarity between the feature representation of the query image, based on feature vectors, graphs or statistical models and the database images. The choice of the feature matching technique is essential for the good performance of the system. As a matter of fact an inappropriate approach may lead to bad results even although the considered feature representation is the more appropriate for that task. When the queries and documents are represented as feature vectors, the matching is carried out comparing the vectors according to some similarity measure. In [17], the query word feature vectors are compared with the corresponding vectors of the database by applying a matching which is constrained by the regions of interest features have been computed in. The similarity measure is in this case based on the Euclidean distance between vectors. The matching by means of the Euclidean distance is proposed also in [16][35][80]. The same approach is proposed in [27] and in [31], but in this case the similarity is evaluated by means of the L1 distance to reduce the computational costs. In [72], Super et al. propose to perform the matching between shapes, considering their normalized contours as vectors. To face the different shape sizes, the similarity between two vectors is evaluated by the Euclidean distance normalized by the squared average of the arc lengths of the two poses. In [84] the matching is performed considering how many common relationships at primitive level the symbols share and choosing the shapes that share the most. Also in [77] the matching is performed comparing feature vectors but in this case the Bhattacharyya distance between two vectors is used. This distance exploits the correlation between vector contents to obtain the similarity measure. A similar approach is proposed by Joutel et al. in [25], where the matching between two document signatures is performed by means of the normalized correlation similarity.

When the document images are representing by means of the vector model, some other techniques can be applied for the matching, in particular a common matching schema is the cosine similarity measure. In [41], Lu et al. propose a word shape coding technique for document retrieval and the vector representation of the strings. The similarity among two document representations is evaluated considering the angle formed by the two vectors. The same matching approach is used in [15][39][51][57].

Different matching techniques have been proposed when documents are represented by means of the word or character shape encoding. In [2][21][60][61], the Dynamic Time Warping is performed to compare two sequences of data points. This method aligns the feature vectors of the query and of the word database using a dynamic programming-based algorithm. The algorithm computes distance scores

**Table 6** Matching

| Representation | Similarity | Method | Ref. |
|---|---|---|---|
| Feature vectors | Euclidean distance | Word feature vectors are compared with the database feature vectors. | [80][35][16] |
| | Constrained euclidean distance | Word feature vectors are compared with the database feature vectors constrained by the regions where features have been computed. | [17] |
| | L1 distance | All word feature vectors are compared with the corresponding feature vectors of database images. | [27][31] |
| | Normalized correlation | Word feature vectors are compared with the database feature vectors. | [25][77] |
| Vector model | Cosine similarity | The similarity is evaluated considering the angle formed by the two document vectors: if it is close to zero, the documents are similar. | [41][39][15] [57][51] |
| Word encoding | DTW | Aligns the query and the word feature vectors using a dynamic programming-based algorithm. | [61][21][60] [18][42] [39] [2] |
| | DTW | DTW-based partial matching technique based on word form variations in the beginning and at the end of words. | [53][3][45] |
| | EDT | Minimum number of operations needed to transform one string into the other. | [28][34] |
| Probabilistic | NSSP-HMM SCHMM | The probabilistic model is used to estimate the recognition scores. | [11][63] |
| Graph | Polynomial bound greedy algorithm | Use a graph matching routine where sub graphs are matched against model graphs. | [59] |

for matching points by means of the Euclidean distance. At the end of the process, similar shapes have a lower distance value than different ones. The same approach is used in [53] and [3], but a DTW-based partial matching technique that takes care of word form variations in the beginning and at the end of the word is also proposed. The DTW technique is also used in [18] to perform matching on layout vector representations while in [45] a modified version on DTW that takes into account both the clusters similarity and the estimated widths of alternative sub-words is proposed. In [42] and [39] the DTW algorithm is considered after a pruning step by means of coarse matching that is used to reduce the number of dataset elements the query code has to be compared to.

Another matching technique is proposed in [28] at word level and in [34]. In these cases the matching among code strings is performed by means of the minimum edit distance. The edit distance between two strings is given by the minimum number of operations needed to transform one string into the other, where an operation is an insertion, deletion, or substitution of a single character.

When the document representation is probabilistic-based, the model obtained in the feature extraction phase is used to evaluate the recognition score [11][63]. On the other hand when the objects are represented in a graph-based approach some ad hoc methods should be used. In particular Qureshi et al. [59] propose to use a graph matching routine where sub graphs are matched against model graphs using polynomial bound greedy algorithm. The output of this process is the score similarity. This technique is error-tolerant and works well in case of under or over segmentation of symbols.

## 8 Conclusions

In this chapter we have described in a comprehensive survey the main document image analysis and recognition techniques which have been proposed in recent years to perform document image retrieval. We have focused our analysis on recognition-free approaches that do not explicitly recognize the document content (e.g. with OCR tools) but work at various levels with a symbolic or sub-symbolic representation of the document image. The comparison of the different techniques has been organizes along the main steps of a general retrieval process.

In the first step the features are extracted from the document images that are represented at various levels, from pixel to primitive, character, word or even zones. In each case different features can be considered according to the task (e.g. wordspotting or trademark retrieval) and to the typology of processed documents (e.g. printed vs handwritten or modern vs historical). In the second processing step the features previously extracted are encoded in suitable representations that depend both on the nature of the features and on the subsequent processing steps. Eventually, the feature representations are used to compute the distance between objects. The distance can be used to cluster indexed objects to speed up the subsequent processing or can be employed in similarity measures to compare query items with indexed ones.

The research on document image retrieval is particularly dynamic in the last few years. To give a measure of it, about half of the papers cited in this chapter have been published after our previous survey on the field appeared in 2006 [44]. The research should now focus on scalable systems that could effectively deal with the large amount of data that is nowadays available in several digital libraries around the world.

## References

1. Alajlan, N., Kamel, M.S., Freeman, G.H.: Geometry-based image retrieval in binary image databases. IEEE Trans. on Pattern Analysis and Machine Intelligence 30(6), 1003–1013 (2008)
2. Bai, S., Li, L., Tan, C.: Keyword spotting in document images through word shape coding. In: Proc. Int'l Conf. on Document Analysis and Recognition, pp. 331–335. IEEE Computer Society Press, Los Alamitos (2009)

3. Balasubramanian, A., Meshesha, M., Jawahar, C.: Retrieval from document image collections. In: Proc. IAPR Int'l Workshop on Document Analysis Systems, pp. 1–12 (2006)
4. Banerjee, S., Harit, G., Chaudhury, S.: Word image based latent semantic indexing for conceptual querying in document image databases. In: Proc. Int'l Conf. on Document Analysis and Recognition, vol. 2, pp. 1208–1212. IEEE Computer Society Press, Los Alamitos (2007)
5. Barbu, E., Héroux, P., Adam, S., Trupin, É.: Using bags of symbols for automatic indexing of graphical document image databases. In: Proc. Int'l Workshop on Graphics Recognition, pp. 195–205 (2005)
6. Belaid, A., Turcan, I., Pierrel, J.M., Belaid, Y., Hadjamar, Y., Hadjamar, H.: Automatic indexing and reformulation of ancient dictionaries. In: Proc. Int'l Workshop on Document Image Analysis for Libraries, pp. 342–354. IEEE Computer Society Press, Washington, DC, USA (2004)
7. Belongie, S., Malik, J., Puzicha, J.: Shape matching and object recognition using shape contexts. IEEE Trans. on Pattern Analysis and Machine Intelligence 24(4), 509–522 (2002)
8. Cao, H., Bhardwaj, A., Govindaraju, V.: Journal of Pattern Recognition 42(12), 3374
9. Cao, H., Govindaraju, V.: Vector model based indexing and retrieval of handwritten medical forms. In: Proc. Int'l Conf. on Document Analysis and Recognition, vol. 1, pp. 88–92 (2007)
10. Chellapilla, K., Piatt, J.: Redundant bit vectors for robust indexing and retrieval of electronic ink. In: Proc. Int'l Conf. on Document Analysis and Recognition, vol. 1, pp. 387–391 (2007)
11. Choisy, C.: Dynamic handwritten keyword spotting based on the NSHP-HMM. In: Proc. Int'l Conf. on Document Analysis and Recognition, pp. 242–246. IEEE Computer Society Press, Washington, DC, USA (2007)
12. Curtis, J.D., Chen, E.: Keyword spotting via word shape recognition. In: Proc. SPIE - Document Recognition II, pp. 270–277 (1995)
13. Delalandre, M., Ogier, J.-M., Lladós, J.: A fast CBIR system of old ornamental letter. In: Liu, W., Lladós, J., Ogier, J.-M. (eds.) GREC 2007. LNCS, vol. 5046, pp. 135–144. Springer, Heidelberg (2008)
14. Doermann, D., Doermann, D.: The indexing and retrieval of document images: A survey. Computer Vision and Image Understanding 70, 287–298 (1998)
15. Fataicha, Y., Cheriet, M., Nie, Y., Suen, Y.: Retrieving poorly degraded OCR documents. International Journal of Document Analysis and Recognition 8(1), 1–9 (2006)
16. Fonseca, M.J., Ferreira, A., Jorge, J.A.: Generic shape classification for retrieval. In: Proc. Int'l Workshop on Graphics Recognition, pp. 291–299 (2005)
17. Gatos, B., Pratikakis, I.: Segmentation-free word spotting in historical printed documents. In: Proc. Int'l Conf. on Document Analysis and Recognition, p. 271. IEEE Computer Society Press, Los Alamitos (2009)
18. Gordo, A., Valveny, E.: A rotation invariant page layout descriptor for document classification and retrieval. In: Proc. Int'l Conf. on Document Analysis and Recognition, pp. 481–485. IEEE Computer Society Press, Los Alamitos (2009)
19. Govindaraju, V., Cao, H., Bhardwaj, A.: Handwritten document retrieval strategies. In: Proc. of Workshop on Analytics for Noisy Unstructured Text Data, pp. 3–7. ACM, New York (2009)
20. Harris, Z.: Distributional structure. Word 10(23), 146–162 (1954)
21. Jain, A.K., Namboodiri, A.M.: Indexing and retrieval of on-line handwritten documents. In: Proc. Int'l Conf. on Document Analysis and Recognition, pp. 655–659. IEEE Computer Society Press, Washington, DC, USA (2003)
22. Hu, J., Kashi, R., Wilfong, G.: Comparison and classification of documents based on layout similarity. Information Retrieval 2(2/3), 227–243 (2000)

23. Jones, G., Foote, J., Sparck Jones, K., Young, S.: Video mail retrieval: the effect of word spotting accuracy on precision. In: Int'l Conf. on Acoustics, Speech, and Signal Processing, vol. 1, pp. 309–312 (1995)

24. Journet, N., Ramel, J.Y., Mullot, R., Eglin, V.: A proposition of retrieval tools for historical document images libraries. In: Proc. Int'l Conf. on Document Analysis and Recognition, pp. 1053–1057. IEEE Computer Society, Washington, DC, USA (2007)

25. Joutel, G., Eglin, V., Bres, S., Emptoz, H.: Curvelets based queries for CBIR application in handwriting collections. In: Proc. Int'l Conf. on Document Analysis and Recognition, pp. 649–653. IEEE Computer Society Press, Washington, DC, USA (2007)

26. Karray, A., Ogier, J.M., Kanoun, S., Alimi, M.A.: An ancient graphic documents indexing method based on spatial similarity. In: Proc. Int'l Workshop on Graphics Recognition, pp. 126–134. Springer, Heidelberg (2008)

27. Kesidis, A., Galiotou, E., Gatos, B., Lampropoulos, A., Pratikakis, I., Manolessou, I., Ralli, A.: Accessing the content of greek historical documents. In: Proc. of Workshop on Analytics for Noisy Unstructured Text Data, pp. 55–62. ACM, New York (2009)

28. Khurshid, K., Faure, C., Vincent, N.: Fusion of word spotting and spatial information for figure caption retrieval in historical document images. In: Proc. Int'l Conf. on Document Analysis and Recognition, pp. 266–270. IEEE Computer Society Press, Los Alamitos (2009)

29. Kise, K., Wuotang, Y., Matsumoto, K.: Document image retrieval based on 2D density distributions of terms with pseudo relevance feedback. In: Proc. Int'l Conf. on Document Analysis and Recognition, pp. 488–492. IEEE Computer Society Press, Washington, DC, USA (2003)

30. Kogler, M., Lux, M.: Bag of visual words revisited: an exploratory study on robust image retrieval exploiting fuzzy codebooks. In: Proc. Int'l Workshop on Multimedia Data Mining, MDMKDD 2010, pp. 3:1–3:6. ACM, USA (2010)

31. Konidaris, T., Gatos, B., Ntzios, K., Pratikakis, I., Theodoridis, S., Perantonis, S.J.: Keyword-guided word spotting in historical printed documents using synthetic data and user feedback. International Journal of Document Analysis and Recognition 9(2), 167–177 (2007)

32. Latecki, L.J., Lakämper, R., Eckhardt, U.: Shape descriptors for non-rigid shapes with a single closed contour. In: IEEE Computer Society Conf. in Computer Vision and Pattern Recognition, pp. 424–429 (2000)

33. Leydier, Y., Lebourgeois, F., Emptoz, H.: Text search for medieval manuscript images. Journal of Pattern Recognition 40(12), 3552–3567 (2007)

34. Li, L., Lu, S.J., Tan, C.L.: A fast keyword-spotting technique. In: Proc. Int'l Conf. on Document Analysis and Recognition, pp. 68–72. IEEE Computer Society, Washington, DC, USA (2007)

35. Liang, S., Sun, Z.: Sketch retrieval and relevance feedback with biased SVM classification. Pattern Recognition Letters 29(12), 1733–1741 (2008)

36. Licata, A., Psarrou, A., Kokla, V.: Revealing the visually unknown in ancient manuscripts with a similarity measure for IR-imaged inks. In: Proc. Int'l Conf. on Document Analysis and Recognition, pp. 818–822. IEEE Computer Society Press, Los Alamitos (2009)

37. Llados, J., Sanchez, G.: Indexing historical documents by word shape signatures. In: Proc. Int'l Conf. on Document Analysis and Recognition, pp. 362–366. IEEE Computer Society Press, Washington, DC, USA (2007)

38. Lowe, D.G.: Distinctive image features from scale-invariant keypoints. International Journal of Computer Vision 60, 91–110 (2004)

39. Lu, S., Li, L., Tan, C.L.: Document image retrieval through word shape coding. IEEE Trans. on Pattern Analysis and Machine Intelligence 30(11), 1913–1918 (2008)

40. Lu, S., Tan, C.: Keyword spotting and retrieval of document images captured by a digital camera. In: Proc. Int'l Conf. on Document Analysis and Recognition, pp. 994–998. IEEE Computer Society Press, Washington, DC, USA (2007)

41. Lu, S., Tan, C.L.: Retrieval of machine-printed latin documents through word shape coding. Journal of Pattern Recognition 41(5), 1816–1826 (2008)
42. Lu, Y., Zhang, L., Tan, C.L.: Retrieving imaged documents in digital libraries based on word image coding. In: Proc. Int'l Workshop on Document Image Analysis for Libraries, pp. 174–187. IEEE Computer Society Press, Washington, DC, USA (2004)
43. Manmatha, R., Han, C., Riseman, E.M.: Word spotting: A new approach to indexing handwriting. In: IEEE Computer Society Conf. in Computer Vision and Pattern Recognition, pp. 631–637. IEEE Computer Society, Los Alamitos (1996)
44. Marinai, S.: A Survey of Document Image Retrieval in Digital Libraries. In: Sulem, L.L. (ed.) Actes du 9ème Colloque International Francophone sur l'Ecrit et le Document, SDN 2006, pp. 193–198 (2006)
45. Marinai, S.: Text retrieval from early printed books. International Journal of Document Analysis and Recognition (2011); doi:10.1007/s10032-010-0146-0
46. Marinai, S., Faini, S., Marino, E., Soda, G.: Efficient word retrieval by means of SOM clustering and PCA. In: Bunke, H., Spitz, A.L. (eds.) DAS 2006. LNCS, vol. 3872, pp. 336–347. Springer, Heidelberg (2006)
47. Marinai, S., Gori, M., Soda, G.: Artificial neural networks for document analysis and recognition. IEEE Trans. on Pattern Analysis and Machine Intelligence 27(1), 23–35 (2005)
48. Marinai, S., Marino, E., Soda, G.: Layout based document image retrieval by means of XY tree reduction. In: Proc. Int'l Conf. on Document Analysis and Recognition, pp. 432–436 (2005)
49. Marinai, S., Marino, E., Soda, G.: Font adaptive word indexing of modern printed documents. IEEE Trans. on Pattern Analysis and Machine Intelligence 28(8) (2006)
50. Marinai, S., Marino, E., Soda, G.: Tree clustering for layout-based document image retrieval. In: Proc. Int'l Workshop on Document Image Analysis for Libraries, pp. 243–251 (2006)
51. Marinai, S., Miotti, B., Soda, G.: Mathematical symbol indexing using topologically ordered clusters of shape contexts. In: Proc. Int'l Conf. on Document Analysis and Recognition, pp. 1041–1045 (2009)
52. Marinai, S., Miotti, B., Soda, G.: Bag of characters and SOM clustering for script recognition and writer identification. In: Proc. Int'l Conf. on Pattern Recognition, pp. 2182–2185 (2010)
53. Meshesha, M., Jawahar, C.V.: Matching word images for content-based retrieval from printed document images. International Journal of Document Analysis and Recognition 11(1), 29–38 (2008)
54. Mitra, M., Chaudhuri, B.: Information retrieval from documents: A survey. Information Retrieval 2(2/3), 141–163 (2000)
55. Moghaddam, R., Cheriet, M.: Application of multi-level classifiers and clustering for automatic word spotting in historical document images. In: Proc. Int'l Conf. on Document Analysis and Recognition, pp. 511–515. IEEE Computer Society Press, Los Alamitos (2009)
56. Nakai, T., Kise, K., Iwamura, M.: Real-time retrieval for images of documents in various languages using a web camera. In: Proc. Int'l Conf. on Document Analysis and Recognition, pp. 146–150. IEEE Computer Society Press, Los Alamitos (2009)
57. Nguyen, T.O., Tabbone, S., Terrades, O.R.: Symbol descriptor based on shape context and vector model of information retrieval. In: Proc. IAPR Int'l Workshop on Document Analysis Systems, pp. 191–197. IEEE Computer Society, Washington, DC, USA (2008)
58. Perronnin, F.: Universal and adapted vocabularies for generic visual categorization. IEEE Trans. on Pattern Analysis and Machine Intelligence 30(7), 1243–1256 (2008)
59. Qureshi, R.J., Ramel, J.-Y., Barret, D., Cardot, H.: Spotting symbols in line drawing images using graph representations. In: Liu, W., Lladós, J., Ogier, J.-M. (eds.) GREC 2007. LNCS, vol. 5046, pp. 91–103. Springer, Heidelberg (2008)

60. Rath, T.M., Manmatha, R.: Features for word spotting in historical manuscripts. In: Proc. Int'l Conf. on Document Analysis and Recognition, pp. 218–222. IEEE Computer Society Press, Washington, DC, USA (2003)
61. Rath, T.M., Manmatha, R.: Word spotting for historical documents. International Journal of Document Analysis and Recognition 9(2), 139–152 (2007)
62. Rodriguez, J.A., Perronnin, F.: Local gradient histogram features for word spotting in unconstrained handwritten documents. In: Proc. Int'l Conf. on Handwriting Recognition (2008)
63. Rodriguez-Serrano, J., Perronnin, F.: Handwritten word-image retrieval with synthesized typed queries. In: Proc. Int'l Conf. on Document Analysis and Recognition, pp. 351–355. IEEE Computer Society Press, Los Alamitos (2009)
64. Rusiñol, M., Lladós, J.: Symbol spotting in technical drawings using vectorial signatures. In: Liu, W., Lladós, J. (eds.) GREC 2005. LNCS, vol. 3926, pp. 35–46. Springer, Heidelberg (2006)
65. Rusiñol, M., Lladós, J.: A region-based hashing approach for symbol spotting in technical documents. In: Liu, W., Lladós, J., Ogier, J.-M. (eds.) GREC 2007. LNCS, vol. 5046, pp. 104–113. Springer, Heidelberg (2008)
66. Rusiñol, M., Lladós, J.: Word and symbol spotting using spatial organization of local descriptors. In: Proc. IAPR Int'l Workshop on Document Analysis Systems, pp. 489–496. IEEE Computer Society Press, Washington, DC, USA (2008)
67. Rusiñol, M., Lladós, J.: Symbol Spotting in Digital Libraries: Focused Retrieval over Graphic-rich Document Collections. Springer, Heidelberg (2010)
68. Salton, G., Wong, A., Yang, C.S.: A vector space model for automatic indexing. Commun. ACM 18, 613–620 (1975)
69. Schomaker, L.: Retrieval of handwritten lines in historical documents. In: Proc. Int'l Conf. on Document Analysis and Recognition, vol. 2, pp. 594–598 (2007)
70. Sivic, J., Zisserman, A.: Video Google: A text retrieval approach to object matching in videos. In: Proc. Int'l Conf. on Computer Vision, vol. 2, pp. 1470–1477. IEEE Computer Society Press, Los Alamitos (2003)
71. Smeaton, A.F., Spitz, A.L.: Using character shape coding for information retrieval. In: Proc. Int'l Conf. on Document Analysis and Recognition, pp. 974–978 (1997)
72. Super, B.J.: Retrieval from shape databases using chance probability functions and fixed correspondence. International Journal of Pattern Recognition and Artificial Intelligence 20(8), 1117–1138 (2006)
73. Tahmasebi, N., Niklas, K., Theuerkauf, T., Risse, T.: Using word sense discrimination on historic document collections. In: Proc. Joint Conf. on Digital Libraries, pp. 89–98. ACM, New York (2010)
74. Tan, G., Viard-Gaudin, C., Kot, A.: Information retrieval model for online handwritten script identification. In: Proc. Int'l Conf. on Document Analysis and Recognition, pp. 336–340. IEEE Computer Society Press, Los Alamitos (2009)
75. Terasawa, K., Nagasaki, T., Kawashima, T.: Eigenspace method for text retrieval in historical documents. In: Proc. Int'l Conf. on Document Analysis and Recognition, pp. 437–441 (2005)
76. Tzacheva, A., El-Sonbaty, Y., El-Kwae, E.A.: Document image matching using a maximal grid approach. In: Proc. SPIE Document Recognition and Retrieval IX, pp. 121–128 (2002)
77. Uttama, S., Loonis, P., Delalandre, M., Ogier, J.M.: Segmentation and retrieval of ancient graphic documents. In: Liu, W., Lladós, J. (eds.) GREC 2005. LNCS, vol. 3926, pp. 88–98. Springer, Heidelberg (2006)
78. Wan, G., Liu, Z.: Content-based information retrieval and digital libraries. Information Technology & Libraries 27, 41–47 (2008)
79. Waters, D., Garrett, J.: Preserving digital information. report of the task force on archiving of digital information. Tech. rep., The Commission on Preservation and Access (1996)

80. Wei, C.H., Li, Y., Chau, W.Y., Li, C.T.: Trademark image retrieval using synthetic features for describing global shape and interior structure. Journal of Pattern Recognition 42(3), 386–394 (2009)
81. Witten, I.H., Bainbridge, D.: How to Build a Digital Library. Elsevier Science Inc., New York (2002)
82. Wong, W.T., Shih, F.Y., Su, T.F.: Shape-based image retrieval using two-level similarity measures. International Journal of Pattern Recognition and Artificial Intelligence 21(6), 995–1015 (2007)
83. Zhang, B., Srihari, S., Huang, C.: Word image retrieval using binary features. In: SPIE, Document Recognition and Retrieval XI, pp. 45–53 (2004)
84. Zhang, W., Liu, W.: A new vectorial signature for quick symbol indexing, filtering and recognition. In: Proc. Int'l Conf. on Document Analysis and Recognition, pp. 536–540. IEEE Computer Society Press, Washington, DC, USA (2007)
85. Zhang, Z., Jin, L., Ding, K., Gao, X.: Character-SIFT: a novel feature for offline handwritten chinese character recognition. In: Proc. Int'l Conf. on Document Analysis and Recognition, pp. 763–767. IEEE Computer Society Press, Los Alamitos (2009)

# Mining Biomedical Text towards Building a Quantitative Food-Disease-Gene Network

Hui Yang, Rajesh Swaminathan, Abhishek Sharma,
Vilas Ketkar, and Jason D'Silva

**Abstract.** Advances in bio-technology and life sciences are leading to an ever-increasing volume of published research data, predominantly in unstructured text. To uncover the underlying knowledge base hidden in such data, text mining techniques have been utilized. Past and current efforts in this area have been largely focusing on recognizing gene and protein names, and identifying binary relationships among genes or proteins. In this chapter, we present an information extraction system that analyzes publications in an emerging discipline–Nutritional Genomics, a discipline that studies the interactions amongst genes, foods and diseases–aiming to build a quantitative food-disease-gene network. To this end, we adopt a host of techniques including natural language processing (NLP) techniques, domain ontology, and machine learning approaches.

Specifically, the proposed system is composed of four main modules: (1) named entity recognition, which extracts five types of entities including foods, chemicals, diseases, proteins and genes; (2) relationship extraction: A verb-centric approach is implemented to extract binary relationships between two entities; (3) relationship polarity and strength analysis: We have constructed novel features to capture the syntactic, semantic and structural aspects of a relationship. A 2-phase Support Vector Machine is then used to classify the polarity, whereas a Support Vector Regression learner is applied to rate the strength level of a relationship; and (4) relationship integration and visualization, which integrates the previously extracted relationships and realizes a preliminary user interface for intuitive observation and exploration.

Empirical evaluations of the first three modules demonstrate the efficacy of this system. The entity recognition module achieved a balanced precision and recall with an average F-score of 0.89. The average F-score of the extracted relationships is 0.905. Finally, an accuracy of 0.91 and 0.96 was achieved in classifying the relationship polarity and rating the relationship strength level, respectively.

Hui Yang · Rajesh Swaminathan · Abhishek Sharma · Vilas Ketkar · Jason D'Silva
Department of Computer Science, San Francisco State University, USA
e-mail: `huiyang@cs.sfsu.edu`

M. Biba and F. Xhafa (Eds.): Learning Structure and Schemas from Documents, SCI 375, pp. 205–225.
springerlink.com &copy; Springer-Verlag Berlin Heidelberg 2011

## 1   Introduction and Motivation

Advances in bio-technology and life sciences are leading to an ever-increasing volume of published research data, predominantly in unstructured text (or natural language). At the time of writing, the MEDLINE database consists of more than 20 million scientific articles with a growth rate of around 400,000 articles per year [29]. This phenomenon becomes even more apparent in nutritional genomics, an emerging new science that studies the relationship between foods (or nutrients), diseases, and genes [20]. For instance, soy products and green tea have been two of the intensively studied foods in this new discipline due to their controversial relationship with cancer. A search to the MEDLINE database on "soy and cancer" renders a total of 1,413 articles, and a search on "green tea and cancer" renders 1,471 articles. Due to the large number of publications every year, it is unrealistic for even the most motivated to manually go through these articles to obtain a full picture of the findings reported to date. Obtaining such a picture however has become ever more important and necessary due to the following reasons: (1) given a pair of entities, e.g., green tea and cancer, different studies might report different findings with respect to their relationship. For example, A comprehensive review in [52] suggests that extracts of green tea  have exhibited inhibitory effects against the formation and development of tumor". Another study, however, concludes that these two are not related [38]. Clearly, the relationship between two given entities can be reported as positive (good), negative (bad), and neutral. We term this as *the relationship polarity* ; and (2) even if different studies agree with each other on the relationship polarity between two entities, they may report it with a different level of certainty. one study suggests that soy intake may protect against breast cancer [27], while another study indicates that soy intake is an essential factor for the incidence of hormone-dependent tumors (e.g., breast cancer) [50]. Obviously, the latter is more decisive than the former. We term the certainty of a relationship as *the relationship strength* .

Specifically, we have classified the relationship polarity into four types: positive, negative, neutral, and no-relationship. The strength feature on the other hand has three values: weak, medium and strong Table 1 lists several relationships with different polarity and strength. Note that the "no-relationship" polarity is highly evident in biomedical articles and has rarely been explored except the work in [32], where it is termed as "no outcome". No-relationship is different from the neutral polarity: a "no-relationship" indicates that no association is found between the biological entities in consideration, whereas for a neutral polarity, the entities are associated but without polar orientation.

It is essential to quantify the polarity and strength of a relationship, since such information will allow one to integrate the extracted relationships and then present a holistic picture to the end user. For example, after having analyzed the 1,413 research articles on the relationship between "soy" and "cancer", a holistic summary of these articles can be: $x\%$ of soy-cancer relationships were reported as positive, $y\%$ reported as negative, and so on. This can be further elaborated based on the strength of a relationship.

**Table 1** Exemplar relationships with their respective polarity and strength

| Example sentence | Polarity | Strength |
|---|---|---|
| Soy food consumption may reduce the risk of fracture in postmenopausal women. | Positive | Weak |
| Consumption of kimchi, and soybean pastes was associated with increased risk of gastric cancer. | Negative | Medium |
| Increased intake of kimchi or soybean pastes was a significant risk factor for the CYP1A1 genotype. | Negative | Strong |
| These data suggest that soy foods do not have an adverse effect on breast cancer survival. | Neutral | Medium |
| No significant associations were found between intake of carotenoids and colorectal cancer risk. | Not-related | Medium |

In this chapter, we present an information extraction system that aims to construct quantitative networks to capture the complex relationships reported among foods, chemical nutrients, diseases, proteins, and genes. To achieve this goal, we have implemented four main functional modules:

1. *Named entity recognition:* A primarily ontology-based approach is adopted to recognize the following types of entities: foods, chemicals (or nutrients), diseases, proteins, and genes.
2. *Relationship extraction:* Unlike previous work that employs heuristics such as co-occurrence patterns and handcrafted syntactic rules, we propose a verb-centric algorithm. This algorithm identifies and extracts the main verb(s) in a sentence. Syntactic patterns and heuristics are adopted to identify the entities involved in a relationship. For example, the sentence *Soy food consumption may reduce the risk of fracture* specifies a relationship between *soy food consumption* and *fracture*, where *may reduce the risk of* depicts their relationship, termed as relationship-depicting phrase (RDP). The goal of this module is to identify all three components.
3. *Relationship polarity and strength analysis:* We have constructed a novel feature space that includes not only the commonly accepted lexical features (e.g., uni-grams), but also features that can capture both the semantic and structural aspects of a relationship. We then build Support Vector Machine (SVM) and Support Vector Regression (SVR) models to learn the polarity and strength of a relationship, respectively.
4. *Relationship integration and visualization:* This module integrates the previously extracted relationships into networks and summarizes them based on polarity and strength. An interactive graphic user interface is also implemented to allow user intuitively view and explore these networks.

The overall architecture of the proposed system is depicted in Figure 1.

The remainder of this chapter will be structured as follows: Section 2 briefly reviews the studies that are most germane to the proposed system; Section 3 details the

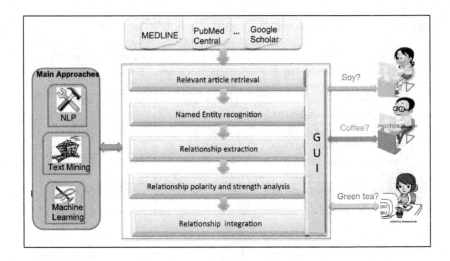

**Fig. 1** System architecture: main functional modules

implementation for named entity recognition; Section 4 describes the verb-centric approach that recognizes and extracts binary relationships; Section 5 first discusses the novel features that are designed to capture both semantic and structural features of a relationship and then presents different feature selection strategies towards building robust SVM and SVR models for polarity and strength prediction; Section 6 describes the visualization tool; Section 7 presents the evaluation results to demonstrate the effectiveness of each module; finally, Section 8 identifies several limitations in the current system and concludes the entire chapter.

## 2  Related Work

The proposed system is related to three areas: named entity recognition, relationship extraction, and opinion mining. We review the most relevant studies in this section.

### 2.1  Named Entity Recognition (NER) in Biomedical Text

Three mainstream approaches have been adopted to recognize entities in biomedical text: (1) dictionary-based; (2) rule-based; and (3) machine learning. The first approach recognizes entities of interest by matching them with entries in a dictionary, such as the Unified Medical Language System (UMLS) MetaThesaurus [1, 3, 46]. This approach often achieves high precision but its recall heavily depends on the quality of the dictionary. The rule-based approach relies on rules that are often hand-crafted using the common orthographical features in a name, e.g., the use of capital letters and digits [5, 9, 16, 45]. Balanced precision and recall can be achieved by this approach but it is a demanding task to construct good and generic rules.

Finally, sequence-based machine learning algorithms such as Hidden Markov Model (HMM) and Conditional Random Field (CRF) have been frequently adopted for the NER task [34, 39, 51]. The advantage of these algorithms is that they can recognize previously unknown entities, but their success is closely linked to the availability and quality of the training data, which is often time consuming to generate. The proposed system adopts a primarily dictionary-based approach using the UMLS MetaThesaurus and the USDA Food Database [49].

## 2.2  Relationship Extraction

Relationship extraction is often considered as a task that takes place after NER. Three main approaches have been applied in the past to extract relationship, namely, the co-occurrence based, rule-based and machine learning approaches. In the co-occurrence approach, if two entities frequently collocate with each other, a relationship is inferred between them [44, 18, 31]. A natural extension to this approach is the link-based approach. Two entities are considered to have a relationship if they co-occur with a common entity [25, 42, 47]. On the other hand, rule-based techniques heavily reply on NLP techniques to identify both syntactic and semantic units in the text. Handcrafted rules are then applied to extract relationships between entities. For instance, Fundel *et al.* construct three rules and apply them to a dependency parse tree to extract the interactions between genes and proteins [10]. Feldman introduces a template in the form of NP1-Verb-NP2 to identify the relation between two entities corresponding to two noun phrases, NP1 and NP2, respectively [7]. Many rule-based studies often predetermine the relationship categories using target verbs such as 'inhibit' and 'activate' or negation words such as 'but not' [17, 37, 48]. Finally, Machine learning approaches such as SVM and CRF have also been used to extract relationships [2, 12, 15]. These approaches are supervised and require manually annotated training dataset, which can be expensive to construct. All these approaches typically assume that the entities of interest are either predetermined or can be satisfactorily recognized by the NER task. The proposed verb-centric approach (Section 4). does not make such an assumption [40]. In addition, it does not require handcrafted rules.

## 2.3  Polarity and Strength Analysis

Although only a couple of existing studies were conducted in the past to analyze the polarity and strength of relationships in biomedical text [32, 43], the concept is not new. Such tasks are commonly referred to as opinion mining or sentiment analysis and have gained increasing attention in the past few years [22, 26]. Existing work however primarily focuses on text created by day-to-day web users such as product reviews and blogs. Both supervised and unsupervised approaches have been proposed in the past [28, 30, 36, 6]. These approaches are mostly lexicon-based. For instance, adjectives and adverbs are routinely employed to detect the polarity of a review [6]. This however does not apply to formally written biomedical articles, as adjectives and adverbs are used sparingly. In addition, semantics-based structure information can play an important role in the latter case.

## 2.4 Relationship Integration and Visualization

Most existing studies in biomedical text analysis stop short of integrating the extracted relations into networks and present them in a user-friendly interface for user interactions. However, there are a few exceptions including web-based systems such as ArrowSmith, iHop, and STITCH [41, 14, 24]. Again, all these systems are primarily concerned with relationships among proteins and genes. They however serve as excellent examples to demonstrate the importance of data integration and intuitive data representation.

## 3 Named Entity Recognition

The UMLS Metathesaurus–the largest vocabulary that contains biomedical and health-related concepts, common terms, and the relationships among them–is used to recognize the following entity types: foods, diseases, chemicals, genes, and proteins. Specifically, we use the companion software program MetaMap [1] to map terms in an abstract to those in the Metathesaurus. Given a phrase such as "gastric cancer", MetaMap often returns multiple mappings. As shown in Table 2, a total of 5 mappings are returned, where each row identifies one mapping. The first two numbers are the mapping and evaluation scores in $[0, 1000]$ (a perfect mapping), and the last field identifies the semantic type (e.g., disease) of the term given a mapping.

**Table 2** The MetaMap results for the term "Gastric Cancer". It is likely of the following semantic types: neoplastic process (neop), invertebrate (invt), and Biomedical Occupation or Discipline (bmod).

| | |
|---|---|
| 1000 | 1000—C0024623—Malignant neoplasm of stomach—Gastric Cancer—neop— |
| 1000 | 1000—C0699791—Stomach Carcinoma—Gastric Cancer—neop— |
| 888 | 861—C0038351—Stomach—Gastric—bpoc— —Cancer Genus—Cancer—invt— |
| 888 | 861—C0038351—Stomach—Gastric—bpoc—Malignant Neoplasms—Cancer—neop— |
| 888 | 861—C0038351—Stomach—Gastric—bpoc—Specialty Typ cancer—Cancer—bmod— |

Among all the returned mappings in Table 2, which one is the most likely? To answer this question, we employ two heuristics: (1) Let us denote the highest mapping score as Sh. We first select all the mappings whose mapping score is in $[S_h - \delta, S_h]$, where $\delta$ is a user-defined parameter and set to 50 in our evaluation; and (2) we then examine these selected mappings and select the most frequently mentioned semantic type to label the term of interest. Take the case in Table 2 as an example and let $\delta = 150$. All five mapping will be selected, among which the semantic type "neap" (neoplastic process) is the most frequent (3 times). The term "gastric cancer" is therefore labeled as "neop", which is a disease. The rationale behind these heuristics is that the mappings returned by MetaMap with the highest score are not necessarily the best mapping due to the stochastic nature of the algorithm. It is possible that one entity might be associated with multiple semantic types using the above approach. Note that we have manually grouped the 135 UMLS semantic types into the five types of interest after removing irrelevant types (e.g., bird).

## 3.1 Improving the Performance of Food Recognition

MetaMap can identify most food-describing phrases. However, it sometimes fails to correctly identify common whole foods, "kiwi" and "grape seeds". To address this issue, we employ the USDA (United States Department of Agriculture) food database [49] as well. Given a sentence, we first build its parse tree using the Penn TreeBank parser [35]. We then focus on the noun phrases in the sentence. For each multi-word noun phrase, we consider all of its subsequences and obtain a mapping score in $[0, 1]$ for each subsequence. This score is determined by the following factors: (1) the relative length and location of a subsequence in the noun phrase: the longer and the closer to the end of the phrase, the better; (2) the location of the match in a USDA food entry: the nearer to the start, the better; and (3) the number of words matched with a USDA food entry: the more, the better. If the highest mapping score of a noun phrase is greater than a threshold (e.g., 0.5), we label it as food.

We next combine the UMLS-based and USDA-based approaches to finalize the list of food entities. A term will be labeled as food if it is labeled by either or both methods. The UMLS-based label is chosen should a conflict arise. This ensemble method proves to be effective and achieves an F-score of 0.95, compared to 0.85 and 0.64 if only UMLS or USDA database is used.

## 3.2 Abbreviations and Co-reference Recognition

The use of abbreviations is a common practice in scientific writing, for instance, SERM for "selective estrogen receptor modulator". Co-reference is another commonly employed linguistic feature where multiple phrases refer to the same referent. We identify abbreviations based on the observation that they are frequently specified within parentheses right after the long form. We have designed an algorithm to identify the following forms of abbreviations: (1) acronyms, e.g., SERM for "selective estrogen receptor modulator"; (2) those that correspond to a non-contiguous subsequence of the long form, e.g. "nucleophosmin-anaplastic lymphoma kinase" as NPM-ALK; and (3) the abbreviations that are constructed using a combination of initials and a whole word, e.g., "estrogen receptor beta" as ERbeta. Co-references are recognized using the OpenNLP toolkit. It takes the entire abstract as input and identifies all the co-references. We then go over these co-references and retain those whose referent is an entity of interest, abbreviated or not.

## 4   Verb-Centric Relationship Extraction

The relationship extraction algorithm identifies and extracts relationships at the sentence level. Given a sentence, the following two criteria are used to determine whether it contains a relationship: 1) it needs to contain at least one recognized entity; and (2) it contains a verb that is semantically similar to one of the 54 verbs listed in the UMLS Semantic Networks. We use WordNet [8] and VerbNet [21] to identify semantically similar verbs. These criteria are drawn from carefully analyzing a collection of 50 randomly selected abstracts downloaded from PUBMED.

Once a sentence is determined to contain a relationship, the following steps are taken to extract all the binary relationships contained within:

- *Step 1.* split a compound sentence into simple sentences if necessary;
- *Step 2.* identify the main verb in each simple sentence, where a main verb is the most important verb in a sentence and states the action of the subject.
- *Step 3.* recognize the two participating entities of each main verb;
- *Step 4.* handle the "between ... and ..." structure;

We next describe each step in detail.

*Step 1.* To partition a compound sentence into multiple semantic units we analyze the sentence-level conjunctive structure using the OpenNLP parser. A list of conjunctions is retrieved from the parse tree of the sentence. The parent sub-tree of each conjunction is then retrieved. If the parent sub-tree corresponds to the entire input sentence the algorithm asserts that it is a sentence level conjunction. It then uses the conjunctions to break the sentence into smaller semantic units.

*Step 2.* To find the main verb(s), the algorithm considers the parse tree of a sentence generated by the OpenNLP parser. It traverses the parse tree using the preorder traversal and continuously checks the tags of each right sub-tree until it reaches the deepest verb node of a verb phrase. This verb is then considered the main verb of the sentence. For instance, using this approach, the verb "associated" is correctly identified as the main verb in the sentence *Phytoestrogens may be associated with a reduced risk of hormone dependent neoplasm such as prostate and breast cancer.* We assume that each main verb corresponds to one action-based relationship (e.g., *A affects B*) and use it to identify the involved entities.

*Step 3.* For each main verb, a proximity-based approach is adopted to identify its two involved entities. Specifically, we take the entities located immediately to the left and to the right of a main verb in the same semantic unit as the two participating entities. For this proximity-based approach to work effectively, we however have to tackle the following issues: (1) the NER module might not recognize all the entities of interest. We term this as the missing entity issue; (2) the incomplete entity issue, where the entity identified by the NER is part of a true entity. For instance, only the word "equol" in the phrase "serum concentration of equol" is labeled as "chemical" and (3) the co-ordinated phrases, for instance, *B, C and D* in *A activates B, C and D*. A co-ordinated phrase essentially defines a one-to-many or many-to-one relationship. To tackle these three issues, we rely on both dependency parse trees and linguistic features.

For the missing entity issue, we first observe that many of these missing entities are either a subject or an object in a sentence. We therefore address this issue by identifying the subject and object in a sentence using the Stanford NLP Parser [4].

To handle the incomplete entity issue, we search the preposition phrase around an entity and merge it with this entity. This is based on our analysis over the incomplete entities. As an example, the NER module in Section 3 only extracts "equol" and "dietary intake" as entities from the sentence ... *serum concentration of equol and dietary intake of tofu and miso soup* .... By recognizing the preposition phrase right before or after these two labeled entities, the program is able to identify the two entities in their full forms, i.e., "serum concentration of equol" and "dietary intake of tofu and miso soup".

Finally, to address the one-to-many or many-to-one relationships as a result of co-ordinated structures, we utilize the dependency parse tree to first recognize such structures. We then merge the co-ordinated entities into one compound entity. This allows us to correctly establish the one-to-many and many-to-one relationship using the main verb identified earlier. The merged compound entity will be split into individual entities as they are when we produce the final binary relationships.

*Step 4.* A special case in describing a relationship is the use of *between ... and ...* template. The approach described earlier cannot handle this structure. We therefore treat such cases separately. Consider the sentence *Few studies showed protective effect between phytoestrogen intake and prostate cancer risk*. We first analyze its parse tree to identify the two prepositions "between" and "and". We then identify the entity after "between" and before "and" as the left entity, the one immediately after "and" as the right entity. The relationship depicting phrase is constructed by using the main verb plus all the words between the main verb and the word "between". Therefore, the binary relationship of the sentence stated above is (phytoestrogen intake, showed protective effect, prostate cancer risk).

## 5 Relationship Polarity and Strength Analysis

As described in Table 1, a relationship can be categorized into four types according to its *polarity*: positive, negative, neutral, and not-related. Furthermore, a relationship of any polarity can be rated at three levels according to its *strength*: weak, medium, and strong. In Section 2.3, we argue that lexical features adopted in traditional opinion mining approaches are not suitable for relationships extracted from biomedical literature articles. In this section, we first describe a list of new features that are designed to capture not only the lexical, but also the semantic, and structural characteristics of a relationship. We then describe a wrapper-based feature selection method to identify the features that are effective at modeling the polarity and strength of relationships.

## 5.1  Feature Space Design

Given a relationship-bearing sentence $s$, at this point, we have identified all the entities and their semantic types contained in $s$ using the algorithm described in Section 3. In addition, the relationship-depicting phrase (RDP) between two related entities has also been recognized by applying the verb-centric algorithm in Section 4. For example, the sentence *Intake of soy products reduces the risk of breast cancer.* has been automatically annotated as {\food *Intake of soy products*} {\RDP *reduces the risk of*} {\disease *breast cancer*}. Such annotated sentences will be used to construct three types of features including unigrams, bigrams, and semantics based sequential features.

*Part-of-speech based unigrams:* Given an annotated sentence, we consider the following two regions to identify its unigrams: (1) the RDP only, and (2) the RDP-centered neighborhood, which centers at the RDP and spans from the left boundary of the entity immediately preceding the RDP to the right boundary of the entity immediately succeeding the RDP. In our empirical evaluation, the RDP-only region delivers better performance. The unigram features are identified as follows: We first remove all the non-content words (e.g. the, and). We then normalize the unigrams such that different grammatical variations of the same root word (e.g., increase, increasing, and increases) will be treated as a single unigram. We next augment our list by obtaining the synonyms of each unigram from WordNet. We also identify all the verbs in our unigram list. For each verb we use VerbNet to identify its semantic class and add all the verbs in the class to the unigram list as well. Note that the unigrams are organized into equivalence classes, each corresponding to a set of synonyms of a given word.

Two sources are used to determine the part-of-speech (POS) tag of a given unigram: (1) Penn Tree Bank Parser: We use it to first generate a parse tree for a relationship-bearing sentence and then extract the POS tag for each individual unigram using the parse tree; and (2) WordNet: The parse tree rendered by the above parser is probabilistic by nature. As a result, sometimes, the POS tag of a unigram can be either missing or incorrect. In such cases we use the POS from WordNet as the default POS value.

*Bigrams:* We observe that bigram features (e.g., *significantly increase*) are common in our dataset and thus hypothesize that they might have positive effect on the final strength prediction. To construct bigram features, we use a public dataset consisting of bigrams that commonly appear in PUBMED. We use all the bigrams that have a term frequency and document frequency of at least 10000 in the dataset.

*Semantics-based sequential features:* This type of features is designed due to our observations that the semantic types of the entities and their order of occurrence involved in a relationship can influence the polarity of a relationship. For instance, the sentence {\food *Soy consumption*} {\RDP *significantly decreased*} {\disease *cancer risk*} has a positive polarity. But the next sentence {\chemical *Phytoestrogens*} {\RDP *are responsible for decreased*} {\protein *CYP24 expression*} is neutral. we construct these features at three different levels: entity, phrase, and sentence.

*Features to capture the Binary-ternary sequential structure:* Let $E$ represent a bio-entity including relationship and $R$ represent a RDP. These features capture the semantic neighborhood of $R$ in the form of $E - R - E$ (ternary), $E - R$ or $R - E$ (binary). In other words, only the immediately preceding or succeeding entity is included in the neighborhood. The $E - R$ or $R - E$ case exists due to either flaws in the entity recognition algorithm or the complexity in the sentence structure. We then examine the semantic types of the involved entities, which can be one of the following: foods, diseases, chemicals, genes and proteins. A total of 35 binary dimensions are created to capture the 35 possible sequential features. For the two sentences in the above paragraph, their respective sequential structures are $food - R - disease$ and $chemical - R - disease$.

*Features to capture the k-ary sequential structure:* These features capture the semantics based sequential structures at the sentence level. Rather than just focusing on the RDP-based neighborhood within the sentence, we take all the biological entities contained within a sentence into account. For example the sentence {\chemical Genistein}, a natural {\chemical isoflavonoid} found in {\food soy products}, {\RDP has been shown to inhibit cell growth and induce apoptosis} in a wide variety of cell lines exhibits the following K-ary sequential structure $chemical - chemical - food - relationship$.

## 5.2 Feature Selection

In this step, we adopt a wrapper-based method to select the features that can accurately predict the polarity and strength of relationships [11]. Specifically, we build multi-stage Support Vector Machine (SVM) models to classify the four types of relationship polarity by using different subsets of the above features; whereas Support Vector Regression (SVR) models are employed to predict the relationship strength at three levels. We then calculate the classification accuracy of these different models by applying them to a test set. The subset of features that deliver the best performance will be selected for polarity and strength prediction.

*Feature selection for polarity analysis:* Since SVM is a binary classifier, we use two different methods to build a multi-stage classifier to handle the four polarity classes: (1) one vs. all (1 vs. A) and (2) Two vs. Two (2 vs. 2). In 1 vs. A, we first build a SVM to separate positive relationships from negative, neutral and no-relationship. We then build a SVM to separate negative from neutral and no-relationship. Finally we separate neutral from no-relationship. Such an ordering is chosen on the basis of both manual observations and empirical evaluation. In the 2 vs. 2 method, we build a SVM to separate neutral and no-relationships first from positive and negative ones. We then build two other SVMs to separate between neutral and no-relationship and between positive and negative relationships, respectively. The above combination strategy is based on analyses of our dataset, which shows that positive and negative relationships often exhibit similar characteristics when compared against the other two types.

*Feature selection for strength analysis:* We build the SVRs using different features without distinguishing the polarity of the training records. We also tried a variation of this approach by building individual SVRs for each polarity value but found that polarity has negligible effect on strength analysis.

*Kernel selection:* We employ four kernel functions when building a SVM or a SVR, including linear, sigmoid, polynomial and radial-bias function (RBF). We explore a variety of kernel combinations for building the multi-stage classifiers for polarity analysis as reported in Section 7.

## 6  Relationship Integration and Visualization

In this module, all the extracted entities, their relationships, and their polarity and strength are stored into a relational database (MySQL). They are then joined together to construct a set of networks. The Prefuse visualization toolkit [13] is utilized to render these networks with flexible user interactivity. Figure 2 shows a screenshot of a visualized network centered at "soy". Each node in this figure represents a named entity, of which the semantic type is indicated by the color of a node (e.g., green for foods). An edge between two nodes indicates the presence of a relationship. The color of an edge varies according to a relationship's polarity and the thickness of an edge varies according to strength. A user can click any entity in the network to highlight its immediate neighbors. He can also search this network. The matched nodes will be highlighted and the entire network will be re-centered.

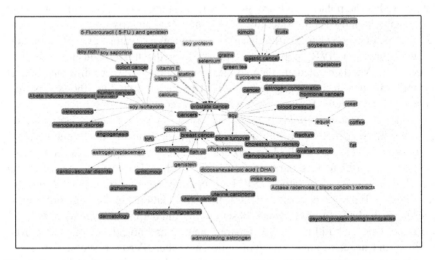

**Fig. 2** A screenshot of a food-disease network drawn from 50 PUBMED abstracts that contain "soy" and "cancer"

# 7  Evaluation

In this section, we evaluate the performance of the Named Entity Recognition module (Section 3), the verb-centric relationship extraction algorithm (Section 4), and the new features designed for relationship polarity and strength classification (Section 5). Three datasets are used for the evaluation. As shown in Table 3, each dataset consists of a number of abstracts downloaded from PUBMED using the following keywords respectively: *soy and cancer, betacarotene and cancer,* and *benzyl and cancer.* These three datasets altogether consist of around 1400 sentences, of which 800 contain a binary relationship. The distribution of these sentences according to polarity and strength is shown in Table 4.

**Table 3** Evaluation datasets

| ID | Keywords | #(Abs.) |
|---|---|---|
| 1 | soy and cancer | 50 |
| 1 | betacarotene and cancer | 80 |
| 1 | benzyl and cancer | 40 |

**Table 4** Distribution of the relationships by polarity and strength, respectively

| #(Positive) | #(Negative) | #(Neutral) | #(Not-related) | #(Weak) | #(Medium) | #(Strong) |
|---|---|---|---|---|---|---|
| 447 | 143 | 117 | 93 | 135 | 558 | 107 |

To evaluate our algorithm, we first manually annotated all the entities of interest (e.g., foods), relationships, the polarity and strength of each relationship in all the datasets. The following is an example of an annotated sentence: {\\*chemical soy isoflavones*} {\\*RDP 1+ are likely to be protective of*} {\\*disease colon cancer*} *and to be well tolerated,* where "1+" indicates that this relationship is positive (+) with a weak (1) strength. The manual annotation was done independently by a group of five members. All members are graduate students who have been analyzing biomedical text for at least one year. To reduce potential subjectivity in the manual annotation, we combine the five versions of annotations from the five members into one version using majority rule policy. Manual intervention was required for this task as we could not find a majority vote for some sentences. In this case, a group discussion was invoked to reach an agreement. Note that we did not evaluate our algorithm using existing public datasets such as the BioCreative datasets [23] because such datasets often focus on genes and proteins, whereas we are concerned with also foods, chemicals, diseases, and their relationships.

## 7.1  Evaluation of the Named Entity Recognition Module

We evaluate the entity extraction module using three measurements: precision (p), recall (r) and the F-score (2pr/(p+r)). Table 5 presents the evaluation results of this module using Dataset 1. We observe that food entities achieve the best results. This is due to the usage of both the UMLS and USDA food databases. The recognition of co-references and abbreviations increases the combined F-score by 1.5Overall, our ER module achieves a balanced precision and recall.

**Table 5** Performance of the named entity recognition module

| Type | #entities | Precision | Recall | F-score |
|------|-----------|-----------|--------|---------|
| Food | 313 | 0.9657 | 0.9306 | 0.945 |
| Chemical | 788 | 0.7718 | 0.8753 | 0.82 |
| Disease | 265 | 0.8148 | 0.9000 | 0.855 |
| Protein-Gene | 364 | 0.8796 | 0.8769 | 0.878 |
| Combined | 1730 | 0.8980 | 0.8896 | 0.8940 |

We also evaluate the NER module using the GENIA V3.02 corpus [33]. This corpus consists of 2000 abstracts from MEDLINE extracted using the terms "human", "blood cells" and "transcription factors". Following a similar protocol in the JNLPBA entity recognition contest [34], we simplify the 36 entities types in GENIA to Proteins, Genes, and Chemicals. We then run our ERE module over 500 abstracts randomly selected from the corpus. Table 6 lists the evaluation results of our algorithm and the top three performers in the JNLPBA contest. Note that all these algorithms use supervised methods (e.g. SVM), are trained over 2000 abstracts and tested over 404 abstracts. In contrast, our algorithm is unsupervised and does not require training time. This demonstrates that our method is comparable with these programs in precision, but with a significantly higher recall.

**Table 6** Performance comparison of the NER module with other leading methods

| Algorithm | Precision | Recall | F-score |
|-----------|-----------|--------|---------|
| GENIA System 1 (Zhou) | 0.694 | 0.760 | 0.726 |
| GENIA System 2 (Fin) | 0.686 | 0.716 | 0.701 |
| GENIA System 3 (Set) | 0.693 | 0.703 | 0.698 |
| Our algorithm | 0.690 | 0.830 | 0.754 |

## 7.2 Evaluation of the Relationship Extraction Algorithm

We use the same three measure–precision, recall and F-score–to evaluate the performance of the verb-centric relationship extraction algorithm. Given a relationship-bearing sentence, it is considered as a true positive if (1) the algorithmically extracted relationship-depicting phrase (RDP) overlaps with the manually annotated RDP, or (2) these two RDPs agree on the same main verb. Otherwise, the sentence is considered as a false positive. A false negative corresponds to a sentence that is manually annotated as relationship-bearing but not by our algorithm.

To first demonstrate the advantage of the proposed verb-centric approach, we also compare it with a rule-based approach, where a rule NP1-VP-NP2 is used. According to this rule, a relationship is composed of a verb phrase and two noun phrases. Furthermore, these noun phrases are labeled as entities of interest. We refer to this approach as the NVN method. We also study the effect of the three entity-related issues–missing, incomplete, and co-ordinated using the NVN algorithm. The results are shown in Table 7. One can observe that there is a gradual improvement in the performance using the NVN method. This indicates the necessity and importance of handling these entity-related issues.

**Table 7** Effect of the missing,incomplete, and co-ordinated entity handling using Dataset 1

| Algorithmm | Precision | Recall | F-score |
|---|---|---|---|
| NVN without missing,incomplete, and co-ordinated entity handling | 0.79 | 0.82 | 0.805 |
| NVN wih missing entity handlingn | 0.80 | 0.87 | 0.843 |
| NVN with missing, incomplete, co-ordinated entity handling | 0.84 | 0.89 | 0.855 |
| Verb-centric relationship extraction | 0.96 | 0.90 | 0.931 |

We also consider a more rigid criterion where we not only compare whether the algorithmically executed RDPs match with their manual counterparts, but also consider the entities involved in a relationship. Table 8 reports the results on all three datasets based on this criterion. These results demonstrate the effectiveness of the verb-centric approach, including the strategies we have introduced to handle missing, incomplete and co-ordinated entities.

**Table 8** Evaluation results of the verb-centric approach on all three datasets

| | Precision | Recall | F-score |
|---|---|---|---|
| Dataset 1 | 0.945 | 0.868 | 0.902 |
| Dataset 2 | 0.858 | 0.844 | 0.851 |
| Dataset 3 | 0.905 | 0.861 | 0.883 |

## 7.3   Evaluation of Relationship Polarity and Strength Analysis

We report the feature selection results (Section 5.2) in this section. These results demonstrate that the proposed feature space can effectively predict the polarity and strength of the relationships extracted from biomedical literature. At the same time, they also indicate that not every feature contributes equally to the two problems under study. We perform 10-fold cross validation throughout our evaluation using all three annotated datasets. Classification accuracy is used to measure the performance of each SVM, whereas prediction accuracy is used for each SVR. We use the SVMLight package by [19] for our experiments.

Figure 3 lists the model accuracy for polarity analysis based on various feature combinations and kernel functions. From this figure, one can observe that the positive polarity constantly has high accuracy for both methods as compared to the

| Feature Set | SVM: One vs. All Schema | | | | | | | |
| --- | --- | --- | --- | --- | --- | --- | --- | --- |
| | Kernel | | Average Accuracy | | | | | StdErr |
| | L1 | L2 | + | - | = | ! | OA | |
| 1 | R | R | 0.88 | 0.9 | 0.87 | 0.88 | 0.87 | 0.0289 |
| 1+3 | L | P | 0.83 | 0.9 | 0.73 | 1 | 0.84 | 0.0312 |
| 1+4 | P | P | 0.85 | 0.9 | 0.67 | 1 | 0.84 | 0.0303 |
| 1+5 | R | R | 0.85 | 0.9 | 0.73 | 1 | 0.85 | 0.0145 |
| 2 | R | R | 0.87 | 0.8 | 0.87 | 0.88 | 0.86 | 0.027 |
| 2+3 | L | R | 0.8 | 0.9 | 0.93 | 1 | 0.86 | 0.0245 |
| 2+4 | P | P | 0.87 | 0.9 | 0.73 | 1 | 0.86 | 0.014 |
| 2+5 | R | R | 0.85 | 0.8 | 0.87 | 1 | 0.86 | 0.031 |
| 2+3+5 | R | R | 0.8 | 0.7 | 0.93 | 1 | 0.84 | 0.0204 |
| 2+4+5 | P | P | 0.9 | 0.7 | 0.73 | 1 | 0.85 | 0.0177 |
| Feature Set | SVM: Two vs. Two Schema | | | | | | | |
| | Kernel | | Average Accuracy | | | | | StdErr |
| | L1 | L2 | + | - | = | ! | OA | |
| 1 | L | P | 0.95 | 0.9 | 0.6 | 1 | 0.88 | 0.0189 |
| 1+3 | R | P | 0.98 | 0.9 | 0.4 | 1 | 0.85 | 0.0261 |
| 1+4 | L | P | 0.95 | 0.9 | 0.73 | 1 | 0.91 | 0.0234 |
| 1+5 | R | P | 0.92 | 0.8 | 0.6 | 1 | 0.85 | 0.014 |
| 2 | R | R | 0.98 | 0.9 | 0.53 | 1 | 0.88 | 0.0239 |
| 2+3 | L | P | 0.95 | 0.8 | 0.4 | 1 | 0.82 | 0.0144 |
| 2+4 | L | P | 0.95 | 0.9 | 0.73 | 1 | 0.91 | 0.0118 |
| 2+5 | R | P | 0.97 | 0.7 | 0.53 | 1 | 0.85 | 0.0189 |
| 2+3+5 | R | L | 0.93 | 0.9 | 0.53 | 1 | 0.85 | 0.0203 |
| 2+4+5 | L | L | 0.9 | 0.9 | 0.73 | 1 | 0.88 | 0.017 |

**Fig. 3** Polarity classification results using the 2 SVM schemas and different feature sets. Column notations: L1–level 1 of the SVM, L2–level 2 of the SVM, L–linear kernel, P–polynomial kernel), R–RBF kernel, OA–overall accuracy, + (Positive), - (Negative), = (Neutral), ! (No-relationship). Feature notations: 1–Penn Treebank based Unigram, 2–unigrams with WordNet based POS correction, 3–Binary semantics-based features), 4–K-ary semantics-based features, 5–bigrams. The top three models are highlighted. Note that the standard error is that of the overall accuracy.

other polarities. One main reason we believe is that the annotated corpus has a large number of positive examples (Table 4). We also observe that the "not-related" has a constantly high accuracy and is not influenced by other feature combinations. The reason behind this is that the unigrams found in the relationships that have "not-related" polarity often contain unique negation terms such as "no" and "not". Therefore unigrams alone are often sufficient. The accuracy of neutral relationships is low because neutral and positive relationships tend to contain identical unigrams and exhibit similar semantics-based sequential structures. We can also observe that the (1 vs. All) SVM schema behaves differently than the (2 vs. 2). The highest overall accuracy of (1 vs. all) is 0.89, generated by using unigrams only. For (1 vs. All), we cannot however find any particular combination of kernels that constantly produce a high accuracy. In contrast, in the (2 vs. 2) schema, the linear and polynomial kernel combination tends to produce good results. In addition, (2 vs. 2) in general outperforms (1 vs. all) regardless of the feature set under use. We hence conclude that the (2 vs. 2) scheme is preferred.

Regarding an optimal feature subset for polarity analysis, one can observe from Figure 3 that the highest overall accuracy is 0.91 and is generated by a feature combination of POS-based unigrams and K-ary semantics-based sequential features, regardless of whether WordNet is used to correct the POS of a unigram or not. In addition, we notice that the inclusion of bigrams does not improve accuracy but rather reduce the accuracy. This agrees with results reported in previous studies.

| Feature Set | Highest Average Accuracy | | | | |
|---|---|---|---|---|---|
| | Accuracy (Weak) | Accuracy (Medium) | Accuracy (Strong) | Overall Accuracy | Standard Error |
| 1 | 0.83 | 0.98 | 1 | 0.96 | 0.0063 |
| 1+3 | 0.83 | 0.98 | 1 | 0.96 | 0.0063 |
| 1+4 | 0.83 | 0.98 | 1 | 0.96 | 0.0062 |
| 1+5 | 0.83 | 0.98 | 0.89 | 0.95 | 0.0062 |
| 2 | 0.83 | 0.98 | 1 | 0.96 | 0.0061 |
| 2+3 | 0.83 | 0.98 | 1 | 0.96 | 0.0061 |
| 2+4 | 0.83 | 0.98 | 1 | 0.96 | 0.0061 |
| 2+5 | 0.92 | 0.98 | 0.89 | 0.96 | 0.0062 |
| 2+3+5 | 0.92 | 0.98 | 0.89 | 0.96 | 0.0062 |
| 2+4+5 | 0.83 | 0.98 | 0.89 | 0.95 | 0.0065 |

**Fig. 4** Strength prediction results using different feature sets. Feature notations: 1–Penn Treebank based Unigram, 2–unigrams with WordNet based POS correction, 3–Binary semantics-based features), 4–K-ary semantics-based features, 5–bigrams. A linear kernel is used to general all the results. Note that the standard error is that of the overall accuracy.

We have also built various SVR models using different feature combinations for strength analysis. The average accuracy from 10-fold cross validation is shown in Figure 4. Note that all the results are generated using a linear kernel based SVR. From this figure, we observe that all the feature combinations deliver approximately similar high-quality results. In other words, the addition of bigrams and semantics-based structural features does not improve the overall performance. This indicates that using unigrams alone is sufficient for the strength prediction task. The main reason behind this phenomenon is that unlike polarity of a relationship, the strength of a relationship is often directly associated with the specific words used in the relationship-depicting phrase. For instance, the sentence *Soy consumption significantly reduces the risk of cancer.* has a strong strength. The word "significantly" carries the most weight for the SVR model to make the correct prediction. This also explains why the addition of other semantics based features does not help in general.

## 8 Discussions and Conclusion

In this chapter, we present an information extraction system to systematically extract relationships from biomedical articles. We specifically focus on articles originated in a relatively new scientific discipline, i.e., nutritional genomics. The primary goal is to build food-disease-gene networks that capture the complex relationships among five types of bio-entities: foods, diseases, chemicals (or nutrients), genes, and proteins. Through interactive visualization techniques, we expect such networks will provide both health professionals and the general public the most up-to-date research results in nutritional genomics. We also expect that these networks can facilitate biomedical researchers to form evidence-based hypotheses.

From a technical point of view, we have proposed an effective verb-centric approach for relationship extract. Furthermore, we have introduced two new concepts–

relationship polarity and strength, which allows one to examine the multi-faceted nature of a relationship. More importantly, the identification of polarity and strength allows one to quantitatively describe the extracted food-disease-gene networks, thereby presenting the general public a much-needed, concise summary of recent research findings.

We empirically evaluate this system on three small datasets. The results are promising. This system, however, exhibits several limitations. First, the current system was only evaluated based on three relatively small datasets. Second, it lacks the ability to recognize synonymous entities. For instance, "soy products" and "soy food" are currently treated as two different entities. Third, it is unable to detect the following relationships between entities of the same semantic type: "is-a", "component-of", "part-of", and "instance-of". It is important to address the above two limitations, since such information will have a significant impact over the connectedness of the resulted food-disease-gene networks. Fourth, the extracted relationships are currently categorized by their polarity and strength. It is also necessary to categorize these relationships based on semantics, for example, based on the actual verbs employed in a relationship-depicting phrase (e.g., "increase" and "inhibit"). Finally, the graphic user interface needs extensive user studies to ensure its usability and user-friendliness. The authors are currently investigating a variety of strategies to overcome these limitations.

**Glossary:** *Binary relationship extraction, Food-disease-gene networks, Nutritional genomics, Relationship polarity and strength, Relationship integration and visualization, Semantics-based sequential features.*

# References

1. Aronson, A.R.: Effective mapping of biomedical text to the umls metathesaurus: The metamap program (2001)
2. Carlson, A., Betteridge, J., Wang, R.C., Hruschka Jr., E.R., Mitchell, T.M.: Coupled semi-supervised learning for information extraction. In: WSDM 2010: Proceedings of the Third ACM International Conference on Web Search and Data Mining, pp. 101–110. ACM, New York (2010)
3. Corney, D.P., Buxton, B.F., Langdon, W.B., Jones, D.T.: Biorat: extracting biological information from full-length papers. Bioinformatics 20(17), 3206–3213 (2004)
4. de Marneffe, M.-C., MacCartney, B., Manning, C. D.: Generating typed dependency parses from phrase structure trees. In: LREC (2006)
5. Denecke, K.: Semantic structuring of and information extraction from medical documents using the umls. Methods of Information in Medicine 47(5), 425–434 (2008)
6. Ding, X., Liu, B., Yu, P.S.: A holistic lexicon-based approach to opinion mining. In: Proceedings of First ACM International Conference on Web Search and Data Mining, WSDM 2008 (2008)
7. Feldman, R., Regev, Y., Finkelstein-Landau, M., Hurvitz, E., Kogan, B.: Mining biomedical literature using information extraction. Current Drug Discovery (2002)
8. Fellbaum, C.: WordNet: An Electronic Lexical Database. MIT Press, Cambridge (1998)

9. Fukuda, K., Tsunoda, T., Tamura, A., Takagi, T.: Toward information extraction: Identifying protein names from biological papers. In: Proceedings of the Pacific Symposium on Biocomputing (1998)

10. Fundel, K., Kffner, R., Zimmer, R.: Relex - relation extraction using dependency parse trees. Bioinformatics 23(3), 365–371 (2007)

11. Guyon, I., Elisseeff, A.: An introduction to variable and feature selection. Journal of Machine Learning Research 3, 1157–1182 (2003)

12. Hakenberg, J., Leaman, R., Vo, N.H., Jonnalagadda, S., Sullivan, R., Miller, C., Tari, L., Baral, C., Gonzalez, G.: Efficient extraction of protein-protein interactions from full-text articles. IEEE/ACM Trans. Comput. Biology Bioinform. 7(3), 481–494 (2010)

13. Heer, J., Card, S.K., Landay, J.A.: Prefuse: a toolkit for interactive information visualization. In: Proc. CHI 2005, Human Factors in Computing Systems (2005)

14. Hoffmann, R., Valencia, A.: A gene network for navigating the literature. Nature Genetics 36, 664 (2004)

15. Hu, X., Wu, D.D.: Data mining and predictive modeling of biomolecular network from biomedical literature databases. IEEE/ACM Trans. Comp. Biol. Bioinf. 4(2), 251–263 (2006)

16. Hur, J., Schuyler, A.D., States, D.J., Feldman, E.L.: Sciminer: web-based literature mining tool for target identification and functional enrichment analysis. Bioinformatics 840, 838–840 (2009)

17. Kim, J.j., Zhang, Z., Park, J.C., Ng, S.-K.: Biocontrasts: Extracting and exploiting protein-protein contrastive relations from biomedical literature. Bioinformatics 22(5), 597–605 (2006)

18. Jensen, L.J., Kuhn, M., Stark, M., Chaffron, S., Creevey, C., Muller, J., Doerks, T., Julien, P., Roth, A., Simonovic, M., Bork, P., von Mering, C.: String 8–a global view on proteins and their functional interactions in 630 organisms. Nucleic acids research 37(database issue), 412–416 (2009)

19. Joachims, T.: Text categorization with support vector machines: learning with many relevant features. In: Nédellec, C., Rouveirol, C. (eds.) ECML 1998. LNCS, vol. 1398, Springer, Heidelberg (1998)

20. Kaput, J., Rodriguez, R.: Nutritional genomics: the next frontier in the postgenomic era. Physiological Genomics 16, 166–177 (2004)

21. Kipper, K., Korhonen, A., Ryant, N., Palmer, M.: Extending verbnet with novel verb classes. In: Proceedings of the Fifth International Conference on Language Resources and Evaluation LREC (January 2006),
http://verbs.colorado.edu/~mpalmer/projects/verbnet.html

22. Kobayashi, N., Iida, R., Inui, K., Matsumoto, Y.: Opinion mining on the web by extracting subject-attribute-value relations. In: Proceedings of AAAI 2006 Spring Sympoia on Computational Approaches to Analyzing Weblogs AAAI-CAAW 2006 (2006)

23. Krallinger, M., Morgan, A., Smith, L., Leitner, F., Tanabe, L., Wilbur, J., Hirschman, L., Valencia, A.: Evaluation of text-mining systems for biology: overview of the second biocreative community challenge. Genome Biol. 9(suppl. 2), S1 (2008)

24. Kuhn, M., Szklarczyk, D., Franceschini, A., Campillos, M., von Mering, C., Jensen, L.J., Beyer, A., Bork, P.: Stitch 2: an interaction network database for small molecules and proteins. Nucleic Acids Research, 1–5 (2009)

25. Lindsay, R.K., Gordon, M.D.: Literature-based discovery by lexical statistics. Journal of the American Society for Information Science 50(7), 574–587 (1999)

26. Liu, B.: Sentiment analysis and subjectivity. In: Handbook of Natural Language Processing, 2nd edn. (2010)

27. Maskarinec, G., Morimoto, Y., Novotny, R., Nordt, F.J., Stanczyk, F.Z., Franke, A.A.: Urinary sex steroid excretion levels during a soy intervention among young girls: a pilot study. Nut. Can. 52(1), 22–28 (2005)
28. McDonald, R., Hannan, K., Neylon, T., Wells, M., Reynar, J.: Structured models for fine-to-coarse sentiment analysis. In: Proceedings of the Association for Computational Linguistics (ACL), pp. 432–439. Association for Computational Linguistics, Prague (2007)
29. MEDLINE. Medline (1999),
    http://www.nlm.nih.gov/databases/databases_medline.html
30. Mei, Q., Ling, X., Wondra, M., Su, H., Zhai, C.X.: Topic sentiment mixture: Modeling facets and opinions in weblogs. In: Proceedings of WWW, pp. 171–180. ACM Press, New York (2007)
31. Müller, H.-M., Kenny, E.E., Sternberg, P.W.: Textpresso: an ontology-based information retrieval and extraction system for biological literature. PLoS Biol. 2(11), e309 (2004)
32. Niu, Y., Zhu, X., Li, J., Hirst, G.: Analysis of polarity information in medical text. In: Proceedings of the American Medical Informatics Association (2005)
33. Ohta, T., Tateisi, Y., Kim, J.D.: The genia corpus: An annotated research abstract corpus in molecular biology domain. In: The Human Language Technology Conference (2002)
34. Ohta, T., Tsuruoka, Y., Tateisi, Y.: Introduction to the bioentity recognition task at jnlpba. In: Kim, J.-D. (ed.) Proc. International Joint Workshop on Natural Language Processing in Biomedicine its Applications (2004)
35. OpenSource. Opennlp (2010), http://opennlp.sourceforge.net/
36. Pang, B., Lee, L., Vaithyanathan, S.: Thumbs up? sentiment classification using machine learning techniques. In: EMNLP 2002: Proceedings of the ACL 2002 Conference on Empirical Methods in Natural Language Processing, pp. 79–86. Association for Computational Linguistics, Morristown (2002)
37. Sahay, S., Mukherjea, S., Agichtein, E., Garcia, E.V., Navathe, S.B., Ram, A.: Discovering semantic biomedical relations utilizing the web. ACM Trans. Knowl. Discov. Data 2(1), 1–15 (2008)
38. Sauvaget, C., Lagarde, F.: Lifestyle factors, radiation and gastric cancer in atomic-bomb survivors in japan. Cancer Causes Control 16(7), 773–780 (2005)
39. Settles, B.: ABNER: An open source tool for automatically tagging genes, proteins, and other entity names in text. Bioinformatics 21(14), 3191–3192 (2005)
40. Sharma, A., Swaminathan, R., Yang, H.: A verb-centric approach for relationship extraction in biomedical text. In: The fourth IEEE International Conference on Semantic Computing, ICSC 2010 (September 2010)
41. Smalheiser, N.R., Swanson, D.R.: Using arrowsmith: a computer-assisted approach to formulating and assessing scientific hypotheses. Computer Methods Programs in Biomedicine 57, 149–153 (1998)
42. Srinivasan, P.: Text mining: Generating hypotheses from medline. J. Amer. Soc. Inf. Sci. Tech. 55(5), 396–413 (2004)
43. Swaminathan, R., Sharma, A., Yang, H.: Opinion mining for biomedical text data: Feature space design and feature selection. In: The Nineth International Workshop on Data Mining in Bioinformatics, BIOKDD 2010 (July 2010)
44. Swanson, D.R.: Fish oil, raynauds syndrome, and undiscovered public knowledge. Perspect. Bio. Med. 30, 7–18 (1986)
45. Tanabe, L.K., John Wilbur, W.: Tagging gene and protein names in full text articles. In: ACL Workshop on Nat. Lang. Proc. in the Biomedical Domain, pp. 9–13 (2002)

46. Tanenblatt, M., Coden, A., Sominsky, I.: The conceptmapper approach to named entity recognition. In: Calzolari, N., Choukri, K., Maegaard, B., Mariani, J., Odijk, J., Piperidis, S., Rosner, M., Tapias, D. (eds.) Proceedings of the Seventh conference on International Language Resources and Evaluation (LREC 2010). European Language Resources Association (ELRA), Valletta (2010)

47. Torvik, V.I., Smalheiser, N.R.: A quantitative model for linking two disparate sets of articles in medline. Bioinformatics 23(13), 1658–1665 (2007)

48. Tsai, R., Chou, W.-C., Su, Y.-S., Lin, Y.-C., Sung, C.-L., Dai, H.-J., Yeh, I., Ku, W., Sung, T.-Y., Hsu, W.-L.: Biosmile: A semantic role labeling system for biomedical verbs using a maximum-entropy model with automatically generated template features. BMC Bioinformatics 8(1), 325 (2007)

49. USDA. Usda national nutrient database for standard reference, release 17 (2006), http://www.nal.usda.gov/fnic/foodcomp/data/

50. Waldschlger, J., Bergemann, C., Ruth, W., Effmert, U., Jeschke, U., Richter, D.U., Kragl, U., Piechulla, B., Briese, V.: Flax-seed extracts with phytoestrogenic effects on a hormone receptor-positive tumour cell line. Anticancer Res. 25(3A), 1817–1822 (2005)

51. Wang, Y., Patrick Cascading, J.: classifiers for named entity recognition in clinical notes. In: Workshop Biomedical Information Extraction, pp. 42–49 (2009)

52. Yang, C.S., Wang, X.: Green tea and cancer prevention. Nutr Cancer 62(7), 931–937 (2010)

# Mining Tinnitus Data Based on Clustering and New Temporal Features

Xin Zhang, Pamela Thompson, Zbigniew W. Raś, and Pawel Jastreboff

**Abstract.** Tinnitus problems affect a significant portion of the population and are difficult to treat. Sound therapy for Tinnitus is a promising, expensive, and complex treatment, where the complete process may span from several months to a couple of years. The goal of this research is to explore different combinations of important factors leading to a significant recovery, and their relationships to different category of Tinnitus problems. Our findings are extracted from the data stored in a clinical database, where confidential information had been stripped off. The domain knowledge spans different disciplines such as otology as well as audiology. Complexities were encountered with temporal data and text data of certain features. New temporal features together with rule generating techniques and clustering methods are presented with a ultimate goal to explore the relationships among the treatment factors and to learn the essence of Tinnitus problems.

Xin Zhang
University of North Carolina at Pembroke, Dept. of Math. Comp. Science,
Pembroke, NC 28372, USA
e-mail: xin.zhang@uncp.edu

Pamela Thompson
University of North Carolina, Dept. of Computer Science, Charlotte, NC 28223, USA
e-mail: pthompso@catawba.edu

Zbigniew W. Raś
University of North Carolina, Dept. of Computer Science, Charlotte, NC 28223, USA &
Warsaw University of Technology, Institute of Comp. Science, 00-665 Warsaw, Poland
e-mail: ras@uncc.edu

Pawel Jastreboff
Emory University School of Medicine, Dept. of Otolaryngology, Atlanta, GA 30322, USA
e-mail: pjastre@emory.edu

M. Biba and F. Xhafa (Eds.): Learning Structure and Schemas from Documents, SCI 375, pp. 227–245.
springerlink.com ⓒ Springer-Verlag Berlin Heidelberg 2011

# 1  Introduction

Tinnitus affects a significant portion of the population. It is rather a symptom than a disease. The definition of Tinnitus based on its individual characters was well discussed by Jastreboff [5]. For many years, Tinnitus was believed impossible to be treated. Not until recently, Tinnitus Retraining Therapy (TRT), which has been developed by Jastreboff (for details, see [5] and [4]) is a promising, complex, and expensive treatment based on the neurophysical model of tinnitus, and is aimed at inducing and sustaining habituation of tinnitus-evoked reactions and tinnitus perception (neurophysiology is a branch of science focusing on the physiological aspect of nervous system function (for details, see [5])). On one hand, TRT has provided relief for many patients as well as generated a high volume of medical data in the format of matrix-in-matrix, which is not suitable for traditional data mining algorithms. On the other hand, due to the fact that objective methods are lacking to detect Tinnitus symptoms, it is also interesting for the clinical doctors to be able to learn the essences of Tinnitus through the audiological evaluation data. Thus, the authors started their initial research by exploring the relationships among the complex factors of the treatment and recovery patterns in different categories of patients for the purpose of optimizing the treatment process as well as learning the essence of the Tinnitus problems.

The rest of this section will focus on the basic domain knowledge necessary to understand TRT and its data collection.

## 1.1  TRT Background

The domain knowledge for tinnitus involves many disciplines, primarily including otology and audiology. Tinnitus appears to be caused by a variety of factors including exposure to loud noises, head trauma, and a variety of diseases. An interesting fact is that Tinnitus can be induced in 94% of the population by a few minutes of sound deprivation [2]. Decreased sound tolerance frequently accompanies tinnitus and can include symptoms of hyperacusis (an abnormal enhancement of signal within the auditory pathways), misophonia (a strong dislike of sound) or phonophobia (a fear of sound) [5]. Past approaches to treatment tend to have been based on anecdotal observations and treatment often focused on tinnitus suppression. Currently a wide variety of approaches are utilized, ranging from sound use to drugs or electrical or magnetically stimulation of the auditory cortex. Jastreboff [4] proposed an important new model (hence treatment) for tinnitus that focuses on the phantom aspects of tinnitus with tinnitus resulting exclusively from activity within the nervous system that is not related to corresponding activity with the cochlea or external stimulation. The model furthermore stresses that in cases of clinically-significant tinnitus, various structures in the brain, particularly the limbic and autonomic nervous system, prefrontal cortex, and reticular formations play a dominant role with the auditory system being secondary. Tinnitus Retraining Therapy (TRT), developed by Jastreboff, is a treatment model with a high rate of success (over 80% of the cases) and is based on the neurophysical model of tinnitus. Tinnitus Retraining Therapy "cures" tinnitus-evoked reactions by retraining its association with specific centers throughout the nervous system, particularly the limbic and autonomic systems. The limbic nervous system (emotions) controls fear, thirst, hunger, joy and happiness and is involved in learning, memory, and stress. The limbic nervous system is connected with all sensory systems. The autonomic nervous system controls functions of the brain and the body over which we have limited control, e.g., heart beating, blood pressure, and release of hormones. The limbic and autonomic nervous systems are involved in stress, annoyance, anxiety etc. When these systems become activated by tinnitus-related neuronal activity (tinnitus signal) negative symptoms are evoked [5]. Unfortunately, many patients seeking treatment other than TRT are often told that nothing can be done about their tinnitus. This can have the negative effect of enhancing the limbic nervous system reactions, which then can cause strengthening of the negative effect of the tinnitus on a patient (see Fig. 1: [5]).

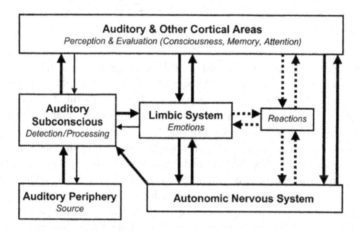

**Fig. 1** Block diagram of the neurophysiological model of tinnitus

TRT is aimed at evoking and sustaining habituation of tinnitus-evoked reactions and tinnitus perception. Degree of habituation determines treatment success, yet greater understanding of why this success occurs and validation of the TRT technique will be useful [1]. The ultimate goal is to lessen or eliminate the impact of tinnitus on the patient's life [5]. Jastreboff has observed statistically significant improvement after three months and treatment lasts approximately 12-18 months.

## 1.2 TRT Data Collection

Tinnitus Retraining Therapy combines medical evaluation, counseling and sound therapy to successfully treat a majority of patients. Based on a questionnaire from the patient as well as an audiological test, a preliminary medical evaluation of patients is required before beginning TRT. Sensitive data from the medical evaluation is not contained in the tinnitus database O presented to the authors (except that Jastreboff maintains this data). Much of this data would contain information subject to privacy concerns, a consideration of all researchers engaged in medical database exploration. Some medical information, however, is included in the tinnitus database such as a list of medications the patient may take and other conditions that might be present, such as diabetes.

| Category | Impact of Life | Tinnitus | Subjective Hearing Loss | Hyperacusis | Prolonged Sound-Induced Exacerbation | Treatment |
|---|---|---|---|---|---|---|
| 0 | Low | Present | - | - | - | Abbreviated counseling |
| 1 | High | Present | - | - | - | Sound generators set at mixing point |
| 2 | High | Present | Present | - | - | Hearing aid with stress on enrichment of the auditory background |
| 3 | High | Not relevant | Not relevant | Present | - | Sound generators set above threshold of hearing |
| 4 | High | Not relevant | Not relevant | Present | Present | Sound generators set at the threshold; very slow increase of sound level |

**Fig. 2** Patient Categories

Patient categorization is performed after the completion of the medical evaluation, the structured interview guided by special forms and audiological evaluation, this interview collects data on many aspects of the patient's tinnitus, sound tolerance, and hearing loss. The interview also helps determine the relative contribution of hyperacusis and misophonia. A set of questions relate to activities prevented or affected (concentration, sleep, work, etc.) for tinnitus and sound tolerance, levels of severity, annoyance, effect on life, and many others. All responses are included in the database. As a part of audiological testing left and right ear pitch, loudness discomfort levels, and suppressibility is determined. Based on all gathered information a patient category is assigned (see Fig. 2). A patient's overall symptom degree is evaluated based on the summation of each individual symptom level, where a higher value means a worse situation. During a medical treatment, comments may be recorded by the doctors or nurses together with a treatment category. Beside surveys for the purpose of symptom evaluation, the TRT process also includes sound therapy, which involves sound parameters for the ear(s), such as mixing point, suppression mask, stochastic resonance (adding low level signal noises), and various sound levels that are applied to a patient. The TRT emphasizes on working on the principle of differences of the stimuli from the background based on the fact that the perceived strength of a signal has no direct association with the physical strength of a stimulus, using a functional dependence of habituation effectiveness model, in Fig. 3. Therefore, once a partial reversal of hyperacusis is achieved, the sound level can be increased rapidly to address tinnitus directly.

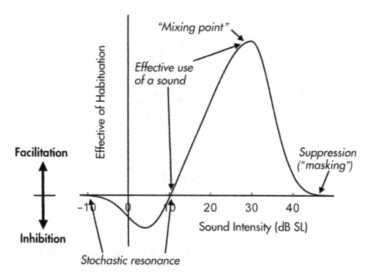

**Fig. 3** Functional dependence of habituation effectiveness on physical intensity of a sound

## 2 Information Retrieval

This paper involves the construction of a database D in the desirable format for classifiers construction based on the database O that was mentioned in the previous section. It uses the term "attribute" to refer to a column in the table from the database O and the term "feature" to refer to a column in the database D. Also, due to the intuition of the process in each visit, recovery of any patient with only one visit cannot be evaluated. Therefore, such records had been removed during the experiments. Features have been developed in this research for the following types of data: text, continuous temporal data, categorical temporal data, among which continuous temporal data have standard statistical features (such as average as well as standard deviation) and new temporal features, as shown in Fig. 4.

Based on subject type, the medical record information for TRT can be grouped into two types: one is to describe properties of a patient, which are relatively stable (e.g., gender, age, occupation, marriage status, etc.); another one is to store properties of a visit of a patient, which may change relatively more frequently than the former type (e.g., total recovery score, drug prescript, clinical doctor's comments, audiological parameters of sound therapy, etc.).

In the light that total visit of each patient during the TRT process varies from several to, sometimes, over fifteen, the authors of this paper developed different formulas and algorithms to capture such subtle changes about visit over time and to transform the sparse data into an ideal data format for traditional classifiers, where each observation describes a patient.

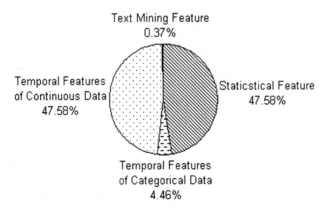

Text Mining Feature
0.37%

Temporal Features
of Continuous Data
47.58%

Staticstical Feature
47.58%

Temporal Features
of Categorical Data
4.46%

**Fig. 4** Features in the Database O

The rest of this section will introduce the features for those three different data types respectively.

## 2.1 Mining Text Data

Many of the attributes in the original database that are stored in text format may contain important information which may have correlation to the new feature tinnitus recovery rate, and to the overall evaluation of treatment method. Some preliminary work has been done in this area with several text fields that indicate the cause or origination of tinnitus.

Text mining (also referred to as text classification) involves identifying the relationship between business categories and the text data (words and phrases). For details, see [14]. This allows the discovery of key terms in text data and facilitates automatic identification of text that is "interesting". The authors designed a new Boolean feature that indicates if tinnitus was induced by exposure to a loud noise. When performing extraction on data entered in text format it is important to recognize that the data may have been entered in a non-systematic manner and may have been entered by multiple individuals. The authors are making the assumption that careful text mining is worthwhile on the tinnitus database in many of the comment style attributes. The following are the text mining steps that were used:

- Term extraction transformation was used which performs such tasks as tokenizing text, tagging words, stemming words, and normalizing words. By this transformation, 60 frequent terms were determined from a text attribute that describes how tinnitus was induced.
- After reviewing these terms, some terms were determined to be inconsequential in the domain. These terms were classified as noise words as they occurred with

high frequency. These terms were then added to the exclusion terms list, which is used as a reference table for the second run of the term extraction transformation.

- The second Term extraction transformation resulted in 14 terms which are related to the Tinnitus induced reason of "noise exposure". The terms included in the dictionary table are "concert", "noise", "acoustic", "exposure", "music", "loud", "gun", "explosion", "pistol", "band", "headphone", "firecracker", "fireworks", and "military".
- Fuzzy-lookup transformation was applied which uses fuzzy matching to return close matches from the dictionary table to extract keywords/phrases into the attribute "keywords". This attribute indicates whether the induced reason for Tinnitus is related to exposure to a noise of some type. As mentioned previously, it is well recognized that noise in general and impulse noise in particular are the most common factors evoking tinnitus.

After adding this new attribute to the table, decision tree algorithms were applied in order to produce relevant rules. Twenty-nine patients have the value of true for the attribute "keywords", where each observation was associated to a patient identification number. The patients identified by this process can be considered clearly to have tinnitus induced by noise exposure. The main purpose of the text mining process is to add extra information to the study in order to improve the accuracy of the prediction model. While expanding the dictionary table will help improve the knowledge gained, relaxation of the rules used for text extraction may serve to enhance the rate of positive responses. This will be addressed in future work, along with more text attributes such as patient occupation, prescription medications, and others.

## 2.2 Temporal Feature Design for Continuous Data

TRT is a complex process with bunches of treatment as well as symptom parameters including not only drug prescriptions, but also audiological therapy, counseling, and evaluations, during which all para-meter values may be subjective to change over time for each individual patient. Therefore, the treatment-record database is sort of matrix-in-matrix format, which is not suitable for traditional data mining algorithms.

Temporal features have been widely used to describe subtle changes of continuous data over time in various research areas, such as stream tracer study [9], music sound classification [11], and business intelligence [6]. It is especially important in the light of the tinnitus treatment process because examining the relationship of patient categorization, the total scores, and audiological test results over time may be beneficial to gaining new understanding of the treatment process. Evolution of sound loudness discomfort level parameters in time is essential for evaluation of treatment outcome for decreased sound tolerance, but irrelevant for tinnitus; therefore it should be reflected in treatment features as well. The discovered temporal patterns may better express treatment process than static features, especially considering that the standard deviation and mean value of the sound loudness discomfort level features can be very similar for sounds representing the same type of Tinnitus treatment category, whereas changeability of sound features with tolerance levels for the same type of patients makes recovery of one type of patients dissimilar. New temporal features include:

Sound level Centroid $C$ is a feature of center of gravity of an audiological therapy parameter over a sequence of visits. It is defined as a visit-weighted centroid.

$$\left\{ V \in \varphi | C = \frac{\sum_{n=1}^{length(T)} n / length(T) \cdot V(n)}{\sum_{n=1}^{length(T)} V(n)} \right) \right\}$$

where $\varphi$ represents the group of audiological features in the therapy, $C$ is the gravity center of the sound audiological feature $V$, $V(n)$ is an audiological therapy feature $V$ in the nth visit, $T$ means the total number of visits.

Sound level Spread $S$ is a feature of the Root of Mean Squared deviation of an audiological therapy feature over a sequence of visits with respect to its center of gravity.

$$\left\{ V \in \varphi | S = \sqrt{\frac{\sum_{n=1}^{length(T)} (n / length(T) - C)^2 \cdot V(n)}{\sum_{n=1}^{length(T)} V(n)}} \right\}$$

where $\varphi$ represents the group of audiological features in the therapy, $S$ is the spread of the audiological feature $V$, $V(n)$ is a sound level feature $V$ in the nth visit, $T$ means the total number of visits. Recovery Rate R describes the recovery over time.

$$\left\{ V \in \varphi | R = \frac{V_0 - V_k}{T_k - T_0}, k \in \min\{V_i\}, i \in [0, N] \right\}$$

where $V$ represents the total score from the Tinnitus Handicap Inventory in a patient visit. $V_0$ is the first score recorded from the Inventory during the patient initial visit. $V_k$ represents the minimum total score which is the best out of the vector of the scores across visits. $V_0$ should be greater meaning the patient is worse based on the Inventory from the first visit. $T_k$ is the date that has the minimum total score, $T_0$ is the date that relates to $V_0$.

A large recovery rate score can mean a greater improvement over a shorter period of time. XY scatter plots were constructed using recovery rate compared to against patient category, and recovery rate compared to against treatment category in order to explore potentially interesting patterns and relationships among those dimensions.

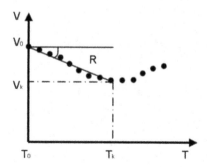

**Fig. 5** Recovery Rate

## 2.3   Temporal Feature Design or Categorical Data

During a period of medical treatment, a doctor may change the treatment from one category to another based on the specifics of recovery of the patients; the symptoms of a patient may vary as a result of the treatment; more so, the category of patient may change over time (e.g., hyperacusis can be totally eliminated and consequently the patient may move from category 3 to 1). Other typical categorical features, which may change over time in the database O include sound-instrument types as well as visiting frequencies. Statistical and econometric approaches to describe categorical data have been well discussed by Powers [7]. In our database, statistical features such as the most frequent pattern, the first pattern and the last pattern were used to describe the changes of categorical data over time.

Most frequent pattern MFP counts the pattern, which occurs most frequently for a particular patient. First and last pattern FP/LP represents the initial and final state of a categorical attribute respectively.

## 2.4   System Overview

Our TRT data mining system consists of five parts: data cleaning, feature extraction, K-means clustering, classification tree construction, and rules generation, as shown in Fig. 6.

A data cleaning procedure filters out inapplicable patient records. For example, patient with only one record is impossible to be analyzed. A feature extraction component transforms the data from the format of matrix-in-matrix to the desirable one with minimum information loss for the purpose of constructing classifiers. A K-means clustering mechanism is applied to produce ideal number of bins for a decision attribute. Decision tree classifiers are then to be constructed for rules extraction.

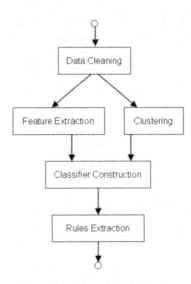

**Fig. 6** Two Clusters

## 3 Experiments and Results

Tools used in this research include Microsoft SQL Server text mining package and SAS clustering functions. Types of learning desired for the sparse data in the database D include classification learning to help with classifying unseen examples, association learning to determine any association among features (largely statistical) and clustering to seek groups of examples that belong together. Decision tree study was performed using C4.5 [8], a system that incorporates improvements to the ID3 algorithm for decision tree induction. C4.5 includes improved methods for handling numeric attributes and missing values, and generates decision rules from the trees [10]. In this study C4.5 was used in order to evaluate the recovery rate by all of the attributes in the research database and to learn patient recovery by loudness discomfort levels, new temporal features and problem types.

We performed experiments for three different tasks: Experiment type I explored Tinnitus treatment records of 253 patients and applied 126 attributes to investigate the association between treatment factors and recovery; Experiment type II explored 229 records and applied 16 attributes to investigate the nature of tinnitus with respect to hearing measurements; Experiment type III compared the recovery rate among different category of patients. All classifiers were 10-fold cross validation with a split of 90% training and 10% testing.

### 3.1 Experiment Type I

Total score was used to represent the difference of overall symptom level between the initial visit and the last visit as a decision attribute, where the data domain contains over two hundred different integer values and therefore not suitable for traditional rule-extraction algorithms. Thus, the authors applied K-means clustering techniques to discretize the domain. Intuitively, total score cannot be applied to patients with only one record, and all such patient records thus were removed from the database D. Due to the limitation of the dataset size, the authors tested up to five bins to maintain a desirable number of supports. In Table 1, we observed a significant gap of the average cluster distance between three-bin clustering and four-bin clustering, which indicated that, regarding to the attribute of total score itself, the four-bin clustering was the most efficient one among these experiments. We also observed that in two-bin clustering, the cutting point is at the score of 20, which coincidences the clinical doctor's empirical cut value for a significant recovery.

**Table 1** Average Distances

| 2 Clusters | 3 Clusters | 4 Clusters | 5 Clusters |
| --- | --- | --- | --- |
| 11.6 | 10.2 | 7.0 | 6.7 |

In Table 2, the cutting points are listed by the ascending order of total score. Tree classifiers had been constructed based on the above discretization methods and produced interesting rules with different semantics of the decision attribute. This paper focuses on discussing the rules with precision and recall close to and above 60% and a support threshold of seven.

**Table 2** Cutting Points

| 2 Clusters | 3 Clusters | 4 Clusters | 5 Clusters |
| --- | --- | --- | --- |
| $(-\infty, 20)$ | $(-\infty, -18)$ | $(-\infty, -18)$ | $(-\infty, -42)$ |
| $[20, +\infty)$ | $[-18, 28)$ | $[-18, 10)$ | $[18, -14)$ |
| — | $[28, +\infty)$ | $[10, 36)$ | $[-14, 10)$ |
| — | — | $[36, +\infty)$ | $[10, 40)$ |
| — | — | — | $[40, +\infty)$ |

Fig. 7 shows the percentage of new temporal feature rules, where the dark regions represent rules having new temporal features and the light pattern regions stand for rules having no new temporal features. We observed that four-bin clustering (with optimal data-driven cutting points) had the highest percentage of new temporal features.

The rest of this section will explain the resultant rules respectively.

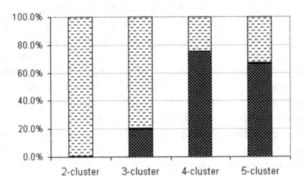

**Fig. 7** Rules Having New Temporal Feature vs. Rules Having no Temporal Features

## 3.2 *Two-Bin Clustering*

Tree classifiers were constructed based on the two-bin clustering method, which resulted in a Boolean decision attribute. The semantic meaning of the decision attribute happened to be the same as empirical one from the clinic doctors.

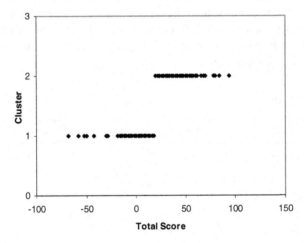

**Fig. 8** Two Clusters

Fig. 8 shows the two-bin clustering, where the total score can be interpreted as either "significant" or "insignificant". In this experiment, we collected the following rules:

**Rule 1.** *When the total visit is not more than two times, the average WN of the right ear is smaller than 30, and the average right ear sensitivity degree is greater than 90, the recovery tends to be insignificant.*

**Rule 2.** *When the instrument model is Tr-COE, the recovery tends to be insignificant.*

**Rule 3.** *When the problem in the last visit is Tinnitus and the standard deviation of R4 of the right ear is smaller than 5.77, the recovery tends to be insignificant.*

**Rule 4.** *When the standard deviation of the right ear Tinnitus loudness is smaller than 5.66, the recovery tends to be insignificant.*

**Rule 5.** *When the instrument model is ITE, the recovery tends to be significant.*

**Table 3** For Rules in the Test by Two-Bin Clustering

| Rule | Precision (%) | Recall (%) | Support |
|------|---------------|------------|---------|
| 1 | 95.8 | 89.4 | 23 |
| 2 | 80.0 | 64.5 | 8 |
| 3 | 70.6 | 63.4 | 12 |
| 4 | 72.7 | 61.1 | 8 |
| 5 | 86.7 | 69.0 | 13 |

## Three-Bin Clustering

Three clusters were generated in this experiment, where the semantic meaning of "significant" is slightly more restricted than the empirical definition by the clinic doctors. The cutting points of clustering pattern were approximately symmetric, which were complied the common sense of "neutral".

Fig. 9 shows the three-bin clustering, where the total score can be interpreted as either "significantly improved", "neutral", or "significantly worse". In this experiment, we collected the following rules:

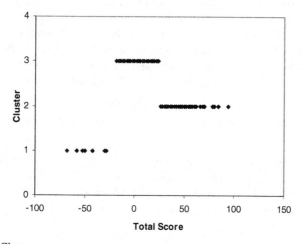

**Fig. 9** Three Clusters

**Rule 6.** *When the average loudness discomfort level-50 of the right ear is less than 74.5, the recovery of a patient tends to be neutral.*

**Rule 7.** *When the weighted spread of the WN of the left ear is less than 0.29, the recovery of a patient tends to be neutral.*

**Rule 8.** *When the instrument model is Tr-COE, the recovery of a patient tends to be neutral* (complying Rule 2).

**Rule 9.** *When the standard deviation of the right ear Tinnitus loudness is smaller than 5.66, the recovery of a patient tends to be neutral* (complying Rule 4).

**Rule 10.** *When the instrument model is ITE, the recovery tends to be significantly improved* (same as Rule 5).

**Table 4** For Rules in the Test by Three-Bin Clustering

| Rule | Precision (%) | Recall (%) | Support |
|------|---------------|------------|---------|
| 6    | 100           | 93.0       | 19      |
| 7    | 73.0          | 59.3       | 92      |
| 8    | 80.0          | 64.5       | 8       |
| 9    | 75.0          | 61.1       | 9       |
| 10   | 87.5          | 69.0       | 14      |

## Four-Bin Clustering

The four-bin clustering had a larger range for non-insignificant recovery and split the range into two categories. The semantics of category three differed from the doctor's empirical definition of significant, which included both significant and neutral. Assuming that there is an approximately even distribution in each cluster, based on the cutting point, the possibility of significant over neutral is about 1:1.

Fig. 10 shows the four-bin clustering, where the total score can be interpreted as either "extensively improved", "significantly improved or neutral", "neutral", or "significantly worse". In this experiment, we collected the following rules:

**Rule 11.** *When the average loudness discomfort level-50 of the right ear is less than 74.5 and total number of visits is not 2, the recovery of a patient tends to be neutral* (similar to Rule 6).

**Rule 12.** *When the symptoms in last visit include both Tinnitus and hearing loss, the weighted centroid of the Tinnitus pitch matched from the left ear is greater than 0.26, and the weighted spread of the threshold of hearing for matching sound for the left ear is not greater than 0.25, the recovery of a patient tends to be significantly improved or neutral.*

**Rule 13.** *When the symptoms in last visit include Tinnitus, hyperacusis, and hearing loss, the weighted centroid of the Tinnitus pitch matched from the left ear is not*

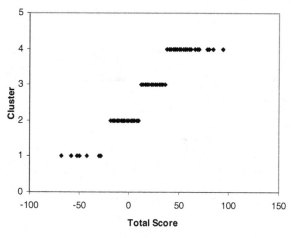

**Fig. 10** Four Clusters

*greater than 0.125, and the weighted centroid of the loudness discomfort level-8000 of the right ear is greater than 0.64, the recovery of a patient tends to be extensively improved.*

**Rule 14.** *When the instrument type is ITE, and the weighted spread of the discrimination level of the left ear is not greater than 0.36, the recovery of a patient tends to be extensively improved* (complying Rule 5 and Rule 10).

**Table 5** For Rules in the Test by Four-Bin Clustering

| Rule | Precision (%) | Recall (%) | Support |
|------|---------------|------------|---------|
| 11 | 81.2 | 70.2 | 13 |
| 12 | 100.0 | 88.2 | 11 |
| 13 | 100.0 | 82.0 | 7 |
| 14 | 73.3 | 53.4 | 11 |

**Five-bin Clustering**

Like the four-bin clustering, the five-bin clustering had a larger range for non-insignificant recovery and split the range into two categories. The semantics of category four differed from the doctor's empirical definition of significance, which included both significant and neutral. Assuming that there is an approximately even distribution in each cluster, based on the cutting point, the possibility of significant over neutral is about 2:1.

Fig. 11 shows the five-bin clustering, where the total score can be interpreted as either "extensively improved", "significantly improved or neutral", "neutral",

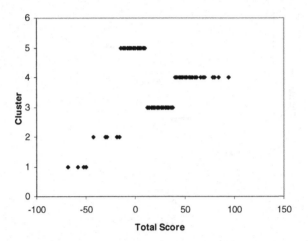

**Fig. 11** Five Clusters

"significantly worse", or "extensively worse". In this experiment, we collected the following rules:

**Rule 15.** *When the symptoms in the last visit include both Tinnitus and hearing loss, the total number of visits is not 2, and the average loudness discomfort level-8000 of the right ear is greater than 101, the recovery of a patient tends to be significantly improved or neutral.*

**Rule 16.** *When the total number of visits is not 2 and average R2 of the right ear is not more than 25, the recovery of a patient tends to be significantly improved or neutral.*

**Rule 17.** *When the symptoms in last visit include both Tinnitus and hearing loss, the weighted centroid of the Tinnitus pitch matched from the left ear is greater than 0.26, and the weighted spread of the threshold of hearing for matching sound for the left ear is not greater than 0.19, the recovery of a patient tends to be significantly improved or neutral.*

**Rule 18.** *When the total number of visits is greater than 4, the symptoms in last visit include both Tinnitus and hearing loss, and the weighted centroid of the L12 of the left ear is greater than 0.59, the recovery of a patient tends to be significantly improved.*

**Rule 19.** *When the symptoms in last visit include Tinnitus, hyperacusis, and hearing loss, the weighted centroid of the tinnitus loudness of the left ear over the sound level threshold of the right ear is not positive, and the weighted spread of the discrimination of the left ear is not greater than 0.15, the recovery of a patient tends to be extensively improved.*

**Rule 20.** *When the symptoms in last visit include both Tinnitus, and hearing loss, the weighted centroid of L12 of the left ear is not greater than 0.59, and the weighted* *centroid of the loudness discomfort level for Tinnitus pitch of the left ear is not greater than 0.11, the recovery of a patient tends to be neutral.*

**Table 6** For Rules in the Test by Five-Bin Clustering

| Rule | Precision (%) | Recall (%) | Support |
|------|---------------|------------|---------|
| 15   | 86.7          | 75.7       | 13      |
| 16   | 83.3          | 70.0       | 10      |
| 17   | 100.0         | 88.2       | 11      |
| 18   | 91.7          | 79.4       | 11      |
| 19   | 88.9          | 73.1       | 8       |
| 20   | 78.6          | 66.2       | 11      |

## 3.3 Experiment Type II

The standard deviation of audiological testing features related to loudness discomfort levels was derived and stored in various attributes in the analysis table. Loudness discomfort level is related to decreased sound tolerance as indicated by hyperacusis or dislike of sound-misophonia, and phonophobia - fear of sound. Loudness discomfort levels change with treatment and patient improvement, unlike other audiological features. For this reason the audiological data related to loudness discomfort levels is included in analysis. Decreased values of loudness discomfort level parameters are not necessary for decreased sound tolerance but they are always lower in case of hyperacusis. In this research the relationship between loudness discomfort level parameters and decreased sound tolerance was investigated. The temporal feature of problems was used as the decision attribute to represent the most often symptoms in all visits of a patient. The value of this feature may be a symptom or a combination of symptoms, where the order in the combination implies the importance of individual symptoms. For example, "THL" means that Tinnitus (T) is the most important symptom and that Hyperacusis (H) is the second most important one and that Hearing Loss (L) is the least important one for a patient. Based on the same thresholds as those in the Experiment I, we collected the following rule:

**Rule 21.** *When the loudness discomfort level-50 of the right ear is between 19 and 40, Tinnitus tends to be a minor symptom for a patient most time.*
    *The support of the rules is 27, the precision is 100.0%.*

## 3.4 Experiment Type III

Scatter plot analysis in Fig. 12 shows that when recovery rate R is compared to patient category, patient category 4 has a smaller rate of recovery value possibly indicating slower or reduced treatment success than that of other patient categories.

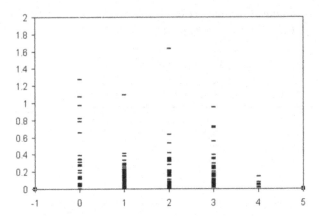

**Fig. 12** Scatter plot for the LDL parameters.

## 4  Conclusion

TRT is a complex treatment process, which generates high volume of matrix data over time: some attributes have relatively stable values while others may be subjective to change as the doctors are tuning the treatment parameters as well as the symptoms of patients are altering. Understanding the relationships between and patterns among treatment factors helps to optimize the treatment process. The authors investigated new features to describe the subtle changes of data, whose domain spans from different disciplines, such as otology as well as audiology for the purpose of classification tree construction. Different clustering methods have been applied to quantize scores into decision attributes. The cutting point of two-bin clustering that was found in the experiment supported the empirical opinion of the clinical doctors about significant recovery. We observed that greater number of bins tended to increase both the precision and recall of the same resultant rules and that rules based on different clustering methods had no conflictions to each other.

Interesting rules about the relationship recovery against symptoms, audiological therapy parameters, and other factors were revealed during the experiments. Also, during the experiments we observed that the new temporal features for continuous data gave more interesting rules especially for extensively improved recovery cases, which confirmed our assumption that carefully designed temporal features may contribute to a better representation of the characteristics of subtle changes over time. Due to the fact that over 77.36% of the patient body in the database has positive recovery results, we may not have enough examples to learn rules about negative cases.

The paper presented an initial research on data mining of medical data about TRT. Expending the patient records will increase the precision as well as the total number of rules. More temporal features for both categorical attributes and continuous attributes shall be explored. It is also interesting to explore the efficient, economic, and independent group of treatment factors to reduce the high expense of the TRT treatment.

# References

1. Baguley, D.M.: What Progress Have We Made With Tinnitus. Acta Oto-Laryngologica 556, 4–8 (2006)
2. Heller, M.F., Bergman, M.: Tinnitus in normally hearing persons. Ann. Otol. 62, 73–83 (1953)
3. Henry, J.A., Jastreboff, M.M., Jastreboff, P.J., Schechter, M.A., Fausti, S.: Guide to Conducting Tinnitus Retraining Therapy Initial and Follow-Up Interviews. Journal of Rehabilitation Research and Development 40(2), 159–160 (2003)
4. Jastreboff, P.J.: Tinnitus as a phantom perception: theories and clinical implications. In: Vernon, J., Moller, A.R. (eds.) Mechanisms of Tinnitus, pp. 73–94. Allyn and Bacon, Boston (1995)
5. Jastreboff, P.J., Hazell, J.W.P.: Tinnitus Retraining Therapy - Implementing the Neurophysiological Model. Cambridge University Press, Cambridge (2004)
6. Povinelli, R.J., Xin Feng, X.: Temporal Pattern Identification of Time Series Data using Pattern Wavelets and Genetic Algorithms. In: Proceedings of Artificial Neural Networks in Engineering, pp. 691–696 (1998)
7. Powers, D., Yu, X.: Statistical Methods for Categorical Data Analysis. Academic Press, London (1999)
8. Quinlan, J.R.: C4.5: Programs for Machine Learning. Morgan Kaufmann, San Francisco (1993)
9. Waldon, M.G.: Estimation of Average Stream Velocity. Journal of Hydraulic Engineering 130(11), 1119–1122 (2004)
10. Witten, I.H., Eibe, F.: Data Mining: Practical Machine Learning Tools and Techniques. Morgan Kaufmann Publishers, San Francisco (2005)
11. Zhang, X., Ras, Z.W.: Differentiated Harmonic Feature Analysis on Music Information Retrieval for Instrument Recognition. In: Proceedings of IEEE International Conference on Granular Computing, Atlanta, GA, May 10-12, pp. 578–581 (2006)
12. Zhang, X., Ras, Z.W., Jastreboff, P.J., Thompson, P.L.: From Tinnitus Data to Action Rules and Tinnitus Treatment. In: Proceedings of 2010 IEEE Conference on Granular Computing, pp. 620–625. IEEE Computer Society, Silicon Valley (2010)
13. Thompson, P.L., Zhang, X., Jiang, W., Ras, Z.W., Jastreboff, P.J.: Mining Tinnitus Database for Knowledge. In: Berka, P., Rauch, J., Zighed, D. (eds.) Data Mining and Medical Knowledge Management: Cases and Applications, pp. 293–306. IGI Global (2009)
14. Nahm, U.Y.: Text Mining with Information Extraction, Ph.D. thesis, Department of Computer Sciences, University of Texas at Austin (August 2004)

# DTW-GO Based Microarray Time Series Data Analysis for Gene-Gene Regulation Prediction

Andy C. Yang and Hui-Huang Hsu

**Abstract.** Microarray technology provides an opportunity for scientists to analyze thousands of gene expression profiles simultaneously. Due to the widely use of microarray technology, several research issues are discussed and analyzed such as missing value imputation or gene-gene regulation prediction. Microarray gene expression data often contain multiple missing expression values due to many reasons. Effective methods for missing value imputation in gene expression data are needed since many algorithms for gene analysis require a complete matrix of gene array values. In addition, selecting informative genes from microarray gene expression data is essential while performing data analysis on the large amount of data. To fit this need, a number of methods were proposed from various points of view. However, most existing methods have their limitations and disadvantages.

To estimate similarity between gene pairs effectively, we propose a novel distance measurement based on the well-defined ontology structure for genes or proteins: the gene ontology (GO). GO is a definition and annotation for genes that describe the biological meanings of them. The structure of GO can be described as a directed acyclic graph (DAG), where each GO term is a node, and the relationships between each term pair are arcs. With GO annotations, we can hence acquire the relations for the genes involved in the experiment. The semantic similarity of two genes within biological aspect can be identified if we perform some quantitative assessments on the gene pairs with their GO annotations.

In this chapter, we first provide the reader with fundamental knowledge about microarray technology in Section 1. A brief introduction for microarray experiments will be given. We then discuss and analyze essential research issues about microarray in Section 2. We also present a novel method based on k-nearest neighbor (KNN), dynamic time warping (DTW) and gene ontology (GO) for the analysis of microarray time series data in Section 3. With our approach, missing value imputation and gene regulation prediction can be achieved efficiently. Section 4 introduces a real microarray time-series dataset. Effectiveness of our method is shown with various experimental results in Section 5. A brief conclusion is made in Section 6.

Andy C. Yang · Hui-Huang Hsu
Department of Computer Science & Information Engineering,
Tamkang University, Taipei, 25137, Taiwan R.O.C.

M. Biba and F. Xhafa (Eds.): Learning Structure and Schemas from Documents, SCI 375, pp. 247–274.
springerlink.com            © Springer-Verlag Berlin Heidelberg 2011

# 1  Introduction

Content of this section tends to bring essential knowledge for the reader to understand the process of microarray technology. The importance of this technology is also mentioned. This section ends with the description of microarray data processing and its relation to the ontology structure: the gene ontology (GO).

## 1.1  What Is Microarray?

Microarray is a widely-used biological experimental approach in this decade. It makes it possible to perform large amounts of gene or protein data experimental operations at the same time. The concept of microarray is based on the differential reactions acted by each sample on the microarray gene chip relative to the experimental conditions. Generally, microarray technology is divided into two aspects: cDNA microarray and Affymetrix microarray. In cDNA microarray, controlled and experimental samples are dyed with two different colors and then hybridized to generate various experimental results. Affymetrix microarray is chosen while biologists or associations need to perform tests on huge amounts of data that are previously cloned and manufactured by Affymetrix microarray producers. Applications of cDNA microarray are more common in several biological research laboratories because one can produce cDNA microarray chips with data of interest more easily. On the other hand, costs of Affymetrix microarray are much more expensive than cDNA microarray. Therefore, we focus on cDNA microarray in this chapter.

## 1.2  Importance of Microarray Technology

Traditionally, biologists need to perform the same operation for biological experiments due to the limitation of instruments. For example, if we want to experiment on one gene sample and observe its reactions, we have to prepare the sample for several copies. This pre-processing task is critically time-consuming, not to mention much more time needed during the process of molecular or biological experiments. With the development of microarray technology, performing experiments on genes of interest becomes much easier because biologists can now retrieve enough amounts of data they need with little time required. Due to this high throughput biological technology, numerous gene expression data are generated simultaneously. In the meanwhile, the large amounts of data provide us great challenges of analysis. Retrieving meaningful information hidden in these data is essential to facilitate the development of drugs, or the discovery of diseases.

## 1.3  Microarray Data Processing

Procedures of microarray experiments can be described as following steps:

-   Prepare cDNA data for a certain gene which is going to be experimented.
-   Print the cDNA data of interest onto microarray chips.

- Design the suitable probes consisting of two cDNA or mRNA samples: One controlled sample and one experimented sample.
- Label the two different probe samples with red (experimented sample) and green (controlled sample) fluorescent dyes.
- Hybridize probes to the microarray chip, and clean up the chip.
- Scan the hybridized microarray chip with computer instruments and save the quantified data for subsequent analyses.

As listed above, quantified data are generated after scanning the hybridized microarray chip. These data represent different degrees of reactions for each gene sample. In other words, we can identify whether a gene tends to act as controlled or treated samples by calculating the ratio of red and green colors in the quantified data for this gene. For example, if the quantified data of two genes are with four and two for their red-green ratio respectively, it means the gene with larger red-green ratio acts like the treated sample than the other gene. For observation convenience, these data are usually transformed into the logarithm format with base two. These logarithmic data for genes on a gene chip are called microarray gene expression data.

Microarray time series data are matrix-like collections of gene expression values that represent reactions for each gene at different time slots as shown in Table 1. Each row in the microarray time-series data stands for a gene ORF profile, while each column in the matrix represents the specific time point. Different kinds of microarray time series data are with different time slots due to distinct gene sampling time and frequency. Gene expression values in the microarray time-series data may be positive or negative numbers. Positive gene expression values of some gene samples on the chip show that these genes are induced with treated sample, and negative values mean repressed reactions. The task is to analyze these gene expression values in different time slots and find the correlations between genes for the inferring of gene-gene interactions.

**Table 1** Microarray time series data

| Gene | Time Slot 1 | Time Slot 2 | ... | Time Slot n |
|---|---|---|---|---|
| **Gene #1** | 0.56 | 0.80 | | 0.90 |
| **Gene #2** | -0.24 | -0.1 | | 0.60 |
| **Gene #3** | 0.12 | 0.24 | | 0.50 |
| **...** | ... | ... | | ... |
| **Gene #n** | 0.78 | -0.14 | | -0.56 |

## 1.4 Microarray and Gene Ontology

Gene ontology (GO) is a biological definition and annotation for genes that describes the biological meanings of each gene. Generally, most known genes have specific annotations (terms) in GO structure within three independent domains: molecular function (MF), biological process (BP), and cellular component (CC). Terms within three above domains record and represent various molecular or biological meanings for each annotated gene from different aspects respectively. Molecular function considers the biological or biochemical activity at the molecular level. Biological process consists of many molecular functions that are involved in a related biological activity or reaction. It denotes a biological objective which genes contribute to. Cellular component records the place in cells where a gene product is active. One gene may have more than one annotation in each domain. These annotations provide hidden information for corresponding genes from the biological aspect.

The main task in microarray gene expression data analysis is to identify the gene pairs or groups that are highly co-expressed under individual experimental conditions. Usually, various distance measurements or classification / clustering operations are performed on gene expression values in microarray data. However, these kinds of procedures only take gene expression values into consideration so that they lack biological explanations and are not effective, either. With proper usage of gene ontology, this task can be done more efficiently and accurately. Detail descriptions about gene ontology and its importance in microarray data analysis are shown in Section 3.2.

## 2   Research Issues in Microarray Time-Series Data

Microarray technology is getting more and more popular due to its high throughput for biological data. Research on microarray technology or relative data analysis can be categorized into various aspects. In this section, we discuss three major research issues about microarray including missing value imputation, gene regulation prediction, and gene clustering or statistical operations. Brief descriptions and literature review are presented for these three issues.

### 2.1   Missing Value Imputation

Before further analysis of microarray data, one critical issue must be addressed: missing value imputation. Microarray time series data usually consist of multiple missing values. Certain portions of gene expression values that do not exist in microarray gene expression raw data are called missing values. It is necessary to effectively estimate and impute these missing values for subsequent analysis of microarray gene expression data. Acuna and Rodriguez discuss the reason why missing values occur in [1]. These values possibly resulted from inaccuracy of experimental operations, or unobvious reaction at several time slot points of certain genes. Fig. 1 illustrates the missing value problem in microarray time series data. If there is a particular gene I with one missing value at time slot J, then YIJ is used to represent the target missing value. For example, $G_{3,3}$ in Fig. 1 stands for a missing value of gene 3 at the third time point of that gene.

Ouyang et al. find that there are about 5% to 90% missing values existing in various available microarray gene expression time series datasets respectively [2]. Studies also argue that simply ignoring or removing missing values from the raw data could lose meaningful information of these genes [3][4]. For these large amounts of data, it is required to first impute these missing values with effective methods. Without imputation for missing values, further analysis cannot be performed. To date, many imputation methods for handling missing values in microarray time series data have been developed. Troyanskaya et al. summarize and implement three methods: singular value decomposition based method (SVD-impute), weighted k-nearest neighbor (KNN-impute), and row average imputation [5]. The results in the paper show that the KNN imputation outperforms SVD-impute and naive methods such as zero or row average imputation. The most suitable number of parameter k in KNN method is also proved to be set between 10 and 20 in the paper. Afterward, several imputation methods are proposed based on KNN. For example, Kim et al. develop a new cluster-based imputation method called sequential k-nearest neighbor (SKNN) method [6]. The method imputes the missing values sequentially from the gene having least missing values, and uses the imputed values for the later imputation. The study is typically an example showing the effectiveness of KNN with some improvements on it.

|        | E1   | E2         | E3         | ···       | ···          | E5        |
|--------|------|------------|------------|-----------|--------------|-----------|
| Gene1  | -0.3 | 0.5        | 0.1        | 0.4       | -0.6         | 0.1       |
| Gene2  | 0.4  | $G_{2,2}$  | -0.4       | $G_{2,4}$ | -1.1         | 0.9       |
| Gene3  | -0.2 | 0.3        | $G_{3,3}$  | 0.5       | -0.7         | 0.2       |
| ···    | 0.6  | 0.5        | 0.1        | $G_{4,4}$ | $G_{4,N-1}$  | $G_{4,N}$ |
| ···    | -0.5 | $G_{N-1,2}$| 0.3        | 0.4       | -0.6         | 0.1       |
| GeneN  | 0.7  | 0.1        | $G_{N,3}$  | -0.3      | 0.2          | 0.5       |

**Fig. 1** Missing values in microarray time series data

In addition to KNN or KNN-based imputation methods, there are still other imputation methods proposed from different standpoints. Oba et al. propose an estimation method for missing values based on Bayesian principal component analysis (BPCA) [7]. The method combines mathematical theorems with parameters that need not to be complicated. The results of BPCA outperform the KNN and SVD imputations according to the authors' evaluations. Moreover, an imputation method based on the local least squares (LLS-impute) formulation is proposed to estimate missing values in the gene expression data [8]. Both KNN and LLS imputations need to find similar genes for a target gene while imputing gene missing values. Other proposed methods take various points of view into consideration. Regression modeling approaches are also used to solve the missing value imputation problem despite it is difficult to determine the parameters used for regression models [9-11].

For existing imputation methods, BPCA is shown to outperform others. But it is not easy to determine the number of principal axes, either [12][13]. Among all related works and published research, existing methods for microarray missing value imputation mainly utilize k-nearest neighbor (KNN) or KNN-like approaches to estimate the missing values. When applying KNN to impute missing values, we have to choose k similar genes without missing entries at the time slot point as the target missing value. This issue is discussed in the following subsection.

## 2.2    Gene Regulation Prediction

In the gene cell cycle or in a biological process, the expression level of one gene is usually regulated by other genes. There might be one-to-one or many-to-one regulatory relations. If one gene regulates other genes, it is called an input gene. On the contrary, if one gene is a regulated target, it is called an output gene [14]. For transcriptional regulations among all genes, there are two sorts of situations, activation and inhibition. In activation regulations, the expression of the output gene is increased with the presence of the input gene, and vice versa. In other words, an activator gene regulates the activatee gene in the biological process so that the gene expression level of the two genes forms the trend of positive correlations. On the contrary, a trend of negative correlations results from the inhibition regulations.

Typically, microarray time series data analysis aims to observe and find out pairs of genes with highly-correlated relations as above-mentioned. This kind of issue is called gene regulation prediction. Research on this issue has been performed for these years, and a variety of approaches have been proposed. The most commonly-used distance measurement is Euclidean distance or statistical calculations such as Pearson correlation coefficient (PCC). However, these kinds of distance measurements have many disadvantages. For example, Euclidean distance of two sequences is very sensitive to the points on the sequences that are far away from other points or the mean. These points are so-called outliers and they often occur in many domains. The existence of outliers influences a lot while measuring the similarity of genes [15]. PCC is a statistical measurement to identify whether two sequences are relative to each other or not. But PCC is not suitable here because for microarray time series data we have to focus on the local similarity but not the global correlation of two genes. The reason is that even genes with known regulations may have reaction time delay and offsets on the time axis [16]. As a result, comparing local similarity is more important than comparing the distance of whole time slot points while identifying similarity of two genes. Moreover, gene pairs in microarray data are often of different length. This reduces the practicability of gene regulation prediction methods requiring time sequences of the same length in real microarray datasets.

Other commonly proposed solutions include similarity analysis [17][18] or Bayesian networks [6][19]. Yeung et al. aim to find potential regulatory gene pairs by finding dominant spectral component of gene pairs [20]. Results of our approach for regulatory gene prediction are compared with Yeung's work because datasets and effectiveness assessment we use are the same. Among above regulatory genes prediction methods, some of them may have success for the analysis of microarray time series data, but their effectiveness is very limited. The most important reason is that these methods take only gene expression values into consideration and they lack external or biological information of genes. External information such as gene ontology for genes themselves is regarded as a hint to increase the accuracy of distance measurements between gene pairs [21][22]. This kind of external information for genes is proved to be helpful. As a result, it is necessary to apply a distance measurement that not only has the capability of pointing out local similarity but is also effective even with certain existing outliers in microarray time-series data. Furthermore, gene ontology information for genes should also be taken into consideration.

## 2.3 Gene Clustering and Statistical Operations

Clustering analysis and statistical operations are also used while dealing with microarray time series data [23][24]. Clustering is grouping similar genes into a finite set of separate clusters. This concept aims to group genes into several sets so that genes falling in the same group tend to have similar reactions to experimental conditions or genes themselves. Hierarchical clustering is the most commonly used clustering approach for microarray time series data. Genes with similar biological functions or reactions are found and collected step by step. Eventually, gene groups are constructed that provide some information for biologist to perform further analysis. However, clustering analysis can be taken as the extension of gene-gene regulation prediction. This is because we still need to identify the distance of gene pairs when we are going to build clusters for genes.

Statistical analysis for microarray time series data is performed from different standpoints. Several techniques such as t-test, p-test, or some hypotheses are used to predict whether genes have similar reactions or not. Nevertheless, statistical analysis usually requires large amounts of data sets and time-consuming calculations. Accordingly, we do not leave space for this issue. We mainly focus on microarray data analysis for gene-gene regulation prediction in this chapter.

## 3 DTW-GO Based Microarray Data Analysis

In order to find the distance between gene pairs for missing value imputation and further gene regulation prediction, we propose a novel and effective approach. Our approach takes both gene expression values and external biological information for genes into account. Dynamic time warping (DTW) algorithm is used in our approach as the substitution for commonly-used Euclidean distance while estimating distance between gene pairs. This is because the importance of finding whether there exist subsequences with highly similar relations is emphasized while analyzing whole microarray time series data [8][16]. We then try our method with several variants of DTW to further improve its efficiency and accuracy.

Moreover, we also add gene ontology (GO) information for genes themselves into our approach to make the distance measurement more accurate. GO is a definition and annotation for genes that describe the biological meanings of them. Each known gene has a specific annotation (term) in GO structure within three independent domains: molecular function (MF), biological process (BP), and cellular component (CC). Terms within three above domains consider different aspects respectively. One gene may have more than one annotation in each domain. These GO terms are quite informative because they provide biological meanings for genes. In our approach, GO terms are taken as the external information for genes while estimating the distance between gene pairs.

Finally, we combine our approach with the k-nearest neighbor (KNN) method to first impute missing values. After missing values are estimated, the prediction of regulatory gene pairs can be done with our approach as the distance measurement. This section briefly describes the DTW algorithm and the GO structure, followed by defining our approach that combines the above algorithm and information for genes with the KNN method for missing value imputation and the eventual prediction of regulatory gene pairs.

## 3.1 Dynamic Time Warping

It has been shown in many domains that dynamic time warping (DTW) algorithm works well on finding the similarity for a pair of time series data [25][26]. In general, DTW is widely-used in voice and pattern recognition because it obtains a precise matching along the temporal axis, and it maximizes the number of point-wise matches between two curves [27][28]. If two series with time slot points are given as input, the DTW algorithm can discover the best possible alignment between them by calculating the minimum sum of whole matched points on the two time series.

DTW is a recursive algorithm that starts with matching each point-to-point pair from the first element to the last element on the two input sequences. In Fig. 2, if we are going to align two sequences that are similar with observation, the application of Euclidean distance or Pearson correlation coefficient on these two sequences may be ineffective because of shifts on time axis. With DTW mapping method, local similarity can be found as the best mapping path within the two sequences to be aligned. As a result, if two genes with similar gene expression values at certain time slot points in microarray time series data are analyzed by DTW, it is more precise to determine the similarity between these two genes. This is because DTW can discover their similarity that cannot be identified with other distance measurements.

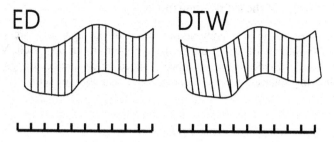

**Fig. 2** Time series sequence similarity measurement

Equations of DTW algorithm are as follows:

**Distance of two time slot points:**

The distance between the elements of the two time series is computed as:

$$dis(i, j) = |x_i - y_i| \tag{1}$$

**Base Conditions:**

$$
\begin{aligned}
&e(0,0) = 0; \\
&e(1,1) = dis(x_1, y_1) * W_D; \\
&e(i,0) = \infty \text{ for } 1 \leqq i \leqq I; \\
&e(0,j) = \infty \text{ for } 1 \leqq j \leqq J;
\end{aligned}
\tag{2}
$$

where $W_D$ is the weighted value for the paths in the diagonal direction.

**Recursive Relation:**

$$e(i, j) = \min \begin{cases} e(i, j-1) + dis(x_i, y_j) * W_V \\ e(i-1, j-1) + dis(x_i, y_j) * W_D \\ e(i-1, j) + dis(x_i, y_j) * W_H \end{cases} \tag{3}$$

where $W_V$, $W_D$, and $W_H$ denote the weighted value for the paths in the vertical, diagonal, and horizontal directions respectively.

**Output: DTW distance for two sequences X and Y:**

$$DTW(X,Y) = \frac{1}{n+m} * e(i, j) \tag{4}$$

where length of X and Y are n and m respectively.

### 3.1.1 Refinement of the DTW Algorithm

To further increase the efficiency and accuracy of our approach, we survey and analyze some variants of DTW and try to add them in our approach. Variants of DTW are usually divided into two aspects: speeding up and accuracy increasing. In the following subsections, we describe these two sorts of refinements for our approach.

#### 3.1.1.1 Computational efficiency of DTW

DTW has a critical disadvantage: high computational cost. Typically, time complexity of the traditional DTW algorithm is $O(n*m)$ for two input sequences with length $n$ and $m$ respectively. As we will show in Section 4, we use the Spellman's dataset to perform missing value imputation with totally 6178 genes in the dataset. If we naively use original DTW algorithm to calculate DTW distance of the whole 6178 genes, the computational time cost is awfully amazing that reduces the practicability of the algorithm. To solve this problem, several methods are proposed to speed up the calculation of DTW. Among all existing methods, we find the most useful one called FastDTW algorithm proposed by Salvador & Chan [29]. The authors propose their algorithm that has only linear time and space complexity. FastDTW uses a multilevel approach with three following operations:

(1) **Coarsening:** Coarsening means that FastDTW shrinks a time series into a smaller one which represents the same curve as accurately as possible with fewer data points.

(2) **Projection:** After FastDTW performs the coarsening step, it will find a minimum-distance warping path at a lower resolution, and use the path to guess another minimum-distance warping path in a higher resolution.

(3) **Refinement:** Finally, FastDTW refines each warping path in every resolution projected from a lower resolution with local adjustments.

If there are 32 points in an original time series, FastDTW cuts the data of points needed from 32 with two-times reducing rate (32->16->8->4->2). However, according to our experiment, we find that coarsening with three-times reducing rate performs better than coarsening with two-times reducing rate in terms of the dataset involved. This is because the datasets we use contain only 18 or 17 time points and need not too many coarsening operations. As a result, we modify the FastDTW algorithm and set the coarsening rate as three-times.

#### 3.1.1.2 Accuracy of DTW

Except computational cost, there's the other attractive issue for the original DTW algorithm called the singularity problem [30]. In some cases, DTW generates unintuitive alignments where a single point on one time series is mapped onto a large subsection of the other time series. This kind of unexpected alignment is the singularity problem. When the two sequences to be aligned are basically similar but with only slightly different amplitude at the peaks or valleys mapped on the two sequences, DTW will perform a one-to-many mapping for the time points. This kind of mapping will easily fail to find obvious and intuitive alignments for sequences. Therefore, it is essential to mitigate the singularity problem of DTW.

We survey and analyze several adjustments aiming to reduce the singularity problem of DTW and choose four of them to implement in our approach. In the following paragraph, we give a brief description about the four adjustments.

(1) **Windowing:** Berndt and Clifford proposed a restricted version of DTW so that the allowable paths for the DTW algorithm are limited with a warping window : $|i-j| \leq w$, where w is a positive value [31]. This constraint may mitigate the seriousness of singularity problem but it is not able to prevent it.

(2) **Slope weighting:** Kruskall and Liberman proposed a modification of DTW so that the recursive equation in original DTW algorithm is replaced by $r(i,j) = d(i,j) + \min\{r(i-1,j-1), X*r(i-1, j), X*r(i, j-1)\}$, where X is a positive real number [32]. With this constraint, the warping path is increasingly biased toward the diagonal if the weighted value X gets larger. This modification of DTW takes the weighted value into consideration and it tries to slightly encourage the warping path to go in the diagonal direction to reduce singularity.

(3) **Step patterns (Slope constraint):** Itakura proposed a permissible step for the warping path with $r(i,j) = d(i,j) + \min\{r(i-1,j-1), r(i-1, j-2), r(i-2, j-1)\}$ [33]. With this constraint, the warping path is forced to move one diagonal step if the previous step goes in the parallel direction to an axis.

(4) **Derivative Dynamic Time Warping:** Keogh and Pazzani introduced a modification of DTW called Derivative Dynamic Time Warping (DDTW) [34]. The authors consider only the estimated local derivatives of gene expression values in sequences instead of using the whole gene expression values themselves. The estimation equation is as follows:

**Distance for two time points in two sequences:**
$dis(i, j) = |E(X_i) - E(Y_i)|^2$
where $E(X_i) = \{ (X_i - X_{i-1}) + [(X_{i+1} - X_{i-1}) / 2] \} / 2$ ,
and $E(Y_i) = \{ (Y_i - Y_{i-1}) + [(Y_{i+1} - Y_{i-1}) / 2] \} / 2$      (5)
DDTW takes moving trends of certain subsequences into account in order to identify the distance of the two sequences.

These methods tend to form some constraints to force the warping path not to go along the horizontal or vertical direction too much. For the four variants of DTW, we consider slope weighting should bring the best results for imputation because it is more flexible and slightly encourages the warping path to go to the diagonal. Forcing the warping path to go to the diagonal too much may mitigate the singularity problem, but it is also at the risk of filtering the most suitable alignment of two genes.

We aim to retrieve suitable modifications of DTW to make our approach the best distance measurement. To fit this need, we implement the four above modifications for DTW in our approach. We also compare the imputation effectiveness resulted from of these modifications of DTW in order to improve the accuracy of our approach while applied in missing value imputation. Experimental results show that performing slope weighting brings the best result. The detail is discussed in Section 5.

## 3.2 Gene Ontology

The structure of GO can be viewed as a directed acyclic graph (DAG), where each GO term is a node, and the relationships between each term pair are arcs. Nodes with parent-children relations imply they are similarly defined within biological functions or reactions, while the children nodes are more specific. In other words, GO is a hierarchical structure, where child terms are more specialized and parent terms are less specialized. Each node in GO can have several parent nodes and several children nodes just in case that relations between each node do not form a cycle. The most commonly-used relations in GO are "is-a" relations and "part of" relations. For example, if the relation "term A is a term B" exists in GO, it means term A is a subtype of term B. By contrast, if the relation "term A is part of term B" stands, it means all children terms of term A with term A itself belong to term B. Each term in GO has one unique GO id for it, but the number of GO id does not represent the similarity between terms. These terms are used to annotate (describe) each gene to identify possible biological functions of it.

Fig. 3 illustrates an example of GO. For instance, GO id 0015749 shown in Fig. 3 denotes a term "monosaccharide transport", which has the relation "is-a" with its parent term (GO id 0008643). Equally, the parent-children relation between terms at consequent levels starting from one specific node can be traced level by level to the root node. If we start from the term GO: 0015749, we can trace the path from the selected node to the root as "GO: 0015749->GO: 0008643->GO: 0006810->GO: 0051234->GO: 0008150". With this directed acyclic graph structure, we can easily query the GO annotation terms of each gene in microarray gene expression time series data to give a general view of the biological activities of the genes involved.

Since each gene may have totally different terms in the three independent domains, deciding which domain we are focusing on is hence very important. Besides, one gene may be annotated by more than one term even in the same domain. Moreover, each term can have more than "one-to-one" relation with its parent term or children term. This forms various complicated reticular relations for annotated genes. Typically, a completed tracing path of GO annotation terms for one gene from the root to the leaf nodes is complex. Therefore, the way how we can use gene ontology differs from the involved data themselves and the algorithm we are applying. Sometimes it can also depend on which kind of analysis we are performing. With GO tern annotation, each gene can have a uniform representation across biological databases. As a result, GO annotations for genes can be taken as their external information while determining distance among them. For more information about gene ontology, please refer to the Gene Ontology website. With GO annotations, we can hence acquire the relations for the genes involved in the experiment. The distance of two genes within biological aspect can be identified if we perform some quantitative assessments on the gene pair with their GO annotations.

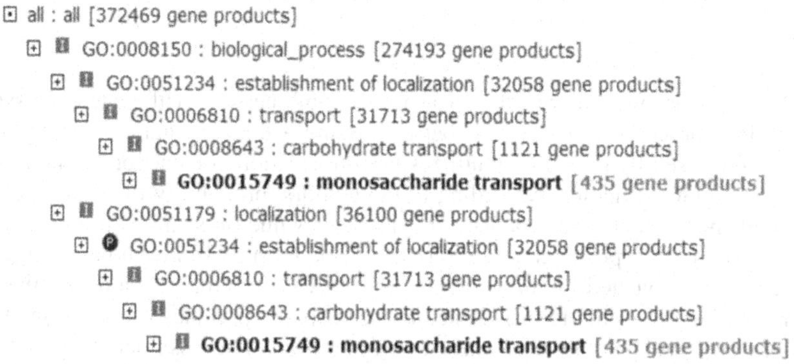

⊡ all : all [372469 gene products]
   ⊞ ◧ GO:0008150 : biological_process [274193 gene products]
      ⊞ ◧ GO:0051234 : establishment of localization [32058 gene products]
         ⊞ ◧ GO:0006810 : transport [31713 gene products]
            ⊞ ◧ GO:0008643 : carbohydrate transport [1121 gene products]
               ⊞ ◧ **GO:0015749 : monosaccharide transport** [435 gene products]
      ⊞ ◧ GO:0051179 : localization [36100 gene products]
         ⊞ ◉ GO:0051234 : establishment of localization [32058 gene products]
            ⊞ ◧ GO:0006810 : transport [31713 gene products]
               ⊞ ◧ GO:0008643 : carbohydrate transport [1121 gene products]
                  ⊞ ◧ **GO:0015749 : monosaccharide transport** [435 gene products]

**Fig. 3** Example of gene ontology

### 3.2.1 Application of Gene Ontology

The main task in microarray gene expression data analysis is to identify the gene pairs or groups that are highly co-expressed under individual experimental conditions. The most common procedure is performing distance measurement or classification and clustering on gene expression values. Nevertheless, with the usage of gene ontology, this task can be done more efficiently and accurately. Lord et al. investigate the validity of using GO information as semantic distance for gens compared with using traditional distance measurement [35]. Effectiveness of taking GO annotations as external information for genes is proved in the work. The authors also recommend choosing the "is_a" relation between gene pairs when determining distance for genes because "is_a" relation occupies almost 90% of all relations recorded. Consequently, we take the "is_a" relation into account in our approach because it is the most used relation in GO structure. Another example of using gene ontology is in [36]. In the study, GO terms are used as the information content. Semantic closeness is defined if the most immediate parent node is shared by two annotation terms. The authors also merge various GO-based distance measurement algorithms that consider intra and inter ontological relations by translating each relative term into a hierarchical relation within a smaller sub-ontology.

To our best knowledge, Tuikkala et al. propose the first method that uses gene ontology [37]. Operations of the algorithm proposed by the authors can be briefly descried as follows:

  –  First find the sets of GO ids for each pair of genes being identified.
  –  Create a table recording the tracing path of all terms annotated for both genes.

- Calculate the probability of the occurrence of each term in the table.
- Estimate all the parent-children relations of each term in the path-tracing table to determine whether the two genes have common ancestors.
- For genes that have shared parent nodes in the GO tracing path, calculate the mean probability of occurrence of all their matched GO term combinations.
- The mean probability of occurrence is taken as the distance between gene pairs.

This study proposes a typical approach that combines GO with gene expression values into processing to retrieve better missing value imputation results. The method proposed by the authors utilizes the information content for each annotation term in GO structure. According to the authors, probability of occurrence of each node has to be first calculated. This numeric value of each GO term is called the p value and it represents how informative this term is. If a term has a larger p value, it is often visited by most tracing paths in GO structure. As a result, the informative degree of this term is hence reduced. If the p values of GO terms that annotate one gene pair are large, these two genes tend to be less related.

Our approach takes the concept in Tuikkala et al.'s work into account. However, the authors in the work only find the minimum p value of shared ancestors of GO terms for two genes. This operation is insufficient because in GO structure two GO terms that are used to annotate different genes may have several relations. Having ancestors in common is only one of the relations for these two GO terms. We have to consider whether GO term pairs that annotate gene pairs form the parent-children relation, or they are even the same one. Theoretically, two genes with GO terms in common tend to be more relative than two genes having GO terms that only have shared ancestors. As a result, our approach gives different weighted values while calculating p values for the three term-term relations: the same terms, parent-children relation, and ancestor- sharing relation. The three relations are marked as case1, case2, and case3 in order as shown in Fig. 4. The star symbol in Fig. 4 illustrates the closest shared ancestors or two GO terms A and B. To find the best weighted values for these three relations, we implement several parameters and the results will be discussed in Section 5. Finally, the mean p value of all GO term pair combinations for two genes is used to the further semantic distance measurement of these two genes.

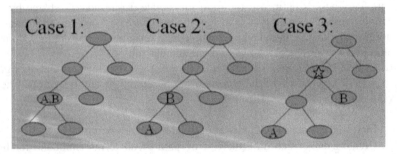

**Fig. 4** Three relations between GO terms

## 3.3 DTW-GO Based Microarray Data Analysis

Our approach aims to provide an accurate distance measurement that takes both gene expression values and external information for genes into account. To fit this need, we survey and analyze existing assessments that provide distance measurements for gene pairs. Tuikkala et al. propose a distance measurement combining both distance for gene expression values and GO information for these genes as the semantic distance. Accuracy for the method is validated in the paper. The equation of the distance measurement for two genes ($g_x$, $g_y$) in Tuikkala et al.'s work is as follows:

$$DIS(g_x, g_y) = D^{GO}(g_x, g_y)^\alpha * D^{EXP}(g_x, g_y) \qquad (6)$$

where $D^{GO}$ is the average p value of all GO term pairs used to annotate $g_x$ and $g_y$, $\alpha$ is a positive weighted parameter that controls how much the semantic dissimilarity value contributes to the combined distance. $D^{EXP}$ is Euclidean distance of ($g_x$, $g_y$).

We then modify equation (6) as follows:

$$DIS(g_x, g_y) = D^{GO\_NEW}(g_x, g_y)^\alpha * D^{DTW}(g_x, g_y) \qquad (7)$$

where $D^{GO\_NEW}$ is our estimation of p values of all GO term pairs used to annotate $g_x$ and $g_y$ as mentioned in Section 3.2.1, $\alpha$ is the positive weighted parameter as shown in equation (6). $D^{DTW}$ is the DTW distance of ($g_x$, $g_y$). In equation (7), we replace Euclidean distance with DTW distance, and replace original p value estimation with our approach. This is because we consider that DTW is more suitable than Euclidean distance while calculating distance between gene expression values. Equally, we use our new estimation for semantic distance between gene pairs to retrieve higher accuracy. After defining our distance measurement for gene pairs, the way we apply our distance measurement in missing value imputation and gene regulation prediction are described in coming subsections.

### 3.3.1 Missing Value Imputation

In this subsection, we propose a novel missing value imputation approach combining our distance measurement with the k-nearest neighbor (KNN) method. The KNN method selects genes with expression values similar to those genes of interest to impute missing values. For example, if we consider gene G that has one missing values at experiment time slot T, KNN would find K other genes that have a value at experiment time slot T, but with expression values most similar to Gene G in experiments time slot points except for T. A weighted average of values at experiment time slot T from the chosen K closest genes is then used as the estimation for the missing value in gene G. The weighted value of each gene in the K closest similar genes is given by the distance of its expression to that of gene G. Euclidean distance is commonly used to determine the k closet genes which are similar to the target gene G with missing values to impute. Here we use our distance measurement as the estimation to determine the closeness of gene pairs. The steps of our approach for missing value imputation are as follows:

1. In order to impute the missing value $G_{IJ}$ for gene I at time slot J, the KNN-impute algorithm chooses k genes that are most similar to the gene I and with the values in position k not missing.
2. The missing value is estimated as the weighted average of the corresponding entries in the selected k expression vectors:

$$G_{IJ} = \sum_{i=1}^{k} W_i \times e_{iJ} \tag{8}$$

3. The weighted value

$$W_i = \frac{1}{DIS(g^*, g_i) \times \Delta} \tag{9}$$

$$\text{where } \Delta = \sum_{i=1}^{k} [1/(Sim(g^*, g_i)] \tag{10}$$

and g* denotes the set of k genes closest to $g_i$, DIS(g*, $g_i$) is our distance measurement as shown in equation (7). Missing values for the target gene are hence imputed with our approach.

When applying the KNN-based method for the imputation of missing values, there are no constant criteria for selecting the best k-value. Choosing a small k value produces poorer performance after imputation. On the contrary, choosing a large neighborhood may include instances that are significantly different from those containing missing values. However, one study shows that setting k-value between 10 and 20 brings the best results for KNN imputation [5]. KNN can be an effective and intuitive imputation method if it works with a proper distance measurement for genes such as our approach.

### 3.3.2 Gene Regulation Prediction

After missing values are imputed with our imputation approach, we will then perform gene regulation prediction. Our approach first calculates and records the distance for all gene pairs with equation (7). The mean of numeric distance for all gene pairs is then calculated, assume DISmean. Gene pairs with distance less then DISmean are retained and recorded as potential regulatory gene pairs. These recorded gene pairs are subsequently compared with the known regulatory gene pairs called Filkov's datasets for validation. Detailed information for Filkov's datasets will be given in Section 4. Afterward, the number of mapping gene pairs between the validation datasets and gene pairs found based on our distance measurement is gathered. Theoretically, potential regulatory gene pairs should have shorter distance compared with the others in all gene pair combinations. The detail algorithm of our approach for gene regulation prediction is described as follows:

**Algorithm for the proposed approach to identify regulatory gene pairs:**

1. For all gene pair combinations, calculate the distance of each gene pair with equation (7).
2. Calculate the mean distance of all gene pair combinations, assume *DISmean*.
3. Record gene pairs with distance less than *DISmean*, assume $S_{SIM}$.
4. Compare $S_{SIM}$ with Filkov's datasets. Count the number of matched gene pairs.

# 4 Datasets and Performance Assessment

In order to evaluate the effectiveness of our approach for missing value imputation and gene regulation prediction, we evaluate it on a real microarray dataset. In this section, we first give a brief description about the dataset used in our experiments. Subsequently, we introduce general performance assessment for missing value imputation and gene regulation prediction respectively.

## 4.1 Real Microarray Dataset

In this chapter, the microarray dataset we used is proposed by Spellman et al. and Cho et al. [38][39]. The data were obtained for genes of Yeast Saccharomyces cerevisiae cells with four synchronization methods: alpha-factor, cdc15, cdc28, and elutriation. Spellman's dataset is widely used as the real dataset in microarray research [5][7][8]. These four subsets of the dataset contain totally 6178 gene ORF profiles with their expression values across various amounts of time slots. In the dataset, the alpha sub-dataset contains 18 time points with seven minutes as the time interval, while the cdc28 sub-dataset contains 17 time points with ten minutes as the time interval. Here we choose alpha and cdc28 sub-datasets in Spellman's microarray datasets as the testing data because these two sub-datasets contain more non-missing gene expression values. Alpha sub-dataset contains missing values with nearly uniform distribution, while cdc28 sub-dataset contains a great portion of missing values occurring almost at some time points. These four kinds

| H | I | J | K | L | M | N | O |
|---|---|---|---|---|---|---|---|
| alpha0 | alpha7 | alpha14 | alpha21 | alpha28 | alpha35 | alpha42 | alpha49 |
| -0.15 | -0.15 | -0.21 | 0.17 | -0.42 | -0.44 | -0.15 | 0.24 |
| -0.11 | 0.1 | 0.01 | 0.06 | 0.04 | -0.26 | 0.04 | 0.19 |
| -0.14 | -0.71 | 0.1 | -0.32 | -0.4 | -0.58 | 0.11 | 0.21 |
| -0.02 | -0.48 | -0.11 | 0.12 | -0.03 | 0.19 | 0.13 | 0.76 |
| -0.05 | -0.53 | -0.47 | -0.06 | 0.11 | -0.07 | 0.25 | 0.46 |
| -0.6 | -0.45 | -0.13 | 0.35 | -0.01 | 0.49 | 0.18 | 0.43 |
| -0.28 | -0.22 | -0.06 | 0.22 | 0.25 | 0.13 | 0.34 | 0.44 |
| -0.03 | -0.27 | 0.17 | -0.12 | -0.27 | 0.06 | 0.23 | 0.11 |
| -0.05 | 0.13 | 0.13 | -0.21 | -0.45 | -0.21 | 0.06 | 0.32 |
| -0.31 | -0.43 | -0.3 | -0.23 | -0.13 | -0.07 | 0.08 | 0.12 |
| 0.02 | -0.33 | -0.49 | -0.3 | -0.15 | -0.24 | 0.4 | 0.53 |
| -0.36 | -0.19 | 0 | -0.32 | -0.27 | -0.12 | 0.04 | 0.17 |
| -0.1 | -0.15 | -0.01 | -0.25 | -0.16 | -0.13 | 0.04 | 0.19 |
| 0 | -0.01 | 0.12 | -0.23 | -0.13 | 0.25 | 0.06 | -0.27 |
| 0.06 | 0.01 | 0.17 | -0.14 | 0.01 | -0.24 | 0.3 | -1.34 |
| -0.4 | -0.22 | 0.19 | -0.2 | -0.09 | 0.41 | 0.15 | -0.05 |
| 0.46 | 0.28 | 0.16 | -1.72 | 0.33 | 0.05 | 0.22 | 0.3 |
| -0.24 | -0.95 | -0.23 | -0.12 | -0.02 | 0.23 | -0.11 | 0.11 |
| -0.02 | -0.29 | -0.07 | -0.22 | -0.06 | -0.07 | 0.2 | 0.2 |
| -0.11 | -0.17 | -0.16 | 0.04 | 0.1 | -0.02 | 0.08 | 0.13 |
| -0.36 | -0.42 | 0.29 | -0.14 | -0.19 | -0.52 | 0.04 | 0.04 |

**Fig. 5** Spellman's yeast dataset

of sub-datasets record the gene expression reactions during different phases in cell cycle. With in these sub-datasets, empty values at certain time slot points are the missing values that we are going to impute and estimate. Fig. 5 illustrates the format of Spellman's dataset.

## 4.2   Assessment of Imputation Accuracy

For assessment of imputation accuracy, genes with missing values in microarray gene expression data are first filtered to generate a complete matrix. There are 3422 and 835 genes in the complete matrix for alpha and cdc28 sub-datasets, respectively. Missing values with different missing rates ranging from 1%, 5%, 10%, 15% and 20% in the complete matrix are deleted at random to create testing datasets. Afterward, we impute missing values in the generated testing datasets with our approach and other methods to recover the deleted missing values for each data set. The estimated values are compared to the original values in the complete matrix. For numeric accuracy assessment of missing value imputation, the commonest way is to calculate the Normalized Root Mean Square (NRMS) error. Equation for NRMS error is as follows:

$$\text{NRMS} = \sqrt{mean[(y_{predict} - y_{known})^2]}\Big/ std[y_{known}] \qquad (11)$$

where $y_{predict}$ and $y_{known}$ are estimated values and known values in the complete matrix respectively, and $std[y_{known}]$ is the standard deviation of known values. An imputation method is said to outperform others if the NRMS error of it is less than that of other imputation methods.

## 4.3   Accuracy of Gene Regulation Prediction

Filkov et al. review related literature and collect all known gene regulations of alpha and cdc28 subsets in Spellman's yeast cell dataset [40]. They also build a database to record all known gene regulations. In our evaluation, the known gene regulations recorded in Filkov's database are taken as the validation datasets. In the database, the number of recorded gene activations and inhibitions for alpha subset is 343 and 96 respectively, while for cdc28 subset is 469 and 155 accordingly. All these regulations are in the format of A (+) B that denotes gene A is an activator that activates gene B. Similarly, C (-) D represents an inhibitor gene C which inhibits gene D. Among these regulations recorded in Filkov's database, one gene could be the activator or inhibitor for more than two other genes. For example, gene ABF1 stands for the activator for totally eight different genes in cdc28 subset. Nevertheless, gene names in Filkov's database are denoted as the gene standard name, while the gene systematic names are used in Spellman's dataset. The systematic and standard names of a gene are like two kinds of aliases for this gene. As a result, a mapping procedure between gene standard name and systematic name is required. For this purpose, we designed a program to perform this operation. The reference database for this phase is the Saccharomyces Genome Database (http://www.yeastgenome.org/) database. The SGD database acts as a platform for biologists to refer and query yeast gene information including the gene standard name and systematic name. During the process of gene name mapping, we find that some of the gene standard name in Filkov's database cannot be found in Spellman's dataset due to the different naming conventions. For example, the mapping systematic name for gene with standard name STA1 cannot be found in the SGD database. Consequently, regulations with gene STA1 are filtered that causes the decrease of gene activations in cdc28 subset from 469 to 466. Therefore, the pre-processing of the raw data is necessary. First, we parse all regulations of alpha and cdc28 sub-datasets in Filkov's database and retrieve unrepeatable involved genes. The parsing result is shown in Table 2. Involved genes in alpha and cdc28 sub-datasets are 295 and 357 respectively.

**Table 2** Parsing result for Gene Regulations

| Dataset | Content | | | |
|---------|---------|---------|---------|---------|
|  | No. of Genes | No. of Activations | No. of Inhibitions | Total |
| alpha | 295 | 343 | 96 | 439 |
| cdc28 | 357 | 466 | 155 | 621 |

Theoretically, the number of pairwise gene combinations for alpha subset is C(295,2) which equals to 43365, and the number of pairwise gene combinations for cdc28 subset is C(357,2) which equals to 63546. Known regulations in Filkov's database are marked as the validation measurement to estimate the accuracy of the gene regulation prediction methods. Finally, we apply our approach on these gene pairwise combinations and count the number of potential regulatory gene pairs found by our approach that are also listed in Filkov's database. Regulations of activations and inhibitions are summed up separately. The results are shown and discussed in Section 5.

## 5 Experimental Results and Discussion

This section presents the way we design our experiments for missing value imputation and gene regulation prediction, following by results of our experiments and discussions about them.

### 5.1 Design of Experiments

To apply our approach, we need to determine several conditions and parameters used in the equations of our approach. These include which DTW adjustment and corresponding parameters that can produce the best results, the weighted value α in equation (7) that controls how much the semantic distance contributes to the combined distance, the selection of the proper GO domain, and the decision of weighted values of the three relations for GO terms. First, we set the weighted value α of our distance measurement as zero to focus on expression values themselves to test the effect of imputation performed by the four adjustments for DTW. We combine KNN method and DTW algorithm modified with FastDTW, along with four adjustments on DTW to impute missing values in alpha and cdc28 testing datasets. NRMS errors are then calculated as the assessment to determine whether an imputation method is effective or not. We choose the adjustment method for DTW that generates the best results as the distance measurement for gene expression values used in our approach. Subsequently, we try different combinations of parameters for weighted value α, GO terms used within the three GO domains, and various weighted values for three relational cases. The parameter set which brings the best results for our distance measurement is chosen. The comparison for NRMS errors of our approach and other methods is made. Due to space limitations, parts of experimental data are not listed. The number of K for KNN is set from 10, 15, 20, 50, and 100. DTW with weighting value ranges from 1.2 to 1.8 because we find that the effectiveness is reduced if the weighting value is larger than 1.8. DTW with windowing parameter ranges from 2 to 5 for the same reason. For each experiment, we run 10 times and calculate the average value to reduce the randomness. Finally, we apply our approach to predict potential regulatory gene pairs and count the number of matched pairs with Filkov's data set as the validation of our prediction approach.

## 5.2 Results and Discussion

In this subsection, we present our experimental results and discussions on the effect of DTW adjustments, effect of parameters used in GO, accuracy of missing value imputation, and practicability of gene regulation prediction in order.

### 5.2.1 Effect of DTW Adjustments

We find that the best result is achieved when we apply our proposed method with FastDTW-based modification and slope weighting with weighted value between 1.5 and 1.8. This indicates that DTW works well with slightly weighted values that force the warping path not to form the "one-to-many" mappings. Only with proper variants of DTW such as slope weighting can the imputation results be further improved. Therefore, we use slope weighting with weight value 1.8 as the adjustment for our approach.

### 5.2.2 Effect of Parameters Used in GO Similarity for Our Approach

After choosing slope weighting with weighted value 1.8 as the adjustment of DTW for our approach, we have to discover the best parameters for conditions and parameters used for the GO part of our approach. For the three GO term-term relations, we try several combinations of the parameters and find that the best parameter for case2 is near the double as the parameter for case1. Similarly, parameter for case3 should be slightly less than the double of parameter for case2. The arrangement for these parameters conforms to the concept that if two GO terms are close or even the same in GO structure, the similarity for these terms is higher. For the validation of choosing the best parameters, we experiment different parameter values. Due to the space limitation, here we only propose the best parameter for the three GO term relations: the same terms, parent-children relation, and ancestor- sharing relation as case1 = 1, case2 = 2.4 and case3 = 4.5.

**Table 3** NRMS values for different parameters of GO similarity in alpha and cdc28 dataset

| GO domain / Value of α | | Molecular Function | Biological Process | ALL |
|---|---|---|---|---|
| 0.25 | alpha | 0.74894 | 0.93786 | 0.63011 |
|      | cdc28 | 0.82003 | 1.00207 | 0.71102 |
| 0.50 | alpha | 0.73745 | 0.92661 | 0.62369 |
|      | cdc28 | 0.81060 | 0.99125 | 0.70331 |
| 0.75 | alpha | 0.74048 | 0.94236 | 0.66014 |
|      | cdc28 | 0.82358 | 1.10559 | 0.74224 |
| 1.00 | alpha | 0.75984 | 0.94713 | 0.69971 |
|      | cdc28 | 0.82276 | 1.13171 | 0.77186 |
| 1.25 | alpha | 0.76688 | 0.95967 | 0.73014 |
|      | cdc28 | 0.82053 | 1.13705 | 0.81677 |
| 1.50 | alpha | 0.78104 | 0.95140 | 0.74144 |
|      | cdc28 | 0.82201 | 1.14055 | 0.82746 |
| 1.75 | alpha | 0.79860 | 0.96237 | 0.75324 |
|      | cdc28 | 0.82555 | 1.14442 | 0.82738 |
| 2.00 | alpha | 0.79925 | 0.97145 | 0.75984 |
|      | cdc28 | 0.83850 | 1.15595 | 0.81154 |
| 2.25 | alpha | 0.80934 | 0.98366 | 0.76658 |
|      | cdc28 | 0.83339 | 1.15542 | 0.81108 |
| 2.50 | alpha | 0.81479 | 0.99147 | 0.77897 |
|      | cdc28 | 0.86462 | 1.16412 | 0.81766 |
| 2.75 | alpha | 0.82369 | 1.00647 | 0.79471 |
|      | cdc28 | 0.86146 | 1.16831 | 0.82707 |
| 3.00 | alpha | 0.83471 | 1.02169 | 0.80036 |
|      | cdc28 | 0.88596 | 1.17059 | 0.82822 |
| 3.25 | alpha | 0.85901 | 1.03004 | 0.80996 |
|      | cdc28 | 0.90022 | 1.18341 | 0.83748 |
| 3.50 | alpha | 0.86971 | 1.05526 | 0.82748 |
|      | cdc28 | 0.91826 | 1.18641 | 0.84589 |

Another task is to determine which GO domain produces the best results. For this experiment, we separate the GO terms for genes within the three domains: biological process (BP), molecular function (MF), and cellular component (CC). We then experiment the imputation results with GO terms within these three domains respectively, compared with the imputation results with the combinations of them. The result of CC is simply removed because the number of GO terms in CC is much less than terms in BP and MF so that it provides very little information for genes. Experimental results show that using all GO terms in the three domains produces the best results. This makes sense because GO information for genes is not sufficient without enough GO terms provided.

Besides, the weighted value $\alpha$ in equation (7) is also needed to be determined. Larger $\alpha$ values mean that the semantic distance is strongly emphasized to our distance measurement. We also try various values set between 0.25 to 3.5 as literatures suggest and record the corresponding imputation results. For experimental convenience, we focus on the testing data set with missing rate = 20% because missing rate of the involved real microarray data is close to the rate. Experimental results for determining these parameters are listed in Table 3. The results show that setting $\alpha$ as 0.5 brings the best imputation result.

### 5.2.3 Accuracy of Missing Value Imputation

After all parameters for our approach are determined, we then perform missing value imputation on alpha and cdc28 sub-datasets with our approach. We also implement several existing methods such as the KNN method, BPCA, and LLS for comparison. Experimental results are shown in Fig. 6 and Fig. 7 for alpha and cdc28 sub-datasets respectively. We observe and compare the results above and hence make some summaries. As shown in Fig. 6, the imputation method that only utilizes KNN with FastDTW achieves better results than using KNN. This proves that taking DTW distance as the distance measurement is more suitable than taking Euclidean distance while handling microarray time series data. BPCA and

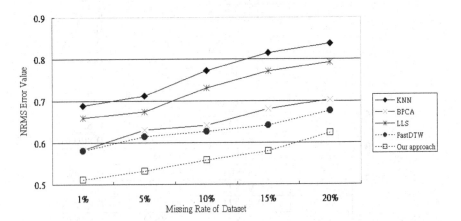

**Fig. 6** Imputation results of alpha dataset

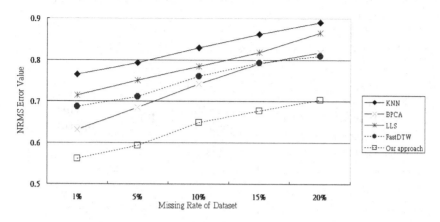

**Fig. 7** Imputation results of cdc28 dataset

LLS seem to outperform KNN. Our approach is the most effective method when using FastDTW with slope weighting and proper parameters for GO distance measurement. Sequences of effectiveness of these imputation methods may change a little in certain percentage of missed data. This may result from the randomness while deciding which values to be removed in the complete matrix. We also experiment on the effectiveness of our semantic distance measurement based on GO with that of Tuikkala et al.'s work. Experimental results show that the NRMS error of our approach is about 0.4 less than that of Tuikkala et al.'s work. We do not list the whole experimental results due to space limitation.

Fig. 7 illustrates almost the same situation as Fig. 6. Basically results of all imputation methods are worse than results in alpha sub-dataset. This is because the cdc28 sub-dataset contains more missing values than the alpha sub-dataset. Theoretically, NRMS error increases when there are many missing values in the dataset. Furthermore, even using FastDTW brings better results than BPCA when the missing rate is larger than 15%. This shows the weakness of BPCA while dealing with microarray time series dataset with a large portion of missing values. To summarize, using our approach with suitable parameters can retrieve the best imputation results. Besides, we find that methods relative to KNN including KNN, FastDTW, and FastDTW with adjustments retrieve the best results when the number of K is set between K =10 and K = 20. This stands for Troyanskaya's research in 2001. As a result, while applying KNN or KNN-liked methods to impute missing values in microarray time series data, setting the number of K between 10 and 20 generates the best result empirically. Assigning the value of K less than 10 or more than 20 will not bring a better result.

### 5.2.4 Practicability of Gene Regulation Prediction

After missing values are imputed with our approach, we then perform gene regulation prediction. Yeung et al. propose their work for similar aim of regulatory gene prediction [20]. Table 4 shows the experimental results of our approach and Yeung et al.'s method.

**Table 4** Number of identified regulatory gene pairs

| Dataset / # of Known gene pairs | Method | | | |
|---|---|---|---|---|
| | PCC | Yeung's method | DTW | Our approach |
| alpha(+)/ 343 | 36 | 223 | 215 | 297 |
| alpha(-)/ 96 | 5 | 55 | 56 | 66 |
| cdc28(+)/ 469 | 66 | N/A | 287 | 380 |
| cdc28(-)/ 155 | 14 | N/A | 87 | 101 |

In Table 4, activation regulations and inhibition regulations from Filkov's database are separated. The four numbers lying in the first column denote the known gene regulations from Filkov's database for alpha and cdc28 sub-datasets. The numbers of mapping gene pairs found by the four methods, including Pearson correlation coefficient (PCC), Yeung et al.'s method, distance measurement with only DTW, and our approach are listed in the corresponding grids of the table. Gene pairs are said to be similar if their PCC values are larger than 0.5 according to Yeung et al.'s work. We can see that PCC can only find very few mapping known regulatory gene pairs, while Yeung et al.'s method than PCC. However, Yeung et al. only experiment alpha sub-dataset. Therefore we mark the result of cdc28 sub-dataset of Yeung et al.'s method with N/A. Obviously, with our method we can find much more known regulatory gene pairs compared with other methods. In alpha activation regulations, we can even find almost $297/343 = 86\%$ of known regulatory gene pairs and $380/469 = 81\%$ of known regulatory gene pairs in cdc28 activation regulations. The results show that our approach is not only accurate for missing value imputation but also effective for regulatory gene prediction.

# 6 Conclusions

In this chapter, we introduce a novel approach that provides an effective distance measurement for genes based on gene ontology (GO) annotations. GO is the structural definition for genes that provides biological information about genes or proteins. With the application of GO terms, external information such as biological functions for genes can be exploited so that the effectiveness of microarray data analysis is improved. We then perform missing value imputation by taking our approach as the distance measurement for gene pairs combined with the KNN method. We also analyze and implement modifications of DTW both for efficiency increasing and accuracy improvement to achieve better imputation results. After missing values are imputed, our approach is then used to predict potential regulatory gene pairs. Experimental results show that our approach with specific adjustments outperforms other methods not only for missing value imputation, but also for gene regulation prediction. Our approach facilitates analysis for microarray time series data.

# References

[1] Acuna, E., Rodriguez, C.: The treatment of missing values and its effect in the classifier accuracy. In: Proceedings of the Classification, Clustering XE Clustering and Data Mining Applications, pp. 639–648 (2004)

[2] Ouyang, M., Welsh, W.J., Georgopoulos, P.: Gaussian mixture clustering and imputation of microarray XE microarray data. Bioinformatics 20(6), 917–923 (2004)

[3] Alizadeh, A.A., Eisen, M.B., Davis, R.E., Ma, C., Lossos, I.S., Rosenwald, A., Boldrick, J.C., Sabet, H., Tran, T., Yu, X., Powell, J.I., Yang, L., Marti, G.E., Moore, T., Hudson, J., Lu, L., Lewis, D.B., Tibshirani, R., Sherlock, G., Chan, W.C., Greiner, T.C., Weisenburger, D.D., Armitage, J.O., Warnke, R., Levy, R., Wilson, W., Grever, M.R., Byrd, J.C., Botstein, D., Brown, P.O., Staudt, L.M.: Distinct types of diffuse large B-cell lymphoma identified by gene expression XE gene expression profiling. Nature 403, 503–511 (2000)

[4] Chen, L.C., Lin, Y.C., Arita, M., Tseng, V.S.: A novel approach for handling missing values in microarray XE microarray data. In: Proceedings of the International Computer Symposium, pp. 45–50 (2008)

[5] Troyanskaya, O., Cantor, M., Sherlock, G., Brown, P., Hastie, T., Tibshirani, R., Botstein, D., Altman, R.B.: Missing value estimation methods for DNA microarray XE microarrays. Bioinformatics 17(6), 520–525 (2001)

[6] Kim, S., Imoto, S., Miyano, S.: Dynamic Bayesian network and nonparametric regression for nonlinear modeling of gene networks from time series gene expression XE gene expression data. Biosystems 75, 57–65 (2004)

[7] Oba, S., Sato, M., Takemasa, I., Monden, M., Matsubara, K., Ishii, S.: A bayesian missing value estimation method for gene expression XE gene expression profile data. Bioinformatics 19(16), 2088–2096 (2003)

[8] Kim, H., Golub, G.H., Park, H.: Missing value estimation for DNA microarray XE gene expression XE gene expressiondata: local least squares XE local least squares imputation. Bioinformatics 21(2), 187–198 (2005)

[9] Choong, M.K., Charbit, M., Yan, H.: Autoregressive-model-based missing value estimation for DNA microarray XE microarray time series data. IEEE Transactions on Information Technology in Biomedicine 13(1), 131–137 (2009)

[10] Choong, M.K., Levy, D., Yang, H.: Study of microarray XE microarray time series data based on forward–backward linear prediction and singular value decomposition XE singular value decomposition. International Journal of Data Mining and Bioinformatics 3(2), 145–159 (2009)

[11] Shan, Y., Deng, G.: Kernel PCA regression for missing data estimation in DNA microarray XE microarray analysis. In: Proceedings of the IEEE International Symposium on Circuits and Systems, pp. 1477–1480 (2009)

[12] Wang, X., Li, A., Jiang, Z., Feng, H.: Missing value estimation for DNA microarray XE microarray gene expression XE gene expression data by support vector regression imputation and orthogonal coding scheme. BMC Bioinformatics 7, 1–10 (2006)

[13] Wong, D.S.V., Wong, F.K., Wood, G.R.: A multi-stage approach to clustering and imputation of gene expression XE gene expression profiles. Bioinformatics 23, 998–1005 (2007)

[14] Liu, J., Ni, B., Dai, C., Wang, N.: A simple method of inferring pairwise gene interactions from microarray XE microarray time series data. In: Proceedings of the Fourth International Conference on Machine Learning and Cybernetics, pp. 3346–3351 (2005)

[15] Yang, A.C., Hsu, H.H., Lu, M.D.: Outlier filtering for identification of gene regulations in microarray XE microarray time-series data XE time-series data. In: Proceedings of the Third International Conference on Complex, Intelligent and Software Intensive System, pp. 854–859 (2009)

[16] Tseng, V.S., Chen, L.C., Chen, J.J.: Gene relation discovery by mining similar subsequences in time-series microarray XE microarray data. In: Proceedings of the IEEE Symposium on Computational Intelligence in Bioinformatics and Computational Biology, pp. 106–112 (2007)

[17] Vlachos, M., Kollios, G., Gunopulos, G.: Discovering similar multidimensional trajectories. In: Proceedings of the Eighteenth International Conference on Data Engineering, pp. 673–684 (2002)

[18] Lee, M.S., Liu, L.Y., Chen, M.Y.: Similarity analysis of time series gene expression XE gene expression using dual-tree wavelet transform. In: Proceedings of the IEEE International Conference on Acoustics, Speech, and Signal Processing, pp. I-413–I-416(2007)

[19] Friedman, N., Linial, M., Nachman, I., Péer, D.: Using Bayesian network to analyze expression data. In: Proceedings of the Fourth Annual International Conference on Computational Molecular Biology, pp. 601–620 (2000)

[20] Yeung, L.K., Yan, H., Liew, A.W.C., Szeto, L.K., Yang, M., Kong, R.: Measuring correlation between microarray XE microarray time series data using dominant spectral component XE dominant spectral component. In: Proceedings of the Second Asia-Pacific Bioinformatics Conference, vol. 29, pp. 309–314 (2004)

[21] Mohammadi, A., Saraee, M.H.: Estimating missing value in microarray XE microarray data using fuzzy clustering and gene ontology XE gene ontology. In: Proceedings of the IEEE International Conference on Bioinformatics and Biomedicine, pp. 382–385 (2008)

[22] Xiang, Q., Dai, X.: Proving missing value imputation in microarray XE microarray data by using gene regulatory information. In: Proceedings of the Second International Conference on Bioinformatics and Biomedical Engineering, pp. 326–329 (2008)

[23] Eisen, M.B., Spellman, P.T., Brown, P.O., Botstein, D.: Cluster analysis and display of genome-wide expression patterns. National Academy of Science 95, 14863–14868 (1998)

[24] Kalpakis, K., Gada, D., Puttagunta, V.: Distance measures for effective clustering of ARIMA time-series. In: Proceedings of the IEEE International Conference on Data Mining, pp. 273–280 (2001)

[25] Myers, C., Rabiner, L., Roseneberg, A.: Performance tradeoffs in dynamic time warping algorithms for isolated word recognition. IEEE Transactions On Acoustics, Speech, and Signal Processing ASSP-28, 623–635 (1980)

[26] Rabiner, L., Rosenberg, A., Levinson, S.: Considerations in dynamic time warping algorithms for discrete word recognition. IEEE Trans. on Acoustics, Speech, and Signal Processing ASSP-26, 575–582 (1978)

[27] Furlanello, C., Merler, S., Jurman, G.: Combining feature selection and DTW for time-varying functional genomics. IEEE Transactions on Signal Processing 54(6), Part 2, 2436–2443 (2006)

[28] Yu, H.M., Tsai, W.H., Wang, H.M.: Query-by-Singing system for retrieving karaoke music. IEEE Transactions on Multimedia 10(8), 1626–1637 (2008)

[29] Salvador, S., Chan, P.: Toward accurate dynamic time warping in linear time and space. Intelligent Data Analysis 11(5), 561–580 (2007)

[30] Sakoe, H., Chiba, S.: Dynamic programming algorithm optimization for spoken word recognition. IEEE Trans. on Acoustics, Speech, and Signal Processing ASSP-26, 43–49 (1978)

[31] Berndt, D., Clifford, J.: Using dynamic time warping to find patterns in time series. In: Proceedings of the Workshop on Knowledge Discovery in Databases (1994)

[32] Kruskall, J.B., Liberman, M.: The symmetric time warping algorithm: from continuous to discrete. Time Warps, String Edits, and Macromolecules: The theory and Practice of String Comparison (1983)

[33] Itakura, F.: Minimum prediction residual principle applied to speech recognition. IEEE Transactions on Acoustics, Speech, and Signal Processing ASSP-23, 52–72 (1975)

[34] Keogh, E., Pazzani, M.: Derivative dynamic time warping. In: Proceedings of the First SIAM International Conference on Data Mining, Chicag, Illinois (2001)

[35] Lord, P.W., Stevens, R.D., Brass, A., Goble, C.A.: Investigating semantic similarity measures across the Gene Ontology: the relationship between sequence and annotation. Bioinformatics 19, 1275–1283 (2003)

[36] Sanfilippo, A., Baddeley, B., Beagley, N., Gopalan, B.: Enhancing automatic biological pathway generation with GO-based gene similarity. In: Proceedings of the International Joint Conference on Bioinformatics, Systems Biology and Intelligent Computing, pp. 448–453 (2009)

[37] Tuikkala, J., Elo, L., Nevalainen, O.S., Aittokallio, T.: Improving missing value estimation in microarray XE microarray data with gene ontology XE gene ontology. Bioinformatics 22, 566–572 (2006)

[38] Spellman, P.T., Sherlock, G., Zhang, M.Q., Iyer, V.R., Anders, K.M., Eisen, B., Brown, P.O., Botstein, D., Futcher, B.: Comprehensive identification of cell cycle-regulated genes of the yeast saccharomyces cerevisiae by microarray XE microarray hybridization. Molecular Biology of the Cell 9, 3273–3297 (1998)

[39] Cho, R., Campbell, M., Winzeler, E., Steinmetz, L., Conway, A., Wodicka, L., Wolfsberg, T., Gabrielian, A., Landsman, D., Lockhart, D.: A genome-wide transcriptional analysis of the mitotic cell cycle. Molecular Cell 2, 65–73 (1998)

[40] Filkov, V., Skiena, S., Zhi, J.: Analysis techniques for microarray XE microarray time-series data XE time-series data. In: Proceedings of the Fifth Annual International Conference on Computational Molecular Biology, pp. 124–131 (2001)

[41] Website: Gene ontology XE Gene ontology website, http://www.geneontology.org/ (last accessed on March 1, 2011)

[42] Website: Saccharomyces Genome Database XE Saccharomyces Genome Database, http://www.yeastgenome.org/ (last accessed on March 1, 2011)

# Integrating Content and Structure into a Comprehensive Framework for XML Document Similarity Represented in 3D Space

Eric Draken, Tamer N. Jarada, Keivan Kianmehr, and Reda Alhajj

**Abstract.** XML is attractive for data exchange between different platforms, and the number of XML documents is rapidly increasing. This raised the need for techniques capable of investigating the similarity between XML documents to help in classifying them for better organized utilization. In fact, the idea of similarity between documents is not new. However, XML documents are more rich and informative than classical documents in the sense that they encapsulate both structure and content; on the other hand, classical documents are characterized only by the content. According, using both the content and structure of XML documents to assign a similarity metric is relatively new. Of the recent research and algorithms proposed in the literature, the majority assign a similarity metric between 0.0 and 1.0 when comparing two XML documents. The similarity measures between multiple XML documents may be arranged in a matrix whereby data mining may be done to cluster closely related documents. In this chapter the authors have presented a novel way to represent XML document similarity in 3D space. Their approach benefits from the characteristics of the XML documents to produce a measure to be used in clustering and classification techniques, information retrieval and searching methods for the case of XML documents. We mainly derive a three dimensional vector per document by considering two dimensions as the document's structural and content, while the third dimension is a combination of both structure and content characteristics of the document. The outcome from our research allows users to intuitively visualize document similarity.

Eric Draken · Tamer N. Jarada
Computer Science Department, University of Calgary, Calgary, Alberta, Canada

Keivan Kianmehr
Computer Engineering Department, University of Western Ontario,
London, Ontario, Canada

Reda Alhajj
Computer Science Department, University of Calgary, Calgary, Alberta, Canada
Department of Computer Science, Global University, Beirut, Lebanon
e-mail: alhajj@ucalgary.ca

M. Biba and F. Xhafa (Eds.): Learning Structure and Schemas from Documents, SCI 375, pp. 275–287.
springerlink.com                                    © Springer-Verlag Berlin Heidelberg 2011

**Keywords:** Similarity measures, XML, 3D space, visualization, intuitive representation, document similarity, platform independence.

# 1 Introduction

Document clustering and classification is an important research area which has a wide range of applications, e.g., email filtering, news monitoring, plagiarism detection, among others. It has attracted considerable attention in the literature, e.g., [4, 17]. A document is normally characterized by its content which is analyzed to derive a feature vector for each document. Once documents are represented by their corresponding feature vectors, different measures could be applied to compare and then cluster/classify the documents. However, techniques that have been developed for handling classical documents are not as effective when applied to documents represented using the eXtensible Markup Language (XML).

XML has become the de facto standard for semi-structured Web data encoded in a textual format [16, 5]. XML has gained its popularity from its flexibility; it is extensible and platform independent. Not surprisingly offline documents (e.g. Word and Open Office documents) are trending toward structured XML formats [9]. There has been great interest to store large data repositories such as digital libraries, online news feeds and web-logs in XML format. With the trend for both online and offline documents being encoded in XML-like structures, the problem of grouping documents by their relative similarity has received increased attention. As the size of document collections grows, data mining techniques such as clustering and classification become necessary to facilitate organization of documents for effective browsing and efficient searching. In addition, all clustering and classification techniques have a basic and vital requirement, namely the measurement of similarity between individual objects. In the context of XML documents clustering and classification, most of the similarity measures that are applied by the employed algorithm to calculate the similarity either work solely based on information retrieval methods and ignore the structural patterns existing in the document tag structure, or the other way around.

The work described in this book chapter attempts to measure the similarity between XML documents based on both content and structural information of XML documents. Our main goal is to find an effective similarity measure that benefits from the characteristics of the XML documents to produce a comprehensive approach that could be integrated into clustering and classification techniques, information retrieval and searching methods when applied to the XML documents. In order to precisely characterize both content and structure of XML documents for measuring the XML documents similarity, we introduce an intuitive way to represent XML document similarity visually; this turns the investigation of document similarity into an attractive approach for a variety of users ranging from professionals to naive users. While the outcome saves the time of professionals by allowing them to get a better flavor of the comparison result, naive users also enjoy better understanding of the outcome from the similarity comparison of documents by watching that on the screen. We have achieved this target by our work described in this chapter by presenting a prototype for an algorithm to assign relative structural and content similarity metrics between a single XML document and multiple other XML documents in three dimensions. The results of an execution of the algorithm then may be directly mapped onto 3D space for immediate visual inspection. The theory of the algorithm, an evaluation of the approach, and a method for back-compatibility converting a similarity vector measurement into a scalar metric is presented.

The rest of this chapter is organized as follows. Section 2 presents the problem statement. The employed similarity matrix is described in Section 4. Visual presentation and experimental results are included in Section 5. Section 6 is conclusions and future work.

## 2 Problem Statement

### 2.1 XML Background

The eXtensible Markup Language, XML, is a W3C standard for document markup with simple, human-readable tags. Some of the attractions of XML are its extendibility, portability and the fact that it is a meta-markup language because there is no fixed set of tags attached to the XML specification as it is the case with HTML. Users are given the flexibility to define their own tags as they need them. This turns XML into a domain independent markup language; it is in fact used for domains as diverse as web sites, data interchange, vector graphics, object serialization, remote procedure calls, databases, among others. The markup in an XML document describes the structure of the document which encapsulates the content; hence XML documents are classified as semi-structured. Further, XML documents may be flat or nested; however, nesting is a more natural representation as opposed to flat XML documents that almost simulate the relational representation of the content. In other words, the relational model suffers from and has been criticized for not supporting nesting, though it gained its popularity for having a well defined underlying theory. On the other hand, XML covers a wider perspective by facilitating both nested and flat structures, hence allows for almost one to one mapping between XML structures and relational database tables.

Another attractive characteristic of the XML technology is its flexibility in allowing developers to load documents with semantics by choosing the tags to well describe the content. For the sake of interoperability, individuals or organizations may agree to use only certain tags, and only a particular structure. The markup permitted can be documented in an XML Schema. Fortunately the XML schema has been recently adopted by W3C as the standard for XML structure definition as opposed to data type definition (DTD), which is not as descriptive and effectives as the schema. Documents can be compared to the schema and those matching the schema are said to be valid. XML Schema is widely adopted to describe the structure of XML documents, e.g., web data and databases. Knowing that a document is valid for a particular schema, assumptions can be made on the contained data and structure, allowing an efficient querying of the document; and even a better detection of the similarity between documents. To sum up, the basic notion of an XML document (herein referred to as *document*) will start the presentation. A document is characterized by *tags* which are terms enclosed by angled brackets without spaces in the term name. Tags are arranged in a hierarchy which provides the structure and semantics of the data enclosed. Tags can appear in two forms: starting and closing tags, or self-closing tags with no value present.

It is necessary for the proposed algorithm that documents are well-formed. That is, tags are properly nested and each starting tag has a corresponding closing tag at the same level. As such, a document may be represented as an ordered tree wherein tags are referred to as *nodes*. For this reason, regular HTML documents are not well-formed as XML documents because many tags such as <rel>,<meta>,<img>,<script>, etc. do not need a corresponding closing tag and do not need to be self-closing. XHTML documents, on the other hand, are a particular kind of XML document which can be processed by our algorithm proposed in this work. A document which can be represented by an ordered tree can be traversed such that from any given node the children of that node as well as its siblings may be known. This way, any document object model (DOM) parser [20] or Simple API for XML (SAX) parser [20] can be used with the algorithm described in this work.

## 3  Problem Definition

Given two XML documents that satisfy the following requirements:

- Are well-formed XML documents
- Are from any two XML schemas (or the same schema; the same schema case requires less effort for deriving the similarity)
- Have the same or different content sizes

This chapter seeks to answer the following questions:

- How similar are the two compared documents?
- Can the similarity measure be represented as a three-dimensional vector which intuitively and visually depicts how similar the two documents are by investigating how spatially close they are?
- Can the calculated metrics be reused (one-time calculation)?

These inquiries are answered by considering the similarity metric described in the sequel.

### 3.1  Similarity Metric Overview

XML documents are characterized by both structure and content. Accordingly, the similarity measures for XML documents, as described in the literature, include two major techniques: content-based and structure-based similarity measures.

Taken alone, the similarity of content between two documents can be evaluated by established methods using *term frequency-inverse document frequency* (TF-IDF) and cosine similarity [2]. For measuring structural similarity, there exist novel approaches [15]. On the other hand, similarity measures for classical text documents concentrate only on the content as the structure is missing, though classical documents implicitly structured into sections, paragraphs, etc. Alternative approaches for deriving the similarity of XML documents have been also proposed to combine different techniques to improve the efficiency and effectiveness of the similarity measures for XML documents clustering and classification.

Fourier transformation is among techniques that have been applied to measure similarity between XML documents. The motivation behind applying the Fourier transformation to study the similarity of XML documents was originated from the initial use of Fourier transformation to check the similarity among time series data [1, 7]. In the context of time series data, each document is converted into a discrete-time series signal. Then the discrete Fourier transform (DFT) is applied and the transform is used to detect similarity [6]. Of course, the tags of the document could be used in place of actual content for performing cosine similarity analysis.

A combination of elements semantics and the nested structures of XML documents have been used as a method of similarity measure by several groups of researchers. For example, Park et al. [13] introduced similarity metrics that were built by considering the similarities in tags' names, tags' values, and the structure of tags. Based on these metrics, they developed a search system to find documents that match the structure and content of a given XML document. Another method based on elements semantics and the nested structures of XML documents was developed by Lee et al. [10] as well. They used synonyms, compound words, and abbreviations to develop a set of extended-element vectors which were used in turn to build a similarity matrix. Then, they measured similarity between XML elements with the similarity matrix. The above mentioned method was also presented by Ma and Chbeir [11]. However, they only considered measuring the similarity between two XML documents. The similarity between two XML documents was measured based on an overall value which was the aggregation of leaf nodes' values similarities according to the XML structure.

XML documents have tree structures in nature. Several research works have used this property of XML documents to calculate the similarity metric. For example, the similarity metric is calculated while converting a tree representation of an XML document to the tree representation of another document [18, 3]. Tekli and Nierman [14, 12] developed a method to calculate the difference between two XML trees by using the edit distance. In general, the edit distance is used to measure the minimum number of node insertions, deletions, and updates required to convert one tree into another. The edit distance can be converted to a similarity measure once the number of edit operations is normalized by considering the number of nodes in the tree.

In [8] a different technique is presented to measure the similarity between any two XML documents. Documents that are structurally identical or structurally contained in each other are identified by a string matching technique. A weight is assigned to the corresponding trees of two XML documents by considering their nodes' names and the paths in order to evaluate how much the content of the two documents are similar.

We argue that by limiting a similarity metric to a scalar value in the interval [0.0,1.0], the way two documents are similar is lost. For example, suppose a given similarity algorithm determined that two documents are similar by 0.5. The latter similarity may characterize only one of the two aspects of the XML documents, namely structure and content. In other words, the structure might be very different between the two documents and the content is similar, or it could be the other way around. Expanding on this problem, what is the justification for grouping similar documents together if we are ultimately unsure of what dimension they have as similar? The similarity metric proposed in this chapter is a three-dimensional vector which encodes structural similarity and content similarity, as well as a third value which is a hybrid of the two.

## 4  Similarity Metric

### 4.1  Structural Encoding

From the research done by Flesca et al. [6], it is determined that pair-wise encoding of nested tags achieves a higher resolution of structural encoding than simply encoding each tag individually. This way, the relative hierarchical relationship between tags is encoded, and when compared to those of another document, these structures can be compared instead of just textual tags.

Consider the following examples:

1.a) A nested XML structure with content

```
<outer>
    <middle>
          <inner>text</inner>
    </middle>
</outer>
```

1.b) Same tags and content as (1.a), but different relationship

```
<outer>
    <middle></middle>
    <inner>text</inner>
</outer>
```

When the two samples shown above are encoded by single tags and these counts, and then compared to each other, both structures appear to be perfectly similar to each other. That is, the two sample contain one <outer>, one <middle> and one <inner> tag set. However, when we take the pair-wise encoding of tags such that we note the frequency of a given tag and its parent tag together, the structural differences become clear as depicted next.

2.a) Pair-wise structural frequency encoding of the structure given above in (1.a)

```
<outer>                 1
<outer><middle>         1
<middle><inner>         1
<inner>                 1
```

2.b) Pairwise structural frequency encoding of the structure given above in (1.b)

```
<outer>                 1
<outer><middle>         1
<middle>                1
<outer><inner>          1
<inner>                 1
```

Without yet explaining how to compare the two encodings shown in (2.a) and (2.b), it can be easily seen that the two structures are not the same using the pair-wise method.

## 4.2  Content Encoding

Content encoding is performed similarly to structural encoding using pair-wise encoding of textual terms. This is preferred over simply recording how many times each word occurs in a given document because we want to preserve some sense of context. For example, consider these two sentences:

3.a) Colorless green ideas sleep furiously.[1]
3.b) Furiously green colorless ideas sleep.

As with the structural encoding problem, if we take the frequency of each term in (3.a) and compare those to the term frequencies of those in (3.b), the content looks identical. However, if we perform pair-wise encoding of the terms in the sentences, we can capture relative positioning of the terms as depicted next.

```
colorless            1
colorless green      1
green ideas          1
ideas sleep          1
sleep furiously      1
furiously            1
```

---

[1] Favourite sentence of Noam Chomsky, respected linguist.

The goal of encoding adjacent pairs of text terms is to detect content differences on the syntactic level as opposed to just the superficial physical level. Then, for example, if the only difference between two documents is that two words anywhere in the document were swapped, this algorithm will detect that.

## 4.3 Nested Content Encoding

The third dimension of this chapter's proposed metric is a hybrid encoding of both content and structure which forgives both the differences between tag names between the two documents, as well as the relative positioning of content terms. Each term is individually tallied and paired with how deep it occurs in the document. To capture the idea better, consider the following example.

```
<outer>
    <middle>
        <inner>term</inner>
    </middle>
    <inner>term</inner>
</outer>

2 term   1
1 term   1
```

This metric takes into account identical text parts which may occur at different levels in the XML document and distinguishes them. The number before the term indicates how deep the term is located in the document, i.e., the level of nesting.

## 4.4 Difference Operator

Once the three metrics described above (structure, content and hybrid) have been calculated for a given document, they do not need to be calculated anymore as long as the given document does not change. They instead can be saved to disk or stored in memory. Once two documents have been analyzed by the proposed algorithm and the metrics have been assigned to them, the next task is comparing the metrics to assign a similarity vector. We shall return to the XML fragments given in (2.a) and (2.b).

The term *merge-annihilate* is now introduced. The two structural frequency encodings of the XML fragments give in (2.a) and (2.b) are merged in such a way that if one term pair does not appear in the other listing for the other document, it as well as its term count is added to the output. If a term pair exists in both lists, the absolute value of the subtraction of their respective pair counts appears in the output. If the same term pair exists with the same frequency in both lists, they annihilate each other and the term pair does not appear in the output. The difference of (2.a) and (2.b) is depicted next.

```
<middle><inner>      1
<middle>             1
<outer><inner>       1
```

The difference information reported above reveals the structural difference between (1.a) and (1.b). If the above lists were empty, it would mean that the structure of both documents were the same with regard to nesting and tag names. Because there is a remainder in the merge-annihilate process, we can count the number of remaining pairs (3) and assign that as the structural difference. The smaller the sum of the frequencies of the remaining terms, the more similar the structure is between two given documents. If two documents have a very different structure, then the sum from the procedure above will be greater. It is only when we take a collection of documents and compare them all to a given document that we can visualize which are spatially closer to the compared-with document. Because a structure can be completely different (different tag names, different nesting, etc.) but the content can be identical, we proceed to continue the merge-annihilate process on the content metrics between the two given documents. Finally, the relative content and the relative structure of two XML documents may be different; yet they may contain the same text terms. Accordingly, we proceed to the hybrid metric that takes into account individual text terms and their nesting level. The final result is a 3-tuple respective sum of frequencies of remaining terms from all three metrics.

## 5   Visual Representation and Experiments

Given a large collection of documents, once individual metrics have been derived for each document (or retrieved from memory or disk), the merge-annihilate process described above can be iteratively performed on all the document metrics in question. The result will be a collection 3-tuples of difference metrics which can be plotted on a graph as shown in Figure 1 and Figure 2 which report some testing results as described in the remainder of this section. We run two different experiments and the results are in favor of the proposed approach as reported in the next two sections (one per experiment).

### 5.1   MS Word Incremental Saves

For this experiment, we used a series of 20 XML documents generated by Word 2003. The process of creating them went as follows:

1. The origin document contained a single sentence and was saved as an XML file.
2. A moment later some words were added and it was saved as another file.
3. These words were rearranged and saved again to a third file.
4. As more modifications to the document took place, more saves were made to new files.

Figure 1 shows an actual plot of the similarity between the 20 documents; there are two noticeable point clusterings which can be observed. The first clustering

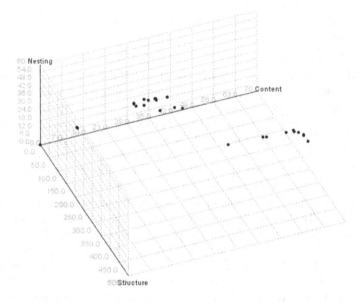

**Fig. 1** Sample 3D visualization of MS Word XML docs

contained a single sentence which was stylized with bolding, different fonts, different font sizes, etc. The second clustering, farthest away from the origin, contained two or more sentences and had further stylizations performed on the text. Relative to the original document, one can already see visually that two major events took place in composing the sample Word document and a host of minor events (styling) took place as well. This first experiment already demonstrates the effectiveness and applicability of the proposed approach.

## 5.2   Random News Documents

For the second experiment, a collection of 200 news XML documents [19] of the same structure but very different contents of similar lengths (all were about 2kb) have been compared and plotted in Figure 2. As expected, all the documents compared to the original document (chosen at random) were far from the origin but all remained along the same structural axis. This was expected because all the documents have identical structures.

In this experiment, the visual clustering does not mean that the documents in the cluster are similar to each other. When compared to the origin document, it shows that they are different in a similar way: different content.

This was a successful test because it clearly shows that, relative to each other, none of the random news entries were more or less similar to the origin document than any other document. There is of course skewing in the plot shown in Figure 2,

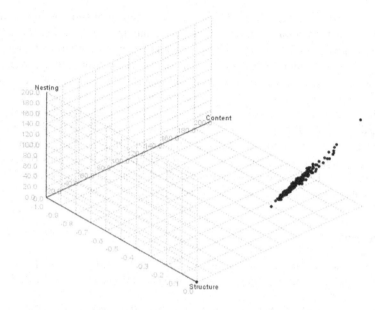

**Fig. 2** Random news documents with same structure

and this is due to common English phrases which contain conjunctions and pronouns which frequently occur in news articles [21].

## 6 Conclusions and Future Work

As demonstrated in the conducted experiments, this line of research is promising. The idea of visually representing the similarity between XML documents in a 3D plot may be extended in the future to search engines which return visual search results on the Web or even internally within a large organization where it is diffi-cult if at all possible for users to realize the similarity between documents as the search space increases. The key features of the experimental algorithm is the use of pairwise encoding to capture structures and textual context within the document, as well a hybrid metric to tie the two together. The merge-annihilate difference opera-tor makes comparisons between these three measurements possible. However, it is worth mentioning that though worked fine for the conducted experiments, we do not want to go far and claim that this is the best algorithm. There is room for improve-ment to the algorithm to utilize weights, statistical modeling, sum-of-squares met-rics, thrice-wise encoding (for example), etc. At this stage of development, we have demonstrated the effectiveness and applicability of the proposed approach which can be used as a conceptual framework in future research. The next step is to com-pare the metrics of all given documents to each other to produce a true 3D plot of all relative similarities in order to see real clustering effects; this has been left as future work. Another step would be to explore modifications to the algorithm to take into account common words and phrases. We anticipate that removing the most common English words [21] from processing and leave content terms with higher intrinsic meaning would further improve the power of the proposed approach. Finally, the algorithm as it stands now in based in Java and makes heavy use of hash tables to keep track of pair-wise terms. These hash tables are based on binary trees which have a logarithmic search and retrieval running time. The process of scanning a single document is linear in terms of the physical lines in the XML file, but every term must be checked and inserted into the hash table in three places. Technically, the scanning algorithm runs in $3(NlogM)$ time, where $N$ is the number of terms in the document and $M$ is size of the current hash table; how this can be improved to take advantage of sorted lists needs to be explored in the future.

# References

1. Agrawal, R., Faloutsos, C., Swami, A.: Efficient Similarity Search in Sequence Databases. In: Proceedings of the Fourth International Conference on Foundations of Data Organization (1993)
2. Aizawa, A.: An information-theoretic perspective of tf-idf measures. Information Processing & Management 39, 45–65 (2003)
3. Chawathe, S., Garcia-Molina, H.: Meaningful change detection in structured data. In: Proceedings of ACM SIGMOD International Conference on Management of Data (1997)
4. Dumais, S., Platt, J., Heckerman, D.: Inductive Learning Algorithms and Representations for Text Categorization. In: Proceedings of ACM International Conference on Information and Knowledge Management (CIKM 1998), Bethesda, MD, pp. 148–155 (1998)
5. Daconta, M.C., Obrst, L.J., Smith, K.T.: The Semantic Web: a guide to the future of XML, Web services, and knowledge management. Wiley, Chichester (2009)
6. Flesca, S., Manco, G., Masciari, E., Ponticri, L., Pugliese, A.: Fast detection of XML structural similarity. IEEE Transactions on Knowledge and Data Engineering 17(2), 60–175 (2005)
7. Goldin, D., Kanellakis, P.: On Similarity Queries for Time Series Data: Constraint Specification and Implementation. In: Proceedings of the International Conference on Constraint Programming (1995)
8. Kim, W.: XML document similarity measure in terms of the structure and contents. In: Proceedings of the International Conference on Computer Engineering and Applications (CEA 2008), pp. 205–212 (2008)
9. Laurent, S.S., Lenz, E., McRae, M.: Office 2003 Xml: Integrating Office with the Rest of the World, 1st edn. O'Reilly & Associates, Sebastopol (2004)
10. Lee, J., Lee, K.: XML Document Analysis based on Similarity. Journal of KISS: Software and Application 29(6) (June 2002)
11. Ma, Y., Chbeir, R.: Content and Structure Based Approach for XML Similarity. In: Proceedings of the International Conference on Computer and Information Technology (September 2005)
12. Nierman, A., Jagadish, H.: Evaluating structural similarity in XML documents. In: Proceedings of the International Workshop on the Web and Databases (2002)
13. Park, U., Seo, Y.: An Implementation of XML Document Searching System based on Structure and Semantics Similarity. Journal of Korean Society for Internet Information 6(2) (April 2005)

14. Tekli, J., Chbeir, R., Yetongnon, K. (eds.): Proceedings of the International Conference on Current Trends in Theory and Practice of Computer Science (January 2007)
15. Wang, L., Cheung, D.W.-I., Mamoulis, N., Yiu, S.-M.: An efficient and scalable algorithm for clustering XML documents by structure. IEEE Transactions on Knowledge and Data Engineering 13(1), 82–96 (2004)
16. Xyleme, L.: Xyleme: A Dynamic Warehouse for XML Data of the Web. In: Proceedings of the International Symposium on Database Engineering and Applications (IDEAS 2001), pp. 3–7 (2001)
17. Yang, Y., Liu, X.: A Re-examination of Text Categorization Methods. In: Proceedings of ACM International Conference on Information Retrieval (SIGIR 1999), Berkley, CA, pp. 42–49 (1999)
18. Zhang, K., Shasha, D., Wang, J.T.L.: Approximate tree matching in the presence of variable length don't cares. Journal of Algorithms 16(1) (1994)
19. AG's corpus of news articles,
    http://www.di.unipi.it/~gulli/newsspace200.xml.bz
20. Xerces Java XML Parser,
    http://xerces.apache.org/xerces-j/
21. Ranks.nl English Stopwords,
    http://www.ranks.nl/resources/stopwords.html

# Modelling User Behaviour on Page Content and Layout in Recommender Systems

Kin Fun Li and Kosuke Takano

**Abstract.** With the exponential growth of information on the Web, recommender systems play an important role in many service applications such as e-commerce and e-learning. Recommender systems are used to assist users in navigating the Web or propose items that the users are likely interested in. Most of the currently prevalent approaches use collaborative filtering based on the preference of a group of similar users. In the past decade, there has been some but rather limited research in personalized recommender systems incorporating an individual user's explicit and implicit feedbacks. In our previous work, a personalized recommender system that extracts an individual user's preference and the associated Web browsing behaviour such as print and bookmark, has been designed and implemented. In this chapter, Web browsing behaviour reflecting a user's preference on layout and design is investigated. We postulate that when a user browses a page, her actions on the content and links could be associated with personal preference on an object's location, icon shape, colour scheme, etc. Furthermore, tags and labels of selected objects contain valuable information to facilitate the recommendation process. Consequently, systematic and automatic analysis of the relationship between information preference and Web browsing behaviour based on structure and schema learning could be exploited to complement recommendation utilizing content similarity. Survey and related work on personal recommender systems that model Web browsing behaviour are presented. A proof-of-concept system is designed with the objective to study whether there is a correlation between browsing behaviour, both in the content and visual aspects of a Web page, and user preference.

Kin Fun Li

Department of Electrical and Computer Engineering, University of Victoria, Canada
Tel.: +1 250 721 8683, Fax: +1 250 721 6052
e-mail: kinli@uvic.ca

Kosuke Takano

Department of Information and Computer Sciences, Kanagawa Institute of Technology, Japan
Tel: +81 (0)46 291 3266, Fax: +81 (0)46 242 8490
e-mail: takano@ic.kanagawa-it.ac.jp

M. Biba and F. Xhafa (Eds.): Learning Structure and Schemas from Documents, SCI 375, pp. 289–313.
springerlink.com                                                    © Springer-Verlag Berlin Heidelberg 2011

# 1  Introduction

The World Wide Web has become an indispensable resource for people to gather information as the Web provides a quick search capability to its rich data and abundant services. To retrieve desired and relevant information effectively, many techniques have been proposed. In particular, recommender systems have been designed to assist a user in navigating the myriad of information on the Web and suggest items that the user is most likely interested in.

Most of the current recommender systems use the collaborative filtering approach [4] that predicts the interest of a user by analyzing preference information collected from a group of similar users. Collaborative filtering has been widely used in many applications such as e-commerce [35], netnews [33], and hobby-sharing in music, movie, etc. [41].

## 1.1  Personalization and Browsing Behaviour

Many search engine users have the experience that the returned results of a query are not exactly what they are looking for. Teevan et al. termed the "large gap between how well search engines could perform if they were to tailor results to individuals, and how well they currently perform by returning a single ranked list of results designed to satisfy everyone" as the potential for personalization [40]. With the advance of monitoring and measuring techniques, implicit personal preference information can be collected easily and harnessed effectively in recommender systems. A user's Web browsing behaviours such as dwell time, mouse click, scroll action, and search query, together with site visit history and personal document collection, are often used in usage and content mining to assist in making personal recommendation.

In the "Stuff I've Seen" system [10], personal contextual items, such as authors and thumbnails from the documents that the user has already seen, are used to search for relevant information. The SEARCHY system [28] filters and re-ranks the Web search results by exploiting the user's profile as obtained from her Web browsing behaviour. Morita et al. proposed an information reminder system [29] where a user's action such as printing, copying and pasting, are recorded during a Web browsing session. This user profile is then utilized to provide personalized information to the user. Chirita et al. proposed a personalized query expansion method to retrieve Web information based on the personal collection of text documents, emails, cached Web pages, etc. [8].

In a previous work [39], we proposed an adaptive personalized recommender system using a preference-thesaurus constructed based on Web browsing behaviour and user feedback. This system is personalized for an individual user by capturing her browsing behaviour into a preference-thesaurus. Moreover, the system can adapt to different users as well as their changing behaviour and interest through direct feedback and continuous update to each individual's preference-thesaurus. Explicit user preference information based on user feedbacks and implicit measures such as browsing history are being used in interest prediction and information filtering. The browsing behaviours captured are the ones associated with actions such as bookmark, print, and save. The contents of the pages associated with these actions are analyzed and used to predict future interest.

## 1.2 Motivation and Objectives

Current personalized information provision systems recommend or navigate to preferable information based on the implicit assumption that a user's preference is strongly correlated to her browsing behaviour. However, to the best of our knowledge, this assumption has only been investigated in a few limited studies. Moreover, recommendation is formulated using only content-based information filter.

Recommender systems based on browsing behaviour have also been used successfully in assistive technology for the mobility or visually impaired [38]. While researching literature in assistive technology for Web browsing, we came upon a recent study conducted by Francisco-Revilla and Crow [13]. They investigated how users interpret the layout of news and shopping pages. Their study reveals that users look for familiar structural elements and use them as references and entry points, before even looking at the main content. Although the target application of their work is in the assistive technology area, this prompts us to postulate that the layout and design of Web pages may also be used as an information filter for establishing user preference.

The objectives and contributions of this chapter are multi-fold. First, an extensive survey on Web browsing behaviour and user behavioural models is presented. Then, previous work on structure and layout of Web pages, though almost all targeted towards the facilitation of Web page design, is discussed. Our proposed recommender system architecture is introduced to reflect how the various modules are integrated. The implementation of the recommender is shown with the important inner working details. Finally, a system design to capture layout and structure information of Web pages, together with how such information can be used for recommendation, are presented.

Specifically, this chapter aims to (1) reinforce the notion that there is a positive correlation between Web browsing behaviour and information preference; (2) strengthen the concept that content-based information filtering is a valid approach for recommendation; (3) promote the novel idea that the layout and design of a Web page is a plausible visual information filter for establishing user preference and thus is useful for recommendation.

## 2    Literature Review and Survey

User Web browsing behaviour is used in many applications including recommender systems and Web page re-design. Much research work has been done in exploiting information derived from various browsing and navigation behaviour to predict and improve the Web search process. By analyzing a user's Web browsing behaviour, her personal preference can be inferred and utilized in recommending information [2, 5, 11, 36]. For example, if a user spends a considerable amount of time on some Web pages, it is reasonable to regard the user is more interested in the contents of these Web pages than other pages. Also, search engine keywords and results are very important factors to detect personal interest [11, 28]. By analyzing such browsing behaviour information, a user's preference can be established and used to recommend information that suits one's individual taste. For instance, if a user browses some Web sites related to Mozart over a long period of time or repeatedly, it can be inferred that the user prefers classical music and she may also be interested in Beethoven and Bach. Likewise, if a user prints Web articles about digital cameras and MP3 music players, it is highly likely that she is also interested in some related electronic devices such as DVD players and mobile phones.

We postulate that some specific actions performed during Web browsing are positively correlated to a user's preference. Also, specific actions could be correlated to particular interest genres. There are many Web browsing behaviours and it is difficult to identify which ones are influential to a specific user's preference, since each user has her peculiar browsing behaviour that may not be universally held by others. For example, although selecting terms on a Web page by mouse-clicks seems to be an important Web browsing behaviour, however, there are some people who click and select items without much thoughts or intentions. In general, bookmarking is a useful resource but old bookmarks may not reflect a user's current interest. It is therefore important to filter out non-influential browsing behaviours in order to make recommendation. We have carried out an extensive literature review on browsing behaviour and found answers to some of the above questions in our survey.

As stated in Section 1, we are interested in the relationship between a user's preference and the layout and design of a Web page, in addition to the correlation between user interest and browsing actions. Therefore, we view a browsing behaviour consisting of two identifiable components: action and visual. Actions are the browsing interaction a user has with the browser such as bookmark and print. Each action also has a visual aspect related to the layout and design of the page, for example, a user most likely prefers a Q & A type of document if she bookmarks such type of pages frequently.

## 2.1 Browsing Behaviour: Action

Browsing behaviour actions are the ones that a user interactively enters into a browser including bookmark, print, save, etc., and the derived ones such as dwell time. These are implicit measures for recommendation effectiveness collected during a browsing session as opposed to explicit measures that require users to state their preference or rank a list of items.

Teevan et al. developed a prototype system that makes use of three sources to improve relevance and search personalization: (1) Explicit ratings; (2) implicit click-through behaviour; and (3) implicit content-based measures including information created, copied, or viewed by an individual [40]. Furthermore, they found that implicit behaviour-based measures are useful in capturing relevance while content-based measures are more suitable for capturing an individual's variation.

Similarly, Seo and Zhang learned a user's preference by observing the user's browsing behaviour implicitly [36]. In their system, a user's implicit feedbacks are profiled including time for reading, bookmarking, scrolling, and following up the hyperlinks in a document.

### 2.1.1    Dwell Time

In one of the early studies in user behaviour and relevance judgment, Morita and Shinoda performed extensive experiments on user behaviour and emphasized that reading time is an important behavioural indicator [30]. Since then, many researchers have established that user browsing time is a major parameter to determine a user's interest of the content [9].

Based on the assumption that the more an object contains the information needed, the longer the viewing time, Liang and Lai [25] presented a time-based approach to determine user interest in news services. In addition, keywords are identified and their position and frequency in the document are analyzed.

Recently, Liu et al. proposed to model the dwell time, the time spent on a document, using the Weibull distribution [27]. They also demonstrated the possibility of predicting dwell time distribution.

### 2.1.2    Other Actions

Many researchers have revealed that click-through is the second most important browsing behaviour behind dwell time [27]. Meanwhile, Claypool et al. established a strong positive correlation between dwell time and mouse scrolling [9]. On the other hand, Seo and Zhang found in their studies of implicit user feedbacks that bookmarked URL reflects a user's strong opinion of relevance [36].

Kumar and Tomkins performed a large-scale study of user online behaviour based on Yahoo toolbar logs [20]. They developed a taxonomy of pageviews consisting of three high-level classes: content, communication, and search; moreover, they found that the ratios of all online pageviews for the three classes are half, one-third, and one-sixth, respectively.

Multiple tabs in browsers have also been a subject of research study. Viermetz et al. investigated the impact of multiple-tab browsing on Web usage mining and its relevance to business applications [43]. Huang et al. examined the effect of parallel browsing sessions on design implication for Web sites, browsers, and search interface [15].

Weinreich et al. conducted an extensive long-term client-side Web usage study [42]. They discovered that users do not use backtracking in Web navigation as frequent as previously thought. One reason for this is due to the usage of multiple windows and tabs. They concluded that Web designers must consider the limited real estate space provided by the browser. This points to the importance and the effect of the layout and design of a Web page on user's browsing behaviour and experience.

### 2.1.3    Classification of Browsing Behaviour

Oard and Kim [32] developed a framework that categorizes observable behaviour into broad classes. Objects at different levels of abstraction, such as a term, paragraph or a document, can be examined, retained, referred, or annotated. The *examination* category consists of the actions view, listen and select. The *retention* category has behaviours, such as bookmarking, that indicate possible future use of an object. Activities that relate two objects, such as linking, form the *reference* category. The last category *annotation* consists of actions, such as highlighting, that intentionally add value to an object. The objective of this framework, however, is for modelling information content using observable browsing behaviour.

Kelly and Teevan later added *create* as a fifth broad category of observable behaviour that includes editing and authoring [19]. They reviewed and classified research work on implicit feedback using this framework of observable behaviour. However, they concluded that "what can be observed does not necessarily reflect the user's underlying intention". This assertion agrees with the fact that the visual aspects of browsing behaviour are implicit feedbacks that cannot be observed directly but can only be estimated statistically.

### 2.1.4    Browsing Behaviour Models

Zheng et al. developed a user interest model based on the following five behaviours: save page, print page, bookmark page, frequency of visit and dwell time on a page [45]. Li and Feng proposed a page interest estimation model based on information found in Web access log, including page size, frequency of access, date of visit and the time spent of each visit [24]. They purposely did not ask for user feedback nor collect any user identifiable information to avoid privacy issues.

Yu and Liu proposed a 'Short-term User Interest Model' for personalized recommendation to accommodate changes in user's interests over time [44]. Using the assertion that a user's interests are related and concentrated in a short period of time, they concluded that Web pages visited are semantically associated. Furthermore, they used a semantic link network to represent these similar pages.

Burklen et al. presented their 'User Centric Walk' algorithm as the basis for modelling browsing behaviour [5]. Their system consists of two models. The Web graph model includes parameters on the structure and the size of the document, while the access behaviour model considers Web page popularity, path length, viewing time, revisiting, link choice, and jump probability.

Sah et al. proposed an architecture to generate dynamic link and personalization using linked data, and the user's browsing strategies [34]. The user strategy model includes search/purposive browsing that looks for specific information, general purpose/explanatory browsing that stemmed from interests, and serendipity/ capricious browsing which is undirected browsing.

## 2.2  Browsing Behaviour: Visual

A Web page's structure includes objects such as text bodies, images, videos, and their associated tags and labels. A Web page's layout includes elements such as background colour, font size, font style, font colour, style sheets, in addition to the locations of objects.

Lerman et al. believed a Web page contains many explicit and implicit structures in the layout and content. They presented an automatic approach for record extraction and segmentation from Web tables [23].

Song et al. asserted that a Web page can be partitioned into several blocks and the importance of those blocks is not equivalent [37]. They found that users do have a consistent view about the importance of blocks on a web page. Using machine learning techniques, they managed to find functions to describe the correlations between page blocks and importance values. Lim et al. described an algorithm for selecting the main content of a Web page automatically [26]. This is done by first segmenting the page into several blocks and then extracting the main content from the important blocks.

Cai et al. argued that traditional link analysis algorithms ignore the fact that a Web page contains multiple semantics [6]. They treated a Web page as a set of blocks and linkages are from blocks to pages rather than from pages to pages.

Layout of a Web page is an important part of the Web site design. Most Web pages are designed using either standardized layout templates or some logical placement based on the nature of the site. Individuals do have their own favourite layouts, therefore, layout is an important factor in capturing user's preference.

Fiala et al. used a component-based XML document format to enable Web contents and adaptive presentations to be automatically adjusted to user's preference [12]. Kawai et al. developed a content fusion system that displays news items in the user's favourite layout format [18].

Lam and Chan proposed a graph mining algorithm to study how and what specific patterns and features of layout can affect advertising click rate [21]. Examining a page's five general areas: header, footer, left sidebar, right sidebar, and body, they investigated how the layout influences click rate, either positively or negatively.

Karreman and Loorback conducted a study to investigate the visual effect of text structure on users' browsing behaviour [17]. Their results showed that users prefer text structured as list than as paragraphs. Moreover, they found that sites with text lists have their pages visited and appreciated more by the users.

### 2.2.1  Multimedia Objects

In addition to semantic information from the text body, the structure and semantics of images and videos are also useful in the modelling of user browsing behaviour.

Lee et al. proposed a keyword extraction method for videos by analyzing the distance of text blocks to a video [22]. This 'layout distance' is an indication of how relevant a text block is to the video, and thus important keywords can be extracted from the relevant text blocks.

Textual and link information such as labels and tags of images can be obtained easily and exploited in modelling user behaviour. He et al. presented a method to segment a Web page into blocks and obtain textual and link information of images extracted from blocks that contain those images [14].

Song and Lim partitioned a page into blocks to extract important contents [26, 37]; we, however, use the partitioned blocks to extract visual information. Citing the fact that a page can have semantics associated with the different areas within it, we exploit this further in our proposed system [6]. These are elaborated in Section 4.

We concur with Karreman and Loorback's finding that emphasizes the visual aspect (i.e., the structure and layout) of a page, is highly relevant to a user's browsing behaviour. Furthermore, we believe a user's preference on the layout of a page could be useful in recommender systems.

## 3   A Recommender System Based on Browsing Behaviour

Our proposed recommender system suggests items that match a user's preference in content, layout and design. A user's preference on the two aspects of browsing behaviour (i.e., action and visual) is monitored, extracted, and stored in a preference-thesaurus which is updated continuously. In this section, we describe the overall architecture and implementation details of our recommender system, based on browsing behaviour of actions. Representation and recommendation of the visual aspects of browsing behaviour are presented in the next section.

### 3.1   Recommender System Architecture

Figure 1 shows the architecture of our system. It consists of three iterative phrases. During the first phase, a user's Web browsing behaviours are monitored and an important term set is extracted for each behaviour. An initial personal preference-thesaurus is constructed based on each behaviour's term set and its term score. In the second phase, Web documents to be recommended are ranked by the similarity between the preference-thesaurus term set and each document. During the final learning phase, the preference-thesaurus is updated based on the user's evaluation feedback on the most recent recommended items.

#### 3.1.1   Web Browsing Behaviour Monitor

The user's Web browsing behaviours, such as the typical ones shown in Table 1, are monitored continuously, and important term set from each behaviour is extracted as shown in Figure 1 (P1). The behaviour term sets and their scores are stored in the

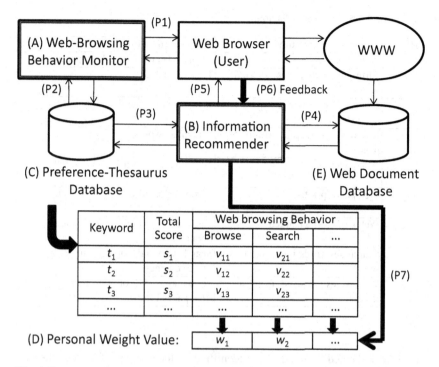

**Fig. 1** Recommender Architecture

preference-thesaurus database (P2). For example, for *Web pages browsed* as shown in Table 1, terms appeared on a Web page are regarded as important terms related to a user's preference. The score of each extracted term is the accumulated browsing time of the Web pages that contain the term. For *clipboard copy*, terms copied onto the clipboard are extracted and their scores are the frequency of copies.

### 3.1.2 Information Recommender

The candidate documents or their URLs for recommendation are stored in the Web document database. These Web documents are collected by the user through various means such as Web crawling, RSS feeds, search engine results, etc. Recommended documents are ranked by calculating the similarity between the preference-thesaurus made up of weighted behaviour term sets and each document in the Web document database (P3, P4), and presented to the user (P5).

### 3.1.3 Evaluation Feedback

In this phase, the user evaluates whether Web documents recommended are relevant or not (P6). The top-*n* Web browsing behaviours associated with the relevant

**Table 1** Typical Web browsing Behaviour

| ID | Web browsing behavior | Term set to be extracted |
|---|---|---|
| $I_1$ | Web pages browsed | Terms appeared on the Web pages |
| $I_2$ | Terms on Web pages selected by mouse-click | Terms selected |
| $I_3$ | Terms on Web pages copied onto the clipboard | Terms copied onto the clipboard |
| $I_4$ | Keywords searched within Web pages | Search keywords |
| $I_5$ | Web pages saved | Terms appeared on the saved Web pages |
| $I_6$ | Web pages printed | Terms appeared on the Web pages printed |
| $I_7$ | Web pages bookmarked | Terms appeared on the Web pages bookmarked |
| $I_8$ | Search keywords input to the Web search engines | Search keywords input to the Web search engines |
| $I_9$ | Web pages browsed from search results | Terms appeared on the returned Web pages browsed |

documents as indicated by the user are identified by the similarity between the behaviours' term sets and the relevant documents. Here, $n$ is the number of behaviours which term set's score is greater than zero, and these $n$ behaviours are deemed to be influential on user's preference. The personal weights associated with the top-$n$ behaviours are increased (P7) to reflect the most recent changes in browsing behaviour and preference.

## 3.2 Recommender Implementation

### 3.2.1 Extraction of Influential Browsing Behaviour

In evaluation feedback, a user evaluates an item on the recommended list, by navigating to the linked page or giving it a score according to her preference. The recommender then associates the specific item with certain influential browsing behaviours, as shown in Figure 2.

For example, Table 2 shows the ranking of browsing behaviours for three explicit feedbacks: EFB-1, EFB-2, and EFB-3. In the table, each $I_x$ corresponds to a Web-browsing behaviour. For instance, $I_5$, $I_6$, $I_7$, and $I_9$ refer to *save*, *print*, *bookmark*, and *browsed from search results*, respectively. In this example, the user prefers a music-related document in EFB-1 and EFB-2, and a politics-related document in EFB-3.

For EFB-1 and EFB-2, $I_9$ (*browsed from search results*), $I_6$ (*print*), and $I_7$ (*bookmark*) are identified as the most influential Web-browsing behaviours. Thus, one

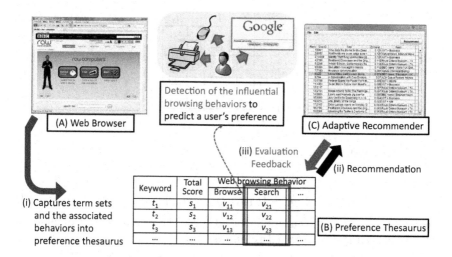

**Fig. 2** Extraction of influential browsing behaviour

**Table 2** Example of Web-browsing behaviour ranking

| Rank | EFB-1 (music) | EFB-2 (music) | EFB-3 (politics) |
|------|---------------|---------------|------------------|
| 1 | $I_9$ | $I_9$ | $I_7$ |
| 2 | $I_6$ | $I_6$ | $I_6$ |
| 3 | $I_7$ | $I_7$ | $I_5$ |
| 4 | $I_2$ | $I_5$ | $I_3$ |
| 5 | $I_1$ | $I_1$ | $I_1$ |

can assert that the specific user searches, prints, and bookmarks music related documents regularly. In addition, based on EFB-3, $I_7$ (*bookmark*), $I_6$ (*print*), and $I_5$ (*save*) are deemed to be strong influential behaviours. This user seems to prefer bookmarking, printing, and saving politics related articles. One can also deduce that in general, this user bookmarks her preferred documents.

The most influential browsing behaviours for each individual user can be extracted from the positive items selected from the recommended list via this evaluation feedback mechanism.

### 3.2.2   Personal Preference-Thesaurus Construction

Typical Web browsing behaviours and their corresponding term sets as shown in Table 1 are extracted by monitoring the user's browsing behaviour and are used to construct a personal preference-thesaurus.

Let a Web browsing behaviour be $I_x$. Let the term set be $T_x$ that includes the $m$ terms extracted from behaviour $I_x$.

|       | $I_1$      | $I_2$      | $I_3$      | $\cdots$ | $I_x$      |
|-------|------------|------------|------------|----------|------------|
| $t_1$ | $s_{11}$   | $s_{21}$   | $s_{31}$   | $\cdots$ | $s_{x1}$   |
| $t_2$ | $s_{12}$   | $s_{22}$   | $s_{32}$   | $\cdots$ | $s_{x2}$   |
| $t_3$ | $s_{13}$   | $s_{23}$   | $s_{33}$   | $\cdots$ | $s_{x3}$   |
| $t_4$ | $s_{14}$   | $s_{24}$   | $s_{34}$   | $\cdots$ | $s_{x4}$   |
| $\vdots$ | $\cdots$ | $\cdots$ | $\cdots$ | $\cdots$ | $\cdots$ |
| $t_{|T|}$ | $s_{1|T|}$ | $s_{2|T|}$ | $s_{3|T|}$ | $c$ | $s_{x|T|}$ |

**Fig. 3** Personal preference-thesaurus matrix

$$T_x = \{t_{x1}, t_{x2}, t_{x3}, \cdots, t_{xm}\} \tag{1}$$

where $t_{xi}$ $(i = 1, 2, \cdots, m)$ is a term included in $T_x$. The total term set T of all terms appeared in each $T_x$ is represented as follows:

$$T = \bigcup_{i=1}^{x} T_i \tag{2}$$

By using each term included in the term set $T$ and each behaviour $I_x$, a term-behaviour matrix is created as shown in Figure 3, which is referred to as the *personal preference-thesaurus* matrix. In Figure 3, each element $s_{ij}$ is the score of a term $t_j$ for a behaviour $I_i$.

Most of the score $s_{ij}$'s are defined as the frequency of a behaviour. For instance, the behaviour *clipboard copy*'s score $s_{ij}$ indicates how many times a user copied the term $t_j$ to the clipboard. Other scores are expressed in different units such as the browsing time in behaviour *Web page browsed*. Since the units of each behaviour may be different (frequency, time, etc.), the score of each behaviour $I_i$ is normalized in the manner of (5), (8), and (11) as described in the next sections.

### 3.2.3 Web Documents Recommendation

Typical Web browsing behaviours and their corresponding term sets as shown in Table 1 are extracted by monitoring the user's browsing behaviour and are used to construct a personal preference-thesaurus. The recommendation of Web documents is based on the similarity between the personal preference-thesaurus and each Web document.

First, a document vector space $S$ using the term set $T$ is created. Each document $d_i$ in the Web document set $D$ is represented as a vector $\mathbf{d}_i$ based on term frequencies appeared in the document as follows:

$$\mathbf{d}_i = \begin{bmatrix} e_1 \\ e_2 \\ \vdots \\ e_{|T|} \end{bmatrix} \tag{3}$$

where, $e_k$ is the term frequency of $t_k$ ($t_k \in T$) in the document $d_i$, and $|T|$ is the number of terms in the term set $T$.

Second, in order to realize personal document retrieval for the document set $D$, personal ranking is performed in the following steps.

**Step-1**: A query based on user's Web browsing behaviours is created, first by representing each behaviour $I_k$ as a vector $\mathbf{I}_k$. Each element $s_{kj}$ is that behaviour's term score in the personal preference-thesaurus matrix.

$$\mathbf{I}_k = \begin{bmatrix} s_{k1} \\ s_{k2} \\ \vdots \\ s_{k|T|} \end{bmatrix} \tag{4}$$

**Step-2**: By summing the behaviour vectors $\mathbf{I}_k$'s, a query vector $\mathbf{q}$ is created as follows:

$$\mathbf{q} = \sum_{k=1}^{x} w_k \cdot \frac{\mathbf{I}_k}{|\mathbf{I}_k|}, \quad \sum_{k=1}^{x} w_k = 1 \tag{5}$$

where, $x$ is the number of Web browsing behaviours, $w_k$ is a weighing value for each behaviour $I_k$ and is normalized with 1-norm. Since the relative importance of each behaviour cannot be pre-determined, therefore the initial values of $w_k$'s are set as follows:

$$w_k^{init} = \frac{1}{x} \tag{6}$$

**Step-3**: The similarity between the query $\mathbf{q}$ and each Web document $d$ in $D$ is calculated. Various similarity measures such as asymmetric measure, Jaccard measure and extended Jaccard measure can be used for this purpose. Here, the commonly used Cosine measure is employed:

$$sim(\mathbf{q}, \mathbf{d}) = \frac{(\mathbf{q} \cdot \mathbf{d})}{|\mathbf{q}||\mathbf{d}|} \tag{7}$$

Then, each document is ranked according to its Cosine similarity score.

### 3.2.4 Document Evaluation Feedback and Re-recommendation

In (5), it is assumed that each weight $w_k$ of behaviour $I_k$ differs for each person due to individual's Web browsing habit. Therefore, it is necessary to set $w_k$ adaptively based on the characteristic of each user's Web browsing behaviour.

When a user selects a document of her interest from the recommended rank list, the behaviour $I_k$ that strongly affect the similarity score of the selected document can be identified. The weight corresponding to the behaviour $I_k$ is then increased, and a new personal query $\mathbf{q}^{new}$ is formed. This feedback process is performed in the following steps.

**Step-1**: From the recommended rank list, a user selects a document $d_f$ ($\in D$) which she is interested in.

**Step-2**: In order to detect the influential behaviour $I_k$ that strongly affects the similarity score between each behaviour $I_k$ and $d_f$, the following Cosine similarity measure is used:

$$score(\mathbf{I}_k, \mathbf{d}_f) = \frac{(\mathbf{I}_k \cdot \mathbf{d}_f)}{|\mathbf{I}_k||\mathbf{d}_f|} \tag{8}$$

The behaviour $I_k$ is ranked according to the similarity score.

**Step-3**: The weight of a behaviour $I_k$ whose score in (8) is greater than 0 is increased:

$$w_k = w_k + \alpha_k \tag{9}$$

where, $\alpha_k$ is an incremental value of $w_k$. Let the rank of the behaviour $I_k$ be $r$. Each $\alpha_k$ is set according to its rank in Step 2 as follows:

$$\alpha_k = w_k^{init} \cdot \frac{1}{r} = \frac{1}{x \cdot r} \tag{10}$$

**Step-4**: The new personal query vector $\mathbf{q}^{new}$ is represented as follows:

$$\mathbf{q}^{new} = \sum_{i=1}^{x} w_i^{new} \cdot \frac{\mathbf{I}_i}{|\mathbf{I}_i|}, \quad \sum_{i=1}^{x} w_i^{new} = 1 \tag{11}$$

where, $w_i^{new}$ is the weight for each behaviour $I_k$, and is normalized with 1-norm after Step-3 when (9) and (10) are processed.

Using the new query $\mathbf{q}^{new}$, the ranking process as described in Section 3.2.3 is performed again. The Web documents with the top-$n$ similarity scores are recommended to the user. This feedback is an iterative process so that even if a user changes her information preference and browsing behaviour over time, appropriate recommendation can still be made with the adaptive capability of our system.

# 4  Structure, Layout, and Schema Learning

## 4.1  Profiling Layout and Design

Similar to content preference as described in the last section, our system profiles personal preference on the visual aspects (i.e., layout and design) of a Web page continuously by monitoring a user's action browsing behaviour such as printing and bookmarking. Common layout and design attributes are extracted from the Web page, analyzed, and stored in the preference-thesaurus. Attributes of interest include background and foreground font style, size, and colour, shape and colour of icons, link position and colour, video and image position, etc.

## 4.2   Formalizing Layout and Design

When a user is browsing a Web page $W$, its layout and design scheme $WS$ is profiled. We define $WS$ as a 4-tuple:

$$WS = \langle A, O, C, S \rangle \tag{12}$$

$A$ is a set of $n \times m$ square-shaped areas $A_{x,y}$ partitioning the page as shown in Figure 4. The numbers $n$ and $m$ depend on, and are adjusted to the width and height of $W$.

$$A = \{A_{1,1}, A_{2,1}, \cdots, A_{n,1}, A_{1,2}, A_{2,2}, \cdots, A_{n,2}, \cdots, A_{1,m}, A_{2,m}, \cdots A_{n,m}\} \tag{13}$$

The second $WS$ element $O$ is a set of object type $ot_p$ for each object appeared in $W$.

$$O = \{ot_1, ot_2, \cdots, ot_p\} \tag{14}$$

For example, table, list, link, image, video, icon, background-area with boundary, and so on are elements of object type $O$. HTML objects such as table $\langle TABLE \rangle \cdots \langle /TABLE \rangle$, list $\langle UL \rangle \cdots \langle /UL \rangle$, link $\langle A \rangle \cdots \langle /A \rangle$, and image $\langle IMG / \rangle$ can be extracted by parsing DOM nodes of a HTML document. Similarly, object node information can be obtained from DOM nodes of an XML document. For icon detection, a small image is recognized as an icon, if its size is smaller than a threshold value.

There are two typical ways to create background-area with boundary: (1) a simple boxed-area using CSS description and (2) a complex-shaped background-area using image files. The first type of bounded area can be detected by parsing HTML tags and their corresponding CSS descriptions. To detect complex-shaped bounded area, some image processing techniques for pattern recognition can be used.

The third element $C$ is a set of colour $c_q$ used in $W$, and the fourth element $S$ is a set of shape feature $s_r$ for each object,

$$C = \{c_1, c_2, \cdots, c_q\} \tag{15}$$

$$S = \{s_1, s_2, \cdots, s_r\} \tag{16}$$

Among many options, 8-bit colours can be used in the colour set $C$. In the shape feature set $S$, basic shapes such as box, rectangular-box, circle, solid line, and dotted line can be used.

### 4.2.1   Representing Layout and Design

In order to represent how many objects belonging to object type $ot_p$ that are located in area $A_{i,j}$, we define a matrix $M_o$,

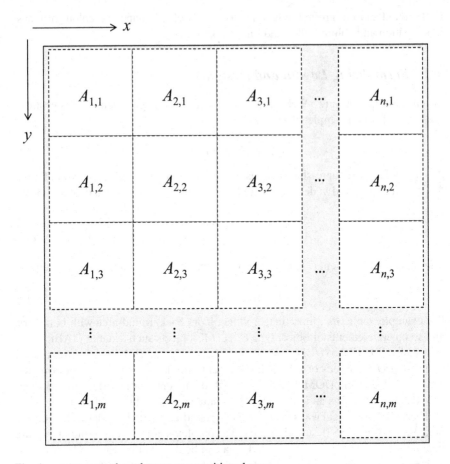

**Fig. 4** $n \times m$ square-shaped areas on a partitioned page

|  | 1 | 2 | 3 | $\cdots$ | $N = (n \times m)$ |
|---|---|---|---|---|---|
|  | $A_{1,1}$ | $A_{1,2}$ | $A_{1,3}$ | $\cdots$ | $A_{n,m}$ |
| $o_{t1}$ | $f_{11}$ | $f_{21}$ | $f_{31}$ | $\cdots$ | $f_{N1}$ |
| $o_{t2}$ | $f_{12}$ | $f_{22}$ | $f_{32}$ | $\cdots$ | $f_{N2}$ |
| $\vdots$ |  |  |  |  |  |
| $o_{tp}$ | $f_{1p}$ | $f_{2p}$ | $f_{3p}$ | $\cdots$ | $f_{Np}$ |

where, $f_{xy}$ is the frequency of $ot_p$ in $A_{i,j}$.

In addition, the colour scheme of objects belonging to $ot_p$ is represented using $c_q$. We define a matrix $M_c$ as follows:

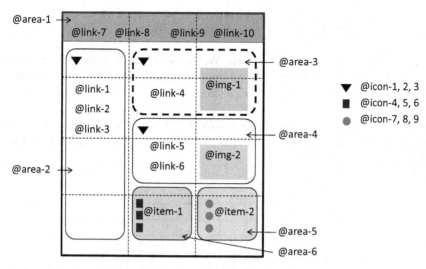

**Fig. 5** Example of the layout and design of a Web page

$$
\begin{array}{c|c|c|c|c|c}
 & c_1 & c_2 & c_3 & \cdots & c_q \\
\hline
o_{t1} & v_{11} & v_{21} & v_{31} & \cdots & v_{N1} \\
o_{t2} & v_{12} & v_{22} & v_{32} & \cdots & v_{N2} \\
\vdots & & & & & \\
o_{tp} & v_{1p} & v_{2p} & v_{3p} & \cdots & v_{Np}
\end{array}
$$

where, $v_{xy}$ is the number of pixels, or the colour histogram using RGB value, for rendering objects belonging to $ot_p$.

We define a matrix $M_s$ to represent the shape feature of objects belonging to $ot_p$ using $s_r$ as follow:

$$
\begin{array}{c|c|c|c|c|c}
 & s_1 & s_2 & s_3 & \cdots & s_r \\
\hline
o_{t1} & u_{11} & u_{21} & u_{31} & \cdots & u_{N1} \\
o_{t2} & u_{12} & u_{22} & u_{32} & \cdots & u_{N2} \\
\vdots & & & & & \\
o_{tp} & u_{1p} & u_{2p} & v_{3p} & \cdots & u_{Np}
\end{array}
$$

where $u_{xy}$ is the number of objects belonging to $ot_y$ that has a shape feature $s_x$.

### 4.2.2   A Layout and Design Example

Figure 5 shows the layout and design of a Web page. Using this example, the following steps illustrate how the three matrices $M_o$, $M_c$, and $M_s$ are generated:

**Step-1**: The Web page is partitioned into 12 square-shaped areas $A_{1,1}$, $A_{2,1}$, $A_{3,1}$, $A_{1,2}, \cdots, A_{3,4}$, proportional to the page's height and width.

**Step-2**: Objects are extracted from the Web page. In this example, five types of objects are extracted: 10 hyperlinks (@link), 2 images (@img), 2 items (@item), 6 boundary areas (@area), and 9 icons (@icon).

**Step-3**: The matrix $M_o$ represents the number of objects within each area. When an object appears in more than one area, it contributes to the frequency count of each of the overlapped areas.

$M_o$:

| | $A_{1,1}$ | $A_{2,1}$ | $A_{3,1}$ | $A_{1,2}$ | $\cdots$ | $A_{1,4}$ | $A_{2,4}$ | $A_{3,4}$ |
|---|---|---|---|---|---|---|---|---|
| @link | 2 | 2 | 2 | 0 | | 0 | 0 | 0 |
| @img | 0 | 0 | 1 | 0 | | 0 | 0 | 0 |
| @item | 0 | 0 | 0 | 0 | | 0 | 1 | 1 |
| @area | 2 | 2 | 2 | 1 | | 1 | 1 | 1 |
| @icon | 1 | 1 | 0 | 0 | | 0 | 3 | 3 |

**Step-4**: The matrix $M_c$ is generated using the number of pixels, or the colour histogram using RGB value, for rendering each object. For this example, let each object use the colours as follows (in practise, the same type of objects could have different colour attributes):

[@link] black ($=c_1$) : 70px, red ($=c_3$) : 20px, blue ($=c_4$) : 10px
[@img] black: 50px, blue: 30px, green ($=c_5$) : 200px, yellow ($=c_6$) : 150px
[@item] black : 80px, red: 80px
[@area] white ($=c_2$) : 1000px, orange ($=c_7$) : 300px, light blue ($=c_8$) : 300px
[@icon] black : 80px, red: 80px, blue: 80px

$M_c$:

| | $c_1$ | $c_2$ | $c_3$ | $c_4$ | $c_5$ | $c_6$ | $c_7$ | $c_8$ | $\cdots$ | $c_{q-1}$ | $c_q$ |
|---|---|---|---|---|---|---|---|---|---|---|---|
| @link | 70 | 0 | 20 | 10 | 0 | 0 | 0 | 0 | | 0 | 0 |
| @img | 50 | 0 | 0 | 30 | 200 | 150 | 0 | 0 | | 0 | 0 |
| @item | 80 | 0 | 80 | 0 | 0 | 0 | 0 | 0 | | 0 | 0 |
| @area | 0 | 1000 | 0 | 0 | 0 | 0 | 300 | 300 | | 0 | 0 |
| @icon | 0 | 0 | 80 | 80 | 0 | 0 | 0 | 0 | | 0 | 0 |

**Step-5**: The matrix $M_s$ is created by counting the number of shape features of each object. In this example, let each object use the shape feature as follows (for simplicity, it is assumed that the same type of objects have the same shape features):

[@img] 2 squares ($=s_1$)
[@area] 1 square and 5 rounded-squares ($=s_2$), 1 dotted line ($=s_5$)
[@icon] 3 squares, 3 circles ($=s_3$), and 3 triangles ($=s_4$)
Typically, the hyperlink objects and item objects have no shape feature.

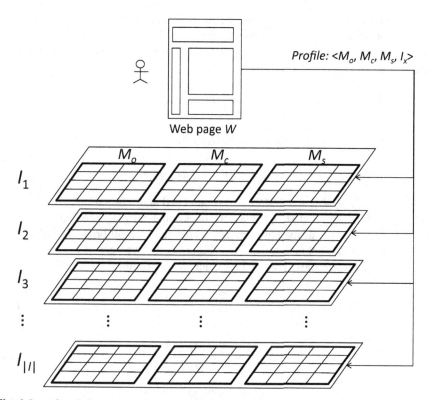

**Fig. 6** Learning design and layout associated with browsing behaviour

$M_s$:

|  | $s_1$ | $s_2$ | $s_3$ | $s_4$ | $s_5$ | $\cdots$ | $s_r$ |
|---|---|---|---|---|---|---|---|
| @link | 0 | 0 | 0 | 0 | 0 | | 0 |
| @img | 2 | 0 | 0 | 0 | 0 | | 0 |
| @item | 0 | 0 | 0 | 0 | 0 | | 0 |
| @area | 1 | 5 | 0 | 0 | 1 | | 0 |
| @icon | 3 | 0 | 3 | 3 | 0 | | 0 |

## 4.3   Layout and Design Learning

When a user exhibits a browsing behaviour $I_x$ on Web page $W$, the layout and design of Web page $W$ is profiled. The three matrices $M_o$, $M_c$, and $M_s$, representing the visual aspects associated with the browsing behaviour $I_x$, are created and stored in the preference-thesaurus.

$$profile_{layout\&design}(W, I_x) \longrightarrow \langle M_o, M_c, M_s, I_x \rangle \qquad (17)$$

When a user, continuously or in separated sessions, browses Web pages and performs a browsing action or behaviour $I_x$, the three matrices $M_o$, $M_c$, and $M_s$, associated with the visual aspects of $I_x$, are updated by summing their current state with the stored state.

$$M_o(I_x) = M_o(I_x) + M_o(I_x)^{cur} \tag{18}$$

$$M_c(I_x) = M_c(I_x) + M_c(I_x)^{cur} \tag{19}$$

$$M_s(I_x) = M_s(I_x) + M_s(I_x)^{cur} \tag{20}$$

where, $M_o(I_x)$, $M_c(I_x)$, and $M_s(I_x)$ are accumulated matrices through profiling, and $M_o(I_x)^{cur}$, $M_c(I_x)^{cur}$, and $M_s(I_x)^{cur}$ are the ones being currently profiled.

In order to represent the entire profile associated with all browsing behaviours $I_x$ ($x = 1, 2, \cdots, |I|$) defined, as shown in Figure 6, the averages of the matrices are used for matching purpose later and are calculated as follows:

$$M_o^* = \frac{\sum_{x=1}^{|I|} M_o(I_x)}{|I|} \tag{21}$$

$$M_c^* = \frac{\sum_{x=1}^{|I|} M_c(I_x)}{|I|} \tag{22}$$

$$M_s^* = \frac{\sum_{x=1}^{|I|} M_s(I_x)}{|I|} \tag{23}$$

Alternatively, the profiling can be obtained by using moving averages to smooth out irregular fluctuations and highlight consistent behaviour, while at the same time taking into account the possible changes in the visual preference of a user. Using moving averages of previous $h$ number of actions or behaviours provides the flexibility of examining behavioural trends simply by changing $h$. Also, the total number of actions profiled, may be dependent on the need of specific application; thus, it is advantageous to establish a profile using moving averages on a per-action basis.

For example, let matrices profiled at $h$ previous actions be $M_o^h(I_x)$, $M_c^h(I_x)$, and $M_s^h(I_x)$, moving averages of the current action and $h$ previous actions for $H$-action matrices are calculated as follows:

$$\overline{M}_o^H(I_x) = \frac{\sum_{h=0}^{H} M_o^h(I_x)}{H} \tag{24}$$

$$\overline{M}_c^H(I_x) = \frac{\sum_{h=0}^{H} M_c^h(I_x)}{H} \tag{25}$$

$$\overline{M}_s^H(I_x) = \frac{\sum_{h=0}^{H} M_s^h(I_x)}{H} \tag{26}$$

The accumulated matrices are then represented as:

$$\overline{M}_o^*(H) = \frac{\sum_{x=1}^{|I|} \overline{M}_o^H(I_x)}{|I|} \tag{27}$$

$$\overline{M}_c^*(H) = \frac{\sum_{x=1}^{|I|} \overline{M}_c^H(I_x)}{|I|} \tag{28}$$

$$\overline{M}_s^*(H) = \frac{\sum_{x=1}^{|I|} \overline{M}_s^H(I_x)}{|I|} \tag{29}$$

## 4.4 Layout and Design Matching

In order to recommend a Web page $W$ with user's preferable layout and design, the similarity between the current visual context $\langle M_o, M_c, M_s \rangle$ of $W$ and the profile $\langle M_o^*, M_c^*, M_s^* \rangle$ has to be calculated. We have chosen a simple Euclidean distance measure for this purpose,

$$
\begin{aligned}
&Preference(W) \\
&= Similarity(\langle M_o, M_c, M_s \rangle, \langle M_o^*, M_c^*, M_s^* \rangle) \\
&= w_1 \times MatrixSim(M_o, M_o^*) + w_2 \times MatrixSim(M_c, M_c^*) + w_3 \times MatrixSim(M_s, M_s^*)
\end{aligned}
\tag{30}
$$

where,

$$
\begin{aligned}
&MaxtrixSim(A, B) \\
&= |a_{11} - b_{11}| + |a_{12} - b_{12}| + \cdots + |a_{21} - b_{21}| + |a_{22} - b_{22}| + \cdots + |a_{mn} - b_{mn}|
\end{aligned}
\tag{31}
$$

and $w_1$, $w_2$, and $w_3$ are the weights assigned to the object, colour, and shape preferences that show a personal order of importance.

When there are multiple pages with similar contents, one can use the layout and design similarity for the purpose of recommendation. The *MaxtrixSim* values of these pages can be used to rank similar-content pages.

If moving averages are employed, the following alternative formula can be used to establish the preference:

$$
\begin{aligned}
&Preference(W) \\
&= Similarity(\langle M_o, M_c, M_s \rangle, \langle \overline{M}_o^*(H), \overline{M}_c^*(H), \overline{M}_s^*(H) \rangle) \\
&= w_1 \times MatrixSim(M_o, \overline{M}_o^*(H)) \\
&\quad + w_2 \times MatrixSim(M_c, \overline{M}_c^*(H)) \\
&\quad + w_3 \times MatrixSim(M_s, \overline{M}_s^*(H))
\end{aligned}
\tag{32}
$$

## 5   Conclusions

In this chapter, we presented a literature survey on Web browsing behaviour, a recommender system based on user browsing behaviour, and the representation and manipulation of attributes associated with the design and layout of Web pages. By modelling and capturing the visual aspects of a Web page, we believe this user preferential information is valuable to complement recommendation utilizing content similarity.

The recommender system that tracks the action aspect of browsing behaviour has been designed and implemented. Preliminary results are positive and warrant further investigation. The design presented here for the visual aspect of a Web page is currently being implemented and incorporated into the recommender.

We plan to have an extensive user study over a long period of time to ascertain the link between information preference and browsing behaviour, and to validate the premise that there is a correlation between user preference and the design and layout of Web pages. This correlation study will be carried out with user experiential interviews and empirical data reviews.

## References

1. BBC, http://www.bbc.co.uk/
2. Bilenko, M., White, R.W.: Mining the Search Trails of Surfing Crowds: Identifying Relevant Websites from User Activity. In: Proceedings of the 17th International World Wide Web Conference, pp. 51–60 (2008)
3. Brafman, R.I., Domshlak, C., Shimony, S.E.: Qualitative Decision Making in Adaptive Presentation of Structured Information. ACM Transactions on Information Systems TOIS Homepage archive 22(4) (2004)
4. Breese, J.S., Heckerman, D., Kadie, C.: Empirical Analysis of Predictive Algorithms for Collaborative Filtering. In: Proceedings of the 14th Conference on Uncertainty in Artificial Intelligence, pp. 43–52 (1998)
5. Burklen, S., Marron, P.J., Fritsch, S., Rothermel, K.: User Centric Walk: An Integrated Approach for Modeling the Browsing Behavior of Users on the Web. In: Proceedings of the 38th Annual Symposium on Simulation, pp. 149–159 (2005)
6. Cai, D., He, X., Wen, J.-R., Ma, W.-Y.: Block-level Link Analysis. In: Proceedings of the 27th Annual International ACM SIGIR Conference on Research and Development in Information Retrieval, pp. 440–447 (2004)
7. Chen, Y., Ma, W.-Y., Zhang, H.-J.: Detecting Web Page Structure for Adaptive Viewing on Small Form Factor Devices. In: Proceedings of the 12th International Conference on World Wide Web, pp. 225–233 (2003)
8. Chirita, P.-A., Firan, C.S., Nejdl, W.: Personalized Query Expansion for the Web. In: Proceedings of the 30th Annual International ACM SIGIR Conference on Research and Development in Information Retrieval, pp. 7–14 (2007)
9. Claypool, M., Brown, D., Le, P., Waseda, M.: Inferring User Interest. IEEE Internet Computing, 32–39 (November/December 2001)
10. Dumais, S., Cutrell, E., Cadiz, J.J., Jancke, G., Sarin, R., Robbins, D.C.: Stuff I've Seen: A System for Personal Information Retrieval and Re-Use. In: Proceedings of the Annual International ACM SIGIR Conference on Research and Development in Information Retrieval, pp. 72–79 (2003)

11. Dupret, G., Piwowarski, B.: A User Browsing Model to Predict Search Engine Click Data from Past Observations. In: Proceedings of the 31st annual international ACM SIGIR Conference on Research and Development in Information Retrieval, pp. 331–338 (2008)
12. Fiala, Z., Hinz, M., Houben, G.-J., Frasincar, F.: Design and Implementation of Componentbased Adaptive Web Presentations. In: Proceedings of the ACM Symposium on Applied Computing, pp. 1698–1704 (2004)
13. Francisco-Revilla, L., Crow, J.: Interpreting the Layout of Web Pages. In: Proceedings of the 20th ACM Conference on Hypertext and Hypermedia, pp. 157–166 (2009)
14. He, X., Cai, D., Wen, J.-R., Ma, W.-Y., Zhang, H.-J.: Clustering and Searching WWW Images Using Link and Page Layout Analysis. Proceedings of the ACM Transactions on Multimedia Computing, Communications, and Applications 3(2) (2007)
15. Huang, J., White, R.W.: Parallel Browsing Behavior on the Web. In: Proceedings of the 21st ACM Conference on Hypertext and Hypermedia, pp. 13–18 (2010)
16. iGoogle, http://www.google.com/ig
17. Karreman, J., Loorbach, N.: Paragraphs or Lists? The Effects of Text Structure on Web Sites. In: Proceedings of the IEEE International Professional Communication Conference IPCC, pp. 1–5 (2007)
18. Kawai, D., Kanjo, D., Tanaka, K.: My Portal Viewer for Content Fusion Based on User's Preferences. In: Proceedings of the IEEE International Conference on Multimedia and Expo., pp. 2163–2166 (2004)
19. Kelly, D., Teevan, J.: Implicit Feedback for Inferring User Preference: A Bibliography. ACM SIGIR Forem 37(2), 18–28 (2003)
20. Kumar, R., Tomkins, A.: Characterization of Online Browsing Behavior. In: Proceedings of the 19th International Conference on World Wide Web, pp. 561–570 (2010)
21. Lam, W.W.M., Chan, K.: Analyzing Web Layout Structures Using Graph Mining. In: IEEE International Conference on Granular Computing, pp. 361–366 (2008)
22. Lee, J., Choi, G., Jang, J., Nang, J.: An Effective Keyword Extraction Method for Videos in Web pages by Analyzing their Layout Structures. In: Proceedings of the IEEE Region 10 Conference TENCON, pp. 1–4 (2007)
23. Lerman, K., Getoor, L., Minton, S., Knoblock, C.: Using the Structure of Web Sites for Automatic Segmentation of Tables. In: Proceedings of the ACM SIGMOD International Conference on Management of Data, pp. 119–130 (2004)
24. Li, Y., Feng, B.-Q.: Page Interest Estimation Model Considering User Interest Drift. In: Proceedings of the 4th International Conference on Computer Science & Education, pp. 1893–1896 (2009)
25. Liang, T.-P., Lai, H.-J.: Discovering User Interests from Web Browsing Behavior: An Application to Internet News Services. In: Proceedings of the 35th Annual Hawaii International Conference on System Sciences, pp. 2718–2727 (2002)
26. Lim, S.H., Zheng, L., Jin, J., Hou, H., Fan, J., Liu, J.: Automatic Selection of Printworthy Content for Enhanced Web Page Printing Experience. In: Proceedings of the 10th ACM Symposium on Document Engineering, pp. 165–168 (2010)
27. Liu, C., White, R.W., Dumais, S.: Understanding Web Browsing Behaviors Through Weibull Analysis of Dwell Time. In: Proceeding of the 33rd International ACM SIGIR Conference on Research and Development in Information Retrieval, pp. 379–386 (2010)
28. Marcialis, I., Vita, E.D.: SEARCHY: An Agent to Personalize Search Results. In: Proceedings of the 3rd International Conference on Internet and Web Applications and Services, pp. 512–517 (2008)
29. Morita, T., Hidaka, T., Tanaka, A., Kato, Y.: System for Reminding a User of Information Obtained Through a Web Browsing Experience. In: Proceedings of the 16th International World Wide Web Conference, pp. 1327–1328 (2007)

30. Morita, M., Shinoda, Y.: Information Filtering Based on User Behavior Analysis and Best Match Text Retrieval. In: Proceeding of the 17th ACM SIGIR, pp. 272–281 (1994)
31. MyYahoo!, http://my.yahoo.com/
32. Oard, D.W., Kim, J.: Modeling Information Content Using Observable Behavior. In: Proceedings of the ASIST Annual Meeting, vol. 38, pp. 481–488 (2001)
33. Resnick, P., Iacovou, N., Sushak, M., Bergstrom, P., Reidl, J.: GroupLens: An Open Architecture for Collaborative Filtering of Netnews. In: Proceedings of the 1994 ACM Conference on Computer Supported Cooperative Work Conference, pp. 175–186 (1994)
34. Sah, M., Hall, W., De Roure, D.C.: Dynamic Linking and Personalization on Web. In: Proceedings of the ACM Symposium on Applied Computing, pp. 1404–1410 (2010)
35. Sarwar, B.M., Karypis, G., Konstan, J.A., Riedl, J.: Analysis of Recommendation Algorithms for E-Commerce. In: Proceedings of ACM Conference on Electronic Commerce, pp. 158–167 (2000)
36. Seo, Y.-W., Zhang, B.-T.: Learning User's Preferences by Analyzing Web-Browsing Behaviors. In: Proceedings of the 4th International Conference on Autonomous Agents, pp. 381–387 (2000)
37. Song, R., Liu, H., Wen, J.-R., Ma, W.-Y.: Learning Important Models for Web Page Blocks Based on Layout and Content Analysis. Proceedings of the SIGKDD Explorations Newsletter 6(2), 14–23 (2004)
38. Spalteholz, L., Li, K.F., Livingston, N.: KeySurf: A Character Controlled Browser for People with Physical Disabilities. In: Proceedings of the 17th International World Wide Web Conference, pp. 31–39 (2008)
39. Takano, K., Li, K.F.: An Adaptive Personalized Recommender Based on Web-Browsing Behaviour Learning. In: Proceedings of the 2009 IEEE International Symposium on Mining and Web, pp. 654–660 (2009)
40. Teevan, J., Dumais, S.T., Horvitz, E.: Potential for Personalization. Proceeding of the ACM Transactions on Computer-Human Interaction 17(1) (2010)
41. Tso-Sutter, K.H.L., Marinho, L.B., Schmidt-Thieme, L.: Tag-aware Recommender Systems by Fusion of Collaborative Filtering Algorithms. In: Proceedings of the 2008 ACM Symposium on Applied Computing, pp. 1995–1999 (2008)
42. Weinreich, H., Obendorf, H., Herder, E., Mayer, M.: Not Quite the Average: An Empirical Study of Web Use. Proceedings of the ACM Transactions on the Web 2(1) (2008)
43. Viermetz, M., Stolz, C., Gedov, V., Skubacz, M.: Relevance and Impact of Tabbed Browsing Behavior on Web Usage Mining. In: Proceedings of the IEEE/WIC/ACM International Conference on Web Intelligence, pp. 262–269 (2006)
44. Yu, X., Liu, F.: A Short-term User Interest Model for Personalized Recommendation. In: Proceedings of the 2nd IEEE International Conference on Information Management and Engineering, pp. 219–222 (2010)
45. Zheng, L., Cui, S., Yue, D., Zhao, X.: User Interest Modeling based on Browsing Behavior. In: Proceedings of the 3rd International Conference on Advanced Computer Theory and Engineering, pp. V5455–V5458 (2010)

## Glossary of Terms and Acronyms

CSS (Cascading Style Sheets)
DOM (Document Object Model)
HTML (HyperText Markup Language)
Q & A (Questions and Answers)
RGB (Red Green Blue Colour Model)
URL (Uniform Resource Locator)
XML (Extensible Markup Language)

# MANENT: An Infrastructure for Integrating, Structuring and Searching Digital Libraries

Angela Locoro, Daniele Grignani, and Viviana Mascardi

**Abstract.** Digital Libraries represent the commitment of research communities to preserve authoritative and well structured sources of knowledge, and to share archival organisations, methods and resources thanks to systems relying on standard metadata formats. This chapter describes some natural language processing techniques exploited for automatically extracting structural information from documents stored in Digital Libraries, based on the exposed metadata. The most prominent results achieved in this area are surveyed and discussed. As an example of an infrastructure for integrating, structuring and searching Digital Libraries based on natural language processing and semantic web techniques, we discuss the MANENT system. MANENT is a working prototype offering services of Digital Library content management and record classification and retrieval. It is hosted on a server at the Computer Science Department of Genova University and, starting from 2011, it will become publicly available. 475,000 records drawn from 138 repositories that all over the world expose OAI-PMH services have been downloaded, stored, and their automatic classification is under way.

## 1 Motivation

Scientific outcomes rely on institutional networks of researchers, leveraged by the Web in their intertwined activity that "help them in criticising and rectifying their findings and preserving the acquired knowledge by transmission to others" [35]. The community does not simply represent an aggregate of individuals based on social

Angela Locoro · Viviana Mascardi
Computer Science Department, University of Genova, Via Dodecaneso 35,
16146 Genova, Italy
e-mail: {locoro,mascardi}@disi.unige.it

Daniele Grignani
Department of Modern and Contemporary History, University of Genova, Via Balbi 6,
16126 Genova, Italy
e-mail: daniele.grignani@gmail.com

M. Biba and F. Xhafa (Eds.): Learning Structure and Schemas from Documents, SCI 375, pp. 315–341.
springerlink.com                                            © Springer-Verlag Berlin Heidelberg 2011

relations through which information flow, but it is instead a community of practice, whose coherent behaviour is defined by the commitments of all its members [44]. In this scenario the presence of even more efficient tools for storing, filtering, sharing and retrieving all the needed theoretical and analytical knowledge becomes crucial. Although the Web seems to represent the ultimate technology for transforming the process of knowledge proliferation and availability, it is more and more clear that this technology is "a source of unprecedented amounts of information. In a content-rich environment where much material is no longer evaluated by traditional gatekeepers such as editors before it has the potential to reach large audiences, the ability to find trustworthy content online is an essential skill" [22].

Organisational criteria for storing resources should reflect the specific goals of structured information. If the goal is that of building an archive, then the most relevant element is the faithful adherence of the documents to their original source, obtained by strictly relating the document to the documental base of origin. In more operative scenarios the organisational criterion is the relational one.

On the one hand, as digitisation is a time consuming and costly effort, a careful analysis of the sources complexity should drive the design of effective devices oriented to a clear separation between the standard archival layer and the relational layer, tailored for specific enquiries. Only in this way changes occurring during activities can be done and remain at the operative level without any impact on the consolidated structure of the documental base.

On the other hand managing fragmented information turns out to cause an irreducible selection of some properties and the loss of other ones, which is typical of a process where the user selects pieces of information and re-contextualises them by creating, de facto, a new source of information as a result of researches [24]. A logical separation between the documental base and the data manipulated by the user can be obtained by setting up a work environment where the researcher may customise and create new metadata structures by following specific research projects criteria. The outcome of this process will generate a new source binded to the documental base on which it depends.

Moreover in the digitisation era every information object is available and accessible worldwide. The dynamic nature of a networked environment where such artifacts are created or their virtual surrogates are placed, outbursts the importance of how and to which extent the information and its context should be delivered to the final user. The role of metadata attached to any information source is then twofold: their schema represents both the high-level document structure and its semantic references to their contextual structure. The first role models the document itself, the second role encompasses the original conditions in which it was created and released as information source. These conditions are characterised by the piece of world knowledge strictly related to the document itself, that is a kind of information that surrounds the document content at one step of inference from all the other objects that it implicitly or explicitly refers to (i.e. authors as people, which are part of a community, with their authoritative power and reputation as well as references that link the document to other documents, and so on).

This chapter surveys some methods, tools, and results relevant for the area of Digital Library integration, structuring, and searching. The approaches we are mainly interested in, are those based on semantic web and natural language processing techniques. Indeed, these are the founding techniques upon which MANENT roots. MANENT performs harvesting of metadata exposed in standard format from real world digital libraries and automatically classifies documents by topic, according to the WordNet Domain structure. This structure has been reproduced into a "WordNet Domain Ontology" that allows MANENT to easily represent and exploit semantic relations among domains. MANENT allows the user to search documents by expressing queries in natural language. It supports a "topic-based" search of documents relevant for the user query, based on the WordNet Domain Ontology. A more sophisticated "text-based" search, usually used for refining the topic-based one, exploits text semantic similarity between the user query and text that appears in documents metadata.

The chapter is organised as follows: Section 2 overviews standards, techniques and tools upon which MANENT roots. Section 3 describes MANENT and Section 4 reports the experiments we conducted with it and the results we obtained. Section 5 analyses the related work. Section 6 concludes and outlines the future developments of our research.

## 2 Background

MANENT and the infrastructures for Digital Library integration and structuring we will describe in this chapter rely on semantic web and natural language processing approaches. In this section we briefly recall the most recent standards, tools, and techniques relevant for designing and implementing such infrastructures. We assume a basic knowledge of XML [48], RDF [47], RDFS [46], and of WordNet [33].

### 2.1 Standards for Digital Library Access and Description

The Open Archives Initiative Protocol for Metadata Harvesting, OAI-PMH [34], provides an application-independent interoperability framework based on metadata harvesting. A OAI-PMH framework involves *data providers*, who administer systems that support the OAI-PMH as a means of exposing metadata, and *service providers* who use metadata harvested via the OAI-PMH as a basis for building value-added services.

OAI-PMH is hence based on a client-server architecture, in which "harvesters" request information on updated records from repositories. Requests for data can be based on a datestamp range, and can be restricted to named sets defined by the provider. Data providers are required to provide XML metadata according to the Open Archives Initiative Protocol for Metadata Harvesting [36] and the Guidelines for Repository Implementers [37].

## 2.2  Ontologies and Related Languages and Tools

According to T. Gruber's definition *"in the context of computer and information sciences, an ontology defines a set of representational primitives with which to model a domain of knowledge or discourse. The representational primitives are typically classes (or sets), attributes (or properties), and relationships (or relations among class members). The definitions of the representational primitives include information about their meaning and constraints on their logically consistent application"* [21].

One of the most widespread languages for describing ontologies is OWL [45], a semantic markup language that extends the vocabulary of RDF.

Protégé is a widely adopted open source ontology editor and knowledge acquisition system developed by Stanford Center for Biomedical Informatics Research at the Stanford University School of Medicine [38].

## 2.3  WordNet Domains and the WordNet Domains Ontology

WordNet Domains[1] [27, 7] is a project developed by the Fondazione Bruno Kessler (FBK), Trento, with the purpose of providing WordNet synsets with a syntagmatic layer beside the existing paradigmatic layer (represented by semantic relations such as for example hyperonymy and meronymy). The project has the scope of better characterising a word meaning within its use in the language and in texts, hence reducing ambiguity.

The theory underlying WordNet Domains is that of Semantic Domains [28], stating that words in the lexicon can be grouped together into sets of strongly associated concepts, determined by the lexical coherence property. Based on such property some set of words tend to highly co-occur in texts, which means that an underlying semantic layer of associations among them exists and thus can be modelled accordingly.

The operation of tagging WordNet synsets with domain labels has been conducted by the FBK team partially by hand (by annotating a subset of synsets) and by automatically extending such labels to related synsets through the WordNet hierarchy, fixing the automatic procedure with corrections where necessary. The domain labels are those of the Dewey Decimal Classification system[2], a standard largely

---

[1] http://wndomains.fbk.eu/index.html.
[2] http://www.oclc.org/dewey/versions/default.htm.

adopted by library systems. The task of labelling synsets has been conducted on WordNet version 2.0. For the more recent versions of WordNet, mappings files from the oldest version to the newest exist and are freely available.

Table 1 shows an excerpt, for the first 32 top-ranked domain labels, of the number of WordNet 3.0 synsets which have been tagged by domain labels. Each synset can be labelled with more than one label and the whole number of synsets upon which this statistics was conducted is 115.248.

**Table 1** WordNet Domains labels and number of synsets tagged with each label. The *facto-tum* label refers to words whose use in the language is not related to any specific domain of discourse.

| Domain | ♯ Syn | Domain | ♯ Syn |
|---|---|---|---|
| biology | 23814 | law | 1887 |
| plants | 17849 | music | 1857 |
| factotum | 16099 | linguistics | 1760 |
| animals | 12265 | metrology | 1673 |
| chemistry | 6495 | administration | 1462 |
| geography | 5169 | physics | 1342 |
| medicine | 4223 | pharmacy | 1339 |
| person | 3790 | psychological_features | 1327 |
| anatomy | 3340 | transport | 1283 |
| religion | 2249 | geology | 1269 |
| gastronomy | 2215 | food | 1197 |
| buildings | 2107 | fashion | 1194 |
| history | 2044 | economy | 1157 |
| military | 2033 | entomology | 1075 |
| politics | 2032 | physiology | 1065 |
| literature | 1986 | industry | 1008 |

As part of our recent research on automatically discovering the WordNet domain of ontology entities and of entire ontologies, we took the WordNet Domains taxonomy[3] and codified it in OWL using Protégé. We called this new ontology WordNet-Domains.owl. It consists of 160 domain labels divided as follows:

- 11 first level classes which represent the upper layer of domains classification. They are: *applied_science, doctrines, factotum, free_time, metrology, number, person, pure _science, quality, social_science, time_period*;
- 42 mid level classes that are used to tag synsets representing concepts used at an intermediate level of generality (e.g. *medicine, economy, sport,* and so on);
- 107 low level classes that are subclasses of one of the 42 mid level concepts or belong to a further level of specialisation and are also used to tag synsets, which are relative to concepts used in more specialised domains (e.g. *psychiatry, banking, athletics* and so on).

---

[3] Available at the official page of the project:
http://wndomains.fbk.eu/hierarchy.html. Last accessed on 30 June 2010.

In [26] we describe how we used the WordNet Domains ontology to successfully tag ontology concepts with WordNet Domains in an automatic fashion. We tagged each concept from one ontology with the label assigned to the WordNet synsets whose lemmas correspond to the concept itself, expanded the domain label tags assigned to each concept with the top domain labels contained in the WordNet Domains ontology through an inference procedure based on its hierarchy, and computed frequencies on the whole set of tags to determine the most frequent domains at ontology level as well as at concept level. Following this approach, we were able to successfully assign the correct domain among the first-level ones to each of the 17 real ontologies we used as testbed, and the correct domain among the mid level classes to 15 of them.

In MANENT, we exploit our WordNet Domains ontology in a very similar way, and with the same satisfactory results (see Section 4).


## 2.4 Text Semantic Similarity

The problem of determining the semantic similarity of sentences has been widely discussed in the literature starting from the late sixties, when two pioneer works [41, 39] were published on concrete applications of text similarity measures.

From then on, significant research results were achieved on this topic. Many recent papers compare similarity measures according to different viewpoints [32, 4, 31].

In MANENT we are experimenting the WordNet-based semantic similarity measurement application by T. Simpson and T. Dao[4], licensed under The Code Project Open License (CPOL).

Given two words $s$ and $t$, Simpson and Dao's algorithm computes *SenseWeight*, a weight calculated according to the frequency of use of the senses assigned to $s$ and $t$ respectively by exploiting the adapted Lesk algorithm [5] and the total of frequency of use of all senses, and *PathLength*, the length of the connection path from $s$ to $t$.

The similarity of $s$ and $t$ is computed as

$$sim(s, t) = SenseWeight(s)*SenseWeight(t)/PathLength$$

and the similarity of two sentences $X$ and $Y$ is computed based on the similarities of $< s,t >$ such as $s \in X$ and $t \in Y$.

Other sentence semantic similarity measures will be tried in the future: MANENT functionality of refining queries by exploiting text similarity measures is still in an early stage. For demonstrating the feasibility of our approach, however, Simpson and Dao's application was powerful enough, and easy to use.

---

[4] http://www.codeproject.com/KB/string/
semanticsimilaritywordnet.aspx.

# 3   MANENT

In this Section we introduce the MANENT architectural and functional features. In order to keep the chapter readable for both specialists and non-specialists, we avoid the technical details and keep our discussion at a high level of abstraction.

The architectural layer we propose in MANENT aims at fostering research communities of practice and encompasses a digital library infrastructure and an OAI-PHM harvesting service that conveys information exchange and retrieval as well as automatic classification of metadata contents by topic. The key characteristic of MANENT is that of modelling, describing, storing and retrieving information objects by means of an OWL ontology derived from EAD (Encoded Archival Description [13]) as a formal and standard conceptualisation of archival rationales maintained by the Library of Congress in partnership with the Society of American Archivists.

Moreover, automatic classification and search services are also provided in MANENT. The WordNet Domains Ontology is exploited for automatically classifying heterogeneous documents owned by hundreds of Digital Libraries, based on the schema mining of the metadata associated with them. Once the metadata harvester retrieves the metadata describing information objects, a service for automatic domain detection of such contents is run in order to provide a classification of both local and remote information based on their topic. A mechanism for extracting the most relevant keywords from metadata annotations (either local or remote) and for tagging them with domain labels has been developed and will provide new structured data for classifying contents, extending existing schemas and ontologies or simply indexing and searching resources based on them.

## 3.1   The MANENT Architecture

The MANENT architecture includes the following components:

- **User Interface Portal** for archive visualisation, browsing, searching, content management and administration tasks.
- **Digital Library Content Manager**, the core content management component in charge of the basic functionalities such as managing contents and requests for visualisation and browsing, managing and updating collections, adding new documents, and so on.
- **Collection Template Composer** for designing and structuring EAD compliant archives.
- **Collections Object Manager**, consisting in all the operations needed to access the knowledge base. Every operation is essentially a SPARQL [42] wrapper.
- **Knowledge Base Component**, the repository with all its Collections and CollectionSets data.

- **Metadata Integration Service**, offering services for integrating different metadata format. This component contains the metadata mapper interface for easy configuration of metadata matching tasks.
- **Metadata Harvester Service**, working as a web service interface for OAI-PMH metadata harvesting.
- **Metadata Classification Service**, providing automatic classification by topics of the repositories metadata records harvested; the service is treated in a dedicated Section (namely Section 3.2) due to its higher relevance with respect to the scope of the chapter.

Besides exploiting the WordNet Domains ontology for offering classification services, MANENT relies on two OWL ontologies that model the hierarchical structure of MANENT archives ("Archives Structure Ontology"), and the basic elements for translating metadata formats into OWL ("Metadata Structure Ontology") respectively. We may consider these two ontologies as the MANENT "meta-model".

### The "Archives Structure Ontology"

Inside this ontology the following classes are defined:

- *Collections*: represent a set of documents ordered and preserved together; collections are created under CollectionSets; collections may be the result of a single process or of a specific activity and are described through a metadata structure derived from EAD and designed by means of the Collection Template Composer. At the time being MANENT digital objects are maintained in the filesystem while the knowledge base keeps a reference to their path through the dao[5] tag. The solution is temporary and the creation of a Digital Object Management System based on OWL is foreseen.
- *CollectionSets*: high level containers for aggregating related or linked Collections. Each CollectionSet may contain either Collections or CollectionSets that, in this case, are called CollectionSubSets. The two elements represent a set of collections created or collected by a single user or institution during their activities. As for a Collection each CollectionSet is also described through a metadata structure derived from EAD, with the difference that, in this case, their structure is fixed a-priori with a set of tags that is not changeable.

### The "Metadata Structure Ontology"

Inside the "Metadata Structure Ontology" the class Element has been defined to represent the set of all XSD elements of the EAD XML Schema [14]. Each of them is translated into an individual of that class. The result is the EAD ontology that is stored in the knowledge base. This procedure is extended to XML Schemas of any

---

[5] Encoded Archival Description Tag Library, Version 2002, EAD Elements, <dao> Digital Archival Object, http://www.loc.gov/ead/tglib/elements/dao.html.

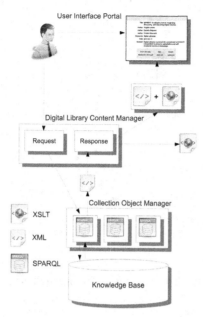

**Fig. 1** A Collection browsing and visualisation workflow in MANENT

other metadata formats (i.e. in this way the MARCXML [29] XML Schema may become MARCXML ontology).

The device responsible for the translation, namely XSD2OWL, is able to parse all the XSD elements and convert them in individuals of the class Element following the "Metadata Structure Ontology". For each XSD element, attributes and relationships are detected and created. At the end of the process the generated ontology is stored in the knowledge base.

## Browsing Contents in MANENT

Figure 1 depicts a typical MANENT browsing operation. The Digital Library Content Manager receives a request from the user, keeps in memory its type and invokes the proper methods of the Collections Object Manager while waiting for results. When data are ready they are organised based on the resource type description (CollectionSet or Collection) and visualised.

## Collection Management

CollectionSets and Collections are created with a request for the creation of a new element sent to the Digital Library Content Manager. The component queries the Collections Object Manager to obtain the proper schema and sends it to the interface

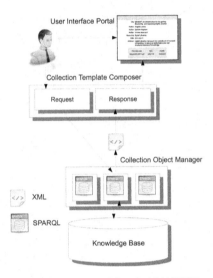

**Fig. 2** An example of Collection creation workflow in MANENT

for user data insertion. Once completed, the two components communicate again in order to store the new data in the knowledge base and visualise it to the user.

The Collection Template Composer is dedicated to the creation of a Collection and may be activated only inside the CollectionSet. It consists of an interface whose purpose is to let the user customise the Collection structure by selecting different combinations of the tags set based on the EAD ontology; as the Collections Template Composer is constrained by the EAD ontology, the insertion of different tags from those expected is not allowed.

The creation procedure for a Collection is depicted in Figure 2 where the user selects a CollectionSet, sends a request for creating a new collection, creates it, and the results are stored in the knowledge base and visualised.

**The Metadata Integration Service**

As already discussed, MANENT represents metadata in an internal format derived from EAD and formalised in the EAD ontology. A matching service is responsible for the mapping of different metadata formats with the EAD ontology provided that they are previously translated in OWL ontologies by means of the XSD2OWL procedure. In order to integrate information expressed in a format $F$ different from EAD, a manual conversion from $F$'s metadata format and the EAD ontology is required. An easy-to-use interface drives the user in the definition of "$F$ to EAD" mappings. Of course, this conversion is needed only once: when mappings from $F$ elements to the EAD ontology have been defined, they are stored into the MANENT knowledge base and will become part of the available representation formats of MANENT collections.

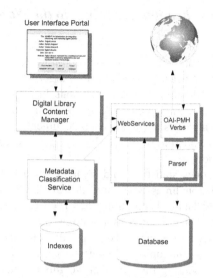

**Fig. 3** The MANENT Metadata Harvester and Metadata Classification Services

## The Metadata Harvester Service

This service is responsible for metadata harvesting and relies on OAI-PMH proto-col release 2.0. The Web Services implemented are integrated with the MANENT interface that accesses data retrieved by means of the harvester itself. Figure 3 ex-emplifies the service.

The harvesting mechanism is built upon URLs belonging to data providers[6], which are under the administration of the Open Archives Initiative[7]. These are pe-riodically read and stored locally.

The service is modelled according to Open Archives Initiative guidelines [37], while the download procedures are those envisioned by the OAI-PMH standard[8]. The service reads one after the other the URLs listed in the data providers page. Two kinds of harvesting are supported:

— Complete, where the repositories description, the list of metadata formats, the repositories structure and, for each metadata format, the list of records are down-loaded.
— Incremental, where only those records added in the period between the last down-load and the current time instant, are downloaded.

---

[6] "Data Providers [are] systems that support the OAI-PMH as a means of exposing meta-data"; The Open Archives Initiative Protocol for Metadata Harvesting, Document Version 2008-12-07T20:42:00Z,
    http://www.openarchives.org/OAI/openarchivesprotocol.html.
[7] http://www.openarchives.org/Register/ListFriends.
[8] http://www.openarchives.org/OAI/openarchivesprotocol.html.

## 3.2 Metadata Classification According to the WordNet Domains Ontology

On the documents' metadata retrieved and stored following the approach described in the previous section, we run our classification algorithm that automatically assigns a set of WordNet domains to each document, based on its metadata. Since concepts in the WordNet Domain Ontology (see Section 2) correspond to WordNet Domains, this activity amounts to linking each document to one or more concepts in the WordNet Domains ontology, hence classifying it according to the WordNet Domains ontology structure.

Our approach integrates statistical and semantic natural language processing techniques. Stemming from an assignment of domain labels to each relevant noun found in document metadata and computing their occurrencies, we extend such domain labels with super-domains ones by reasoning on the WordNet Domains hierarchy. We then propagate frequencies of super-domain labels by summing sub-domains frequencies up to the topmost domain nodes. The list of the most relevant domains is output by a scoring function that weights domain labels according to the metadata type they appear in (i.e., we weight the *dc_subject* field in the Dublin Core [12] metadata schema more than the *dc_description* one, since our goal is to classify documents by topic and we expect that *dc_subject* represents the document topic better than other fields).

A detailed description of the main steps of our procedure is introduced in the following paragraphs, where we depict the pre-processing steps conducted on metadata contents, the tagging stage, and the score computation. To better exemplify each passage we apply the automatic classification methodology to the record shown in Figure 4, described following the Dublin Core metadata schema.

```
dc:title A weather forecast study
dc:subject global warming
dc:description This paper deals with a study on weather
based on forecasts of the last 30 years where the weather
has changed due to climate conditions.
```

**Fig. 4** Example of document metadata

For each record we execute the following steps.

**Tokenisation and POS tagging.** By exploiting the GATE [9] Natural Language Processing tool we tokenise the text in the record, tag contents with a POS (part-of-speech) tagger and retain only noun words.

**Lemmatisation.** We lemmatise each noun in order to gain the canonical word forms from derivationally formed ones. To acquire the lemma from one noun we use the WordNet dictionary.

---

[9] http://gate.ac.uk.

**Lemmas occurrencies count.** We count the occurrencies of each lemma in each metadata content slot. The formula we apply is

$$\forall lemma_i \in metadata_j = F_{metadata_j}(lemma_i) = \mid lemma_i \mid$$

and results in the total number of times the lemma occurs into a metadata slot.

**Filtering.** In case the lemmas occurrencies count results in a huge amount of nouns with low occurrence, which do not characterise the domain of discourse (which may happen with the *dc:description* field in particular), we filter out the least relevant words on this metadata field, by normalising the occurrences of each word and retaining only those words whose frequencies sum amounts to 50% of the total frequencies counts. The frequency computation is as follows:

$$p(w_i) = \frac{freq(w_i)}{\sum_{i=1}^{W} freq(w_i)}$$

where $w$ is a word lemma, $freq$ is the number of occurrencies of that lemma, $W$ is the total number of occurrencies of the whole set of words in the description field. The sum of all $p(w_i)$ amounts to 1.

**Assignment of the right WordNet domain to each lemma.** We look into the "WordNet synsets - WordNet Domains" mapping files (WnToWnD)[10] and assign to each word lemma its domain labels as they result from such file. The result of the mapping operation on our example record is shown in Figure 5.

```
dc:title weather [meteorology] forecast [meteorology] study
[school]
dc:subject warming [meteorology]
dc:description paper [factotum] study [school] weather
[meteorology] forecast [meteorology] year [time_period]
weather [meteorology] climate [meteorology] condition
[factotum]
```

**Fig. 5** Document metadata after text processing: domain labels are in squared brackets

**Extension of WordNet domains to super-domains following the WordNet Domains hierarchy.** Besides the "direct tagging" with the WordNet domain associated with the given lemma (if any), we also tag lemmas with the super-domain labels of the domain labels just assigned. By looking at the WordNet Domains ontology and, hence, at the whole domain hierarchy space, we add all the superclasses, from the

---

[10] The WordNet Domain version we used is 3.1. For conversion from WordNet 2.0 to Word-Net 3.0 synsets we use mappings files available at
http://www.lsi.upc.es/~nlp/tools/mapping.html.

direct superclass up to the root domain label. What we obtain in the end is a set of domain labels associated with each lemma, that we represent as

$$Dom(l_i) = \{d \mid d \in D\}$$

where $l$ stands for the lemma and $d$ stands for a domain label belonging to the WordNet Domains set of labels $D$.

Lemmas that were tagged with no domain label are eliminated.

**Score computation.** We associate the number of occurrencies of each word with the domain labels by means of the following formula

$$\forall d \in Dom(l_i), Dom(l_i) \in metadata_j = F_{metadata_j}(Dom(l_i)) = \mid l_i \mid$$

and we apply a weight to this number, according to the type of metadata field to which the domain labels belong to. In this case we weighted domain labels in *dc:subject* 1, domain labels in *dc:title* 0.5, and domain labels in *dc:description* 0.25. For each domain label we then compute the following score function:

$$s(d_i) = \sum_{j=1}^{3} F_{metadata_j}(d_i) * w_j =$$

$$F_{dc:subject}(d_i) * 1 + F_{dc:title}(d_i) * 0.5 + F_{dc:description}(d_i) * 0.25$$

**Ranking of domain labels.** We rank the domain labels according to the resulting score. Each relevant root domain together with its sub-domains, if present, will appear in the ranking. An example of the final results of our procedure on our sample document is depicted in Figure 6.

```
[pure_science, earth, meteorology] (3)
[social_science, pedagogy, school] (0.75)
[factotum] (0.5)
[time_period] (0.25)
```

**Fig. 6** Document domain classification according to WordNet Domains hierarchy after score computation and ranking

The *Factotum* domain is filtered out. If we want to visit the sub-domains and take the best of them with highest score we can see if the most specific domain has been ranked.

**Querying and Matching Metadata Contents to Discover Semantic Similarities**

Starting from a natural language query expressed by a user, we extract the Word-Net domains associated with them by using the approach discussed above, and we exploit them to retrieve those records whose WordNet domains better match those in the user's query. The matching criterion we adopt is to look for records whose WordNet Domains have the highest overlapping with the user's ones. For refining the query, we use relevant keywords extracted from the metadata and match them with the keywords extracted from the user's query. Moreover, we may use semantic text similarity to further refine queries, when a set of records annotated with the WordNet domains and keywords matching those extracted from the user's query have been found and more fine-grained search must be performed.

## 4  Experiments and Results

We conducted our experiments on a dataset composed by 10 repositories, chosen among 138 randomly picked up from the 1.342 all over the world repositories that expose OAI-PMH services. We downloaded only records within a temporal range (January 2008 to October 2010), able to capture the more recent entries of each harvested provider, for a total of 475.000 records; to begin with, we selected 10 repositories, for a total of 1000 records automatically classified, showing different features in terms of both their content and their structure, and we asked to domain experts in the records' disciplines to manually verify the correctness and completeness of the automatic classification over 100 of them (10 for each repository).

The harvesting procedure is the one depicted in the "complete harvesting" paragraph except for minor details that do not change its behaviour in a substantial way. Since the only format always available for all the repositories is the Dublin Core (*dc* from now on), we harvested records in this format.

The selection of the 10 repositories out of 138 as well as of the 100 records belonging to them to be manually inspected has been based on different qualitative aspects. A discriminating factor was the completeness of the metadata available for each record (i.e., most records of the 10 repositories have at least one *dt:title*, one *dc:subject* and one *dc:description* field). Another relevant factor was the domain of the repository from which the 100 manually evaluated records were drawn from: we considered mono-thematic repositories as well as miscellaneous ones. Notwithstanding an ideal choice would have been to capture at least one repository for each of the WordNet top domains, this choice turned out to penalise completeness of the metadata subset, as most of the 138 repositories lack some of the metadata we choose to exploit for our analysis.

A good compromise was then to select as many metadata complete repositories as possible. For our experiments we performed a further selection on the dataset language, by automatically detecting only metadata records written in English. To operate such selection we used the Java Text Categorisation Library (JTCL)[11] [11], a tool for guessing the language of sentences. A description of the 10 repositories is outlined in Table 2. For each of them we report the institution that owns the repository and the total number of records downloaded. In the rest of this section we will identify such repositories by their number, from r1 to r10. In the sequel, we refer to the 100 records that were manually inspected as "the benchmark".

---

[11] http://textcat.sourceforge.net.

**Table 2** The 10 repositories from which we selected our benchmark

| Institution owning the repository | ♯ rec |
|---|---|
| r1. Centro de information y gestion tecnologica Matanza, Cuba | 154 |
| r2. University of Stirling, Scotland | 2.204 |
| r3. Queensland Dept. of Primary Industries and Fisheries, Australia | 1.502 |
| r4. Nara University of Education Academic Repository, Japan | 1.962 |
| r5. University of Bayreuth, Germany | 250 |
| r6. University of Hohenheim, Germany | 304 |
| r7. University of Fukui, Japan | 1.639 |
| r8. University of Regensburg, Germany | 386 |
| r9. Stellenbosch University, South Africa | 3.393 |
| r10. Indiana University, USA | 831 |

As *dc* standard allows the multiple cardinality of each metadata (i.e. each document may be described by more than one *dc:subject* field and so on), our algorithm processes them in order to obtain a unique entry for each metadata type by concatenating all contents of a *dc* field as they appear in the document metadata record description.

Once the benchmark was set up through such preliminary normalisation steps, we proceeded by applying the classification procedure depicted in the previous section to each record in the benchmark. For each record we obtain a list of relevant keywords as well as the prevailing domain labels arranged hierarchically.

The topic detection results are reported in Table 3 where, for each repository and each domain label whose ranking score was among the first three ones, the number of documents automatically classified according to such domain is shown.

In particular the ranking mechanism adopted for each record works as follows:

— for the top level WordNet Domains (namely, the 11 first level classes that include applied science, doctrines, factotum, free time, metrology, etc; see Section 2) ranked according to the first two higher scores, search their direct sub-domains (the 42 mid level classes that include medicine, economy, sport, and so on) at the same ranking level down to the third ranking score;
— do the same for their leaf domains (classes at the lowest level: psychiatry, banking, athletics, ...), if they exist and if the scoring function has ranked them among the first three ranking scores.

For the final classification shown in Table 3 we used the entire set of domain labels ranked at record level as explained above and we aggregated each domain label by summing up the number of records tagged with them. For conciseness, we only show the first two top domains with higher rank and, for each of them, the first three higher sub-domains.

To evaluate the correctness and completeness of our classification algorithm we asked domain experts to manually check our benchmark.

**Table 3** Final classification of the benchmark, an excerpt

| domain label | r1 | r2 | r3 | r4 | r5 | r6 | r7 | r8 | r9 | r10 |
|---|---|---|---|---|---|---|---|---|---|---|
| social_science | 66 | 70 | 35 | 74 | 32 | 49 | 44 | 63 | 83 | 46 |
| commerce | | | | | 5 | 11 | | | | |
| economy | 28 | 18 | 16 | | 7 | 27 | | 12 | | 11 |
| industry | 13 | | | | | | 10 | | | |
| law | 29 | | | | | | | | | |
| pedagogy | | 20 | | 51 | 2 | | 11 | 16 | 11 | 11 |
| politics | | 24 | 12 | | | 7 | 9 | | 65 | 11 |
| publishing | | | | | | | | | 11 | |
| religion | | | | | | | | 9 | | |
| sociology | | | | 29 | | | | | | |
| telecommunication | | | 6 | 14 | | | | | | |
| applied_science | 43 | | | | | | | | | |
| architecture | 29 | | | | | | | | | |
| computer_science | 12 | | | | | | | | | |
| engineering | 12 | | | | | | | | | |
| pure_science | | 46 | 86 | 39 | 79 | 63 | 81 | 51 | 76 | 83 |
| biology | | 24 | 63 | 19 | 24 | 29 | 35 | 16 | 10 | 37 |
| chemistry | | 11 | 25 | 11 | 40 | 23 | 41 | 12 | 7 | 46 |
| earth | | 14 | 27 | 14 | | 17 | | 16 | 69 | |
| physics | | | | | 26 | | 16 | | | 21 |

The evaluation results are summarised in Table 4 where, for each group of records, we report the presence of the relevant *dc* metadata with their number, the number of correctly detected top level domains (*top lev.* column), sub-domains (*sub-dom.* column) and, if present and properly assigned, the number of leaf domains (*leaf dom.* column). The total number of domain labels assigned (♯ *dom. lab.* column) together with the average number of labels assigned to each record (*avg lab.* column), that we recall are resulting from the ranking mechanism selection above explained, are also shown. A sum of each column value is reported in the *Tot* row and the percentage value (see row *% on Tot*) has been obtained by dividing the total of correctly assigned domain labels for the number of records correctly tagged, while the average number of total labels assigned to each record is reported in the *avg* row and has been obtained by dividing the total number of labels assigned to the benchmark (518) by 10. Moreover, an indication of relevant top level domains and direct sub-domains that are missing, meaning that our procedure was not able to guess them, is reported in the discussion below together with an overview of the results.

The overall results of the evaluation show that about 89% of the records were tagged with a correct top domain label whereas 72% have being assigned a correct sub-domain label, which are very encouraging results. At the leaf level, only 20% of the leaf domain labels were correctly set. Following the preliminary results of [26] we confirm our claim: the top level domains and their direct sub-domains are

**Table 4** Manual evaluation over the benchmark carried out by domain experts

| rep. | title | subj. | descr. | top lev. | sub-dom. | leaf dom. | ♯ dom. lab. | avg lab. |
|------|-------|-------|--------|----------|----------|-----------|-------------|----------|
| r1 | 8 | 4 | 10 | 8 | 5 | 0 | 61 | 6.1 |
| r2 | 5 | 14 | 10 | 8 | 5 | 0 | 56 | 5.6 |
| r3 | 7 | 9 | 10 | 10 | 8 | 0 | 42 | 4.2 |
| r4 | 6 | 10 | 8 | 9 | 9 | 4 | 66 | 6.6 |
| r5 | 7 | 6 | 10 | 9 | 9 | 1 | 42 | 4.2 |
| r6 | 10 | 4 | 10 | 9 | 6 | 2 | 40 | 4 |
| r7 | 6 | 12 | 10 | 7 | 5 | 1 | 55 | 5.5 |
| r8 | 7 | 28 | 12 | 10 | 8 | 2 | 45 | 4.5 |
| r9 | 7 | 16 | 10 | 9 | 5 | 0 | 48 | 4.8 |
| r10 | 7 | 8 | 10 | 9 | 9 | 0 | 63 | 6.3 |
| Tot | 70 | 111 | 100 | 87 | 69 | 10 | 518 | |
| % on Tot | | | | 89% | 72% | 20% | | |
| avg | | | | | | | | 5.18 |

those able to better classify documents, while the leaf domains are still difficult to be reached and correctly detected. In support of this evidence we may observe that 50% of the records examined do not even have a leaf domain in the first three ranking scores and only 10% of the records have been tagged with a sound leaf domain.

For the limited significance of leaf domain labels we exclude them from the following discussion, that provides a qualitative evaluation of the results obtained in each individual repository.

The main lesson that we learned from this evaluation, is that records about medicine, genetics and biology are among the most difficult ones to correctly classify in an automatic way. This is quite surprising, since, instead, they are among the easiest ones to classify for a human being, even without specific skills in the field. Computer science is again almost hard to recognize, probably because of the use of very technical terms.

These results, once confirmed by experiments on larger benchmarks, might provide the basis for suggesting an extension to the WordNet Domains classification in order to keep track of specific terms that strongly characterise a domain, but that are not considered yet.

- In Repository 1, 5 records out of the 10 in the benchmark deal with computer_science but were not correctly tagged with this sub-domain. On the other hand, the 5 remaining records were correctly tagged with top and sub-domain applied_science → computer_science or engineering, which result both correct, and the only one about sociology was correctly labelled with the social_science top domain even if the automatically detected sub-domain was not sociology, but economy and commerce.
- In Repository 2, 9 records out of 10 were assigned the correct top level domain, while only 50% of the records were tagged with a correct sub-domain. The

only record whose top domain classification failed was about applied_science → computer_science. Other relevant missing sub-domains were sociology (for 2 records), biology (1 record) and again computer_science (1 record) despite their top level domains were correctly found.

- For Repository 3, all the top level domains were exactly detected and 8 records out of 10 were tagged with their correct sub-domain. The two records with wrong sub-domains dealt with agricultural systems from the point of view of soil fertilisation and of agricultural models. The correct tagging should have been applied_science → agriculture.
- In Repository 4, 9 records out of 10 were tagged with correct top domains and sub-domains. The only misclassified record should have been tagged with pure_science → mathematics.
- Repository 5 shows a situation similar to Repository 4, with 9 out of 10 records correctly classified at top level as well as sub-domains level. The misclassified record was about medicine and genetics.
- For Repository 6, the 4 misclassified records (including the one with wrong top level) are about biology.
- In Repository 7, only 7 records out of 10 have been classified with a correct top level domain. As a consequence only 50% records were assigned a correct sub-domain. The records with wrong classifications lacked pure_science and biology (1 record) as well as applied_science (2 records), sociology (1 record) and medicine (3 records).
- Repository 8 has only 2 records out of 10 with misclassification at sub-domain level, caused by missing pedagogy and sociology labels respectively.
- For Repository 9, 9 records out of 10 were correctly classified at the top domain level (the misclassified one should have been tagged with social_science → publishing) whereas only 5 records were assigned the right sub-domain. Among these ones, 4 records miss the pedagogy and psychology sub-domains. One record represents a borderline case, since it contains only 2 words and the human expert herself could not assign a discriminating sub-domain.
- In Repository 10 only 1 record was misclassified and its top and sub-domains should have been applied science and medicine respectively.

From the above results we may conclude that the top level domains classification gives very positive results. As shown in Table 4, our procedure tags each record with a limited amount of domains (about 5 on average) in order to limit noise and to provide very synthetic results.

Considering the top level domain classification, in 20% of the repositories (r3, r8), 100% records were correctly classified; in 50% of the repositories (r4, r5, r6, r9, r10), the correct classification applied to 90% of the records; in 20% of the repositories (r1, r2), to 80% of the records. The worst case is represented by Repository 7 where only 70% of the records found a correct top level domain classification.

In 50% of the repositories, sub-domain classification succeeds for more than 80% of the records (r3, r4, r5, r8, r10). In the remaining repositories, the percentage of records assigned a correct sub-domain is between 50% and 60%.

Thanks to these experiments carried out on a large set of records, we observed that many *dc:subject* entries do not affect the classification performance. This confirms our impression that the manual creation of metadata is often error-prone and a service as the one MANENT provides may prove useful in real world scenarios where data provided by users are incomplete.

Despite completeness of our classification procedure (namely, how many domain labels that should have tagged the record, were actually discovered by it) cannot reach impressive values because of the limited number of domain labels we assign to each record, correctness (namely, how many domain labels discovered by the classification procedure, are indeed correct) is very good. Refining techniques such as metadata content relevant keywords, extracted and tagged with domain labels as well as sentence similarity methodologies may be further applied to improve the completeness while keeping the correctness.

## 5 Related Work

In this Section we outline some state of the art Digital Libraries infrastructures as well as approaches that exploit WordNet Domains.

### 5.1 Related Work on Digital Libraries Infrastructures

The consistent amount of investments towards European projects dealing with digital libraries infrastructures for cultural heritage and preservation[12] witnesses the swift growth of a research field becoming crucial for the management of information commitment foreseen in the near future.

Studies on the state of the art of semantic digital libraries architectures [6], [25] emphasise some conclusive aspects for the sustainability of new generation infrastructures. They are:

- easy to use information discovery that may strongly rely on natural language facets;
- design of digital libraries infrastructures and services that should more and more rely on trusted reference models and well grounded standard resource description formats enriched with semantics;
- basic services such as indexing and classification augmented with user-centered annotations for systems customisation and evolution;
- interoperability at different levels of granularity: from document descriptions, through systems architectures and their tight integration.

Projects that have contributed to the fulfillment of this vision towards the integration, structuring and searching of Digital Libraries are, to cite only a few, the TELplus

---

[12] http://cordis.europa.eu/fp7/ict/telearn-digicult/, following the link to "DigiCult".

project[13] and the DELOS project[14]. The main outcomes are the definition of a Reference Model [10] for Digital Libraries systems design and a Digital Library Management System [2]. Formal models for semantic annotations [3] have been studied. A set theoretic model [17] for managing hierarchical structures such as those of the OAI-PMH metadata harvester at resource level and those of the EAD at archival description level has been defined. A search and retrieval component [16] focusing on annotations similarity measures based on boolean operators and hypertext relational features is also provided.

The BRICKS [25] infrastructure joins the flexibility of a service oriented distributed architecture able to orchestrate interoperability among digital library nodes. A metadata manager component that relies on RDF representation of different standard schemas is in charge of providing advanced query search based on SPARQL syntax.

In JeromeDL [25] several ontologies have been integrated under the MarcOnt[15] ontology umbrella, whose design roots from MARC 21 bibliographic standard. MarcOnt is expressed in OWL and encompasses FOAF[18], Dublin Core, BibTeX[16] and S3B Tagging [17] ontologies, each of them representing one aspect of a semantic digital library system, namely people, resources, metadata and users annotations respectively. Based on that some services are built to provide semantic query in form of regular expressions templates from which the users may choose. Moreover search operations are based on full-text and bibliographic search as well as query expansion based on keywords suggested by the users that are saved in their preferences profile for later use and results ranking. To the best of our knowledge topical information in JeromeDL, which can be based on several classification standard such as DDC[18], UDC[19] and LCC[20] has to be manually added by librarians or resources owners and no automatic classification service is foreseen.

In the HarvANA system [23] institutional metadata and users annotations are both harvested and aggregated. An OAI-PMH interface to remote user-annotation servers has been developed for storage and retrieval whereas a mapping procedure from RDF converted annotations to Dublin Core schema is performed. In this way the system is able to index both metadata records and user defined annotation records. The search operation on resources can be then performed on both kind of

---

[13] http://www.theeuropeanlibrary.org/telplus.

[14] http://www.delos.info.

[15] http://www.marcont.org/ontology/2.0.

[16] http://purl.org/net/nknouf/ns/bibtex.

[17] http://s3b.corrib.org/tagging.

[18] Dewey's Decimal Classification.

[19] Universal Decimal Classification, http://www.udcc.org.

[20] Library of Congress Classification
http://www.loc.gov/aba/cataloging/classification.

semantic information. As far as we know the system works mainly with images repositories and annotations are suggested to the users via a controlled vocabularies interface, hence no automatic tagging service is provided.

Starting from the mid nineties, the use of ontologies for the integration of large heterogeneous information sources has been the subject of a very lively research activity [43].

Among the recent systems that exploit ontologies for annotating and classifying documents in knowledge sources, we mention SOBA [9] which processes structured information, text and image captions to extract information and integrate it into a knowledge base whose coherence is ensured by a reference ontology, built at system design time.

The approach discussed in [8] is even closest to ours, since it presents a framework for integrating digital library knowledge sources as well as facts extracted from the content under consideration by means of an ontology-based digital library system. Documents in the knowledge sources are annotated and classified according to the PROTON upper ontology[21] using natural processing techniques, in the same way as we annotate and classify records according to the WordNet Domain Ontology. Bloehdorn et al. allow users to express queries in natural language, as we do. The main differences between our work and Bloehdorn et al.'s one is that in their work, topics extracted from metadata and unstructured documents are instances of the Topic concept in PROTON and can be automatically organised in a subTopic hierarchy, thus allowing the PROTON ontology to grow during time. We use the WordNet Domains ontology instead and with a bottom-up approach such that an expansion of the domain labels assigned to relevant keywords extracted from metadata contents, and hence a subset of WordNet synsets, is applied by attaching super-domain labels to the lemmas considered. Moreover, WordNet has been translated into different languages thus providing multilingual facilities that can be easily integrated with our domains tagging procedure. Also, we use MANENT to harvest metadata records retrieved from more than 1,300 digital libraries spread all over the world. To the best of our understanding, Bloehdorn et al.'s infrastructure was used for annotating documents in the Digital Library of British Telecommunications only.

The PIRATES framework [6] is part of a semantic layer, included in a service-oriented architecture, providing primitive services to the applications built on top of the digital library which communicates with. The framework provides primitive services to automatically classify, annotate and recommend specific content using techniques based on natural language processing. A controlled, ontology-based vocabulary, is used to classify documents as result of the automatic tagging process. The PIRATES framework is still a theoretical model, although a prototype version has been already developed. We are aware neither of a massive use of PIRATES on a large number of digital libraries, nor of an experimental evaluation as the one we performed.

---

[21] http://proton.semanticweb.org.

## 5.2 Related Work on WordNet Domains

Gliozzo and Strapparava [1, 19] have built a Domain Model on top of WordNet Domains and have exploited it in several Natural Language Processing tasks in order to provide a topic similarity measure to documents. Their WordNet Domains Model ($WND_{DM}$) is formed by the set of all WordNet Domains ($WND$) and the set of WordNet synsets in WordNet ($Syns_c$). In this model a function applied to each element of $Syns_c$ results in a subset of $WND$ (that we can call $WND_c$) associated with that element whereas a domain relevance function is able to return a real value for each element of $WND_c$, which represents the relevance ($rel$ from now on) of each domain label in $WND_c$. In this model all the senses for a word $w$ can be viewed as a subset of $Syns_c$ and hence as a union of all those $WND_c$ associated with each element of this subset (and we can call this union $Syns_{cw}$). The domain relevance function for each $w$ in $WND_{DM}$ is computed as the averaged sum of all the $rel$ calculated on $Syns_{cw}$.

The Domain Model explained above has been instantiated for defining the "Domain Driven Disambiguation" (DDD) methodology [28, 19]. This method has the peculiarity to apply disambiguation level to a domain and hence it is interesting in all those tasks where the results do not need to be as fine grained as a word level disambiguation task requires.

This methodology is based on Domain Vectors (DVs) that represent the domain relevance of an object with respect to a Domain Space, which is a vectorial space. Each value of the vector is an estimate of domain (a dimension) in the Domain Space.

Given a target word $w$ to be disambiguated, DDD is computed on every single term of a context window $CW$ surrounding the target term $w$ and gives a score to each possible sense $s$ of the term $w$. The building blocks of the score computation function are the DV for $w$, represented by the relevance values computed by the functions of the $WND_{DM}$ and the DV for $CW$, which are computed according to the Domain Relevance Estimation technique in [20]. Moreover, a prior probability function, relative to a specific distribution of sense $s$ in a reference corpus, is used as a smoothing parameter. The final result undergoes a fixed threshold in order to filter out not relevant outcomes.

A work similar to ours is that of [15] where the DOMINUS framework is described and an approach for tagging documents with WordNet and WordNet Domains is carried out in order to provide text categorisation and extraction of relevant keywords. Stemming from the document structured elements, such as title, abstract, body and bibliography, the authors exploit some natural language processing steps and extract WordNet synsets and the domain labels associated with them. With the aid of a density function based on Naïve Bayes they assign weights to synsets, varying upon the structured element just considered, and hence propagate them to domain labels in order to obtain both a classification of the document by topic and a raked list of the keywords that best represent the document content. The main differences with our work from the point of view of the local classification procedure are that we use metadata contents instead of document contents and we exploit the

WordNet Domains as an ontology in order to augment the domain label associated with WordNed synsets with super domains label, hence automatically structuring a hierarchical classification service. The main differences from the point of view of the global procedure are that we operate on documents metadata of different existing digital libraries all over the world obtained through our metadata harvesting service.

## 6 Conclusions and Future Work

In this chapter we described MANENT, an infrastructure for integrating, structuring and searching Digital Libraries. The technologies and standards enabling the realisation of infrastructures of this kind have been reviewed, as well as the related work.

The experimental results we obtained by evaluating the reliability of MANENT automatic classification of 100 records drawn from libraries spread all over the world are extremely encouraging.

Besides completing and testing the implementation of all the MANENT services, in order to release a working prototype in mid 2011, there are some improvements that will drive our medium-term future work:

- **Integration of different metadata formats**: as anticipated in Section 3, we provide the user the means for defining her own mappings from any metadata format to our EAD ontology. However, we plan to provide a set of already defined, built-in mappings from the most commonly used metadata formats. We would also like to provide a visualisation service for showing our built-in mappings and allowing users to define their own in a graphical and intuitive way.
- **Integration of a community in MANENT**: in the same way as we already foresee an active involvement of users in the definitions of mappings between metadata schemas, we would like to extend the user involvement to any service provided by MANENT. Becoming more and more similar to a community, MANENT, besides containing documents, should also "contain persons". Communities of practice are characterised by homogeneous patterns of experiences. Digital libraries are the results of the interplay between such implicit knowledge and the explicit layer that they incorporate. In a knowledge society a swift and efficient access to all of the knowledge devices, either people or artifacts, is a pressing requirement. Linking people to collections, enabling communities to annotate digital objects in MANENT and providing a working space for users to collect, organise and annotate such digital objects is the social dimension that, besides the organisational one, should become MANENT's most characterising feature.
- **Integration of digital objects in the knowledge base**: the MANENT prototype stores digital objects in the filesystem. In the near future a mechanism for storing them in the knowledge base should be provided. The idea is to exploit the METS [30] metadata format, after a proper conversion in OWL, to describe such resources in the light of the MANENT rationale, which keeps the collections

descriptions separated from that of single resources. This would also enable the straightforward provision of MANENT contents via OAI-PMH.

- **Classification of multilingual resources**: as an extension of our automatic classification service a study for providing classification of multilingual contents is on its way. Our choice of adopting a WordNet based approach has also been driven by its already available multilingual versions provided by worldwide institutions [49].

# References

1. Agirre, E., Edmonds, P.: Word Sense Disambiguation - Algorithms and Applications. Springer, Heidelberg (2007)
2. Agosti, M., Berretti, S., Brettlecker, G., del Bimbo, A., Ferro, N., Fuhr, N., Keim, D., Klas, C.P., Lidy, T., Milano, D., Norrie, M., Ranaldi, P., Rauber, A., Schek, H.J., Schreck, T., Schuldt, H., Signer, B., Springmann, M.: DelosDLMS - the integrated DELOS digital library management system. In: Proceedings of the First International Conference on Digital Libraries: Research and Development, pp. 36–45 (2007)
3. Agosti, M., Ferro, N.: A Formal Model of Annotations of Digital Content. ACM Trans. Inform. Syst., 26(1) (2007)
4. Balasubramanian, N., Allan, J., Croft, W.B.: A comparison of sentence retrieval techniques. In: Proceedings of the Thirtieth Annual International ACM SIGIR Conference on Research and Development in Information Retrieval, pp. 813–814 (2007)
5. Banerjee, S., Pedersen, T.: An Adapted Lesk Algorithm for Word Sense Disambiguation Using WordNet. In: Gelbukh, A. (ed.) CICLing 2002. LNCS, vol. 2276, pp. 136–145. Springer, Heidelberg (2002)
6. Baruzzo, A., Casoto, P., Challapalli, P., Dattolo, A., Pudota, N., Tasso, C.: Toward Semantic Digital Libraries: Exploiting Web2.0 and Semantic Services in Cultural Heritage. Journal of Digital Information 10(6) (2009)
7. Bentivogli, L., Forner, P., Magnini, B., Pianta, E.: Revising WordNet Domains Hierarchy: Semantics, Coverage, and Balancing. In: Proceedings of the Twenty-First International Conference on Computational Linguistics (COLING 2004),, pp. 101–108 (2004)
8. Bloehdorn, S., Cimiano, P., Duke, A., Haase, P., Heizmann, J., Thurlow, I., Völker, J.: Ontology-based question answering for digital libraries. In: Kovács, I., Fuhr, N., Meghini, C. (eds.) ECDL 2007. LNCS, vol. 4675, pp. 14–25. Springer, Heidelberg (2007)
9. Buitelaar, P., Cimiano, P., Frank, A., Hartung, M., Racioppa, S.: Ontology-based information extraction and integration from heterogeneous data sources. Int. J. Hum.-Comput. Stud., 66(11), 759–788 (2008)
10. Candela, L., Castelli, D., Ferro, N., Ioannidis, Y., Koutrika, G., Meghini, C., Pagano, P., Ross, S., Soergel, D., Agosti, M., Dobreva, M., Katifori, V., Schuldt, H.: The DELOS Digital Library Reference Model. Foundations for Digital Libraries. ISTI-CNR, PISA (2007)
11. Cavnar, W.B., Trenkle, J.M.: N-Gram-Based Text Categorization. In: Proceedings of the Third Annual Symposium on Document Analysis and Information Retrieval, pp. 161–175 (1994)
12. Dublin Core Metadata Element Set,
    http://www.dublincore.org/documents/dces/
13. EAD: Encoded Archivial Description,
    http://www.loc.gov/ead/

14. EAD XML Metaschema,
    http://www.loc.gov/ead/ead.xsd
15. Ferilli, S., Biba, M., Basile, T., Esposito, F.: Combining Qualitative and Quantitative
    Keyword Extraction Methods with Document Layout Analysis. In: Proceedings of the
    Fifth Italian Research Conference on Digital Libraries (IRCDL 2009). DELOS: an As-
    sociation for Digital Libraries (2009)
16. Ferro, N.: Annotation search: The FAST way. In: Agosti, M., Borbinha, J., Kapidakis,
    S., Papatheodorou, C., Tsakonas, G. (eds.) ECDL 2009. LNCS, vol. 5714, pp. 15–26.
    Springer, Heidelberg (2009)
17. Ferro, N., Silvello, G.: The NESTOR framework: How to handle hierarchical data struc-
    tures. In: Agosti, M., Borbinha, J., Kapidakis, S., Papatheodorou, C., Tsakonas, G. (eds.)
    ECDL 2009. LNCS, vol. 5714, pp. 215–226. Springer, Heidelberg (2009)
18. FOAF: Friend of a Friend ontology, http://www.foaf-project.org/
19. Gliozzo, A., Strapparava, C.: Semantic Domains in Computational Linguistics. Springer,
    Heidelberg (2009)
20. Gliozzo, A., Strapparava, C., Dagan, I.: Unsupervised and supervised exploitation of
    semantic domains in lexical disambiguation. Computer Speech & Language 18(3), 255–
    299 (2004)
21. Gruber, T.: Definition of Ontology. In: Liu, L., Özsu, M.T. (eds.) Encyclopedia of
    Database Systems, Springer, Heidelberg (2009)
22. Hargittai, E., Fullerton, F., Menchen-Trevino, E., Thomas, K.: Trust Online: Young
    Adults' Evaluation of Web Content. International Journal of Communication 4, 468–494
    (2010)
23. Hunter, J., Khan, I., Gerber, A.: Harvana: harvesting community tags to enrich collec-
    tion metadata. In: Proceedings of the Eighth ACM/IEEE-CS Joint Conference on Digital
    Libraries, pp. 147–156 (2008)
24. Itzcovich, O.: L'uso del calcolatore in storiografia, Milano (1993)
25. Kruk, S.R., McDaniel, B.: Semantic Digital Libraries. Springer, Heidelberg (2009)
26. Locoro, A.: Tagging Domain Ontologies with WordNet Domains: An Approach
    for Fostering Ontology Classification, Engineering and Matching. Technical Report
    DISI-TR-10-10, CS Dept. of Genova University (2010),
    http://www.disi.unige.it/person/LocoroA/
    download/DISI-TR-10-10.pdf
27. Magnini, B., Cavagliá, G.: Integrating Subject Field Codes into WordNet. In: Proceed-
    ings of the Second International Conference on Language Resources and Evaluation
    (LREC 2000), pp. 1413–1414 (2000)
28. Magnini, B., Strapparava, C., Pezzulo, G., Gliozzo, A.: The role of domain information
    in Word Sense Disambiguation. Natural Language Engineering 8, 359–373 (2002)
29. MARCXML, http://www.loc.gov/standards/marcxml/
30. METS: Metadata encoding and Transmission Standard,
    http://www.loc.gov/standards/mets/
31. Metzler, D., Dumais, S.T., Meek, C.: Similarity measures for short segments of text. In:
    Amati, G., Carpineto, C., Romano, G. (eds.) ECiR 2007. LNCS, vol. 4425, pp. 16–27.
    Springer, Heidelberg (2007)
32. Mihalcea, R., Corley, C., Strappavara, C.: Corpus-based and Knowledge-based Measures
    of Text Semantic Similarity. In: Proceedings of the Twenty-First National Conference on
    Artificial Intelligence and Eighteenth Innovative Applications of Artificial Intelligence
    Conference. AAAI Press, Menlo Park (2006)
33. Miller, G.A.: WordNet: A Lexical Database for English. Communications of the
    ACM 38(11), 39–41 (1995)

34. OAI-PMH: Open Archives Initiative Protocol for Metadata Harvesting,
    http://www.openarchives.org/OAI/openarchivesprotocol.html
35. Ortoleva, P.: Persi nella rete? Circolazione del sapere storico. In: Soldani, S., Tomassini, L. (eds.) Storia & Computer, alla ricerca del passato con l'informatica, Milano (1996)
36. The Open Archives Initiative Protocol for Metadata Harvesting: Metadata Prefix and Metadata Schema,
    http://www.openarchives.org/OAI/
    openarchivesprotocol.html#MetadataNamespaces
37. The Open Archives Initiative Protocol for Metadata Harvesting: Guidelines for Repository Implementers,
    http://www.openarchives.org/OAI/2.0/
    guidelines-repository.htm
38. The Protégé Ontology Editor,
    http://protege.stanford.edu/
39. Rocchio, J.: Relevance feedback in information retrieval. In: Salton, G. (ed.) The SMART retrieval system: Experiments in Automatic Document Processing. Prentice-Hall, Englewood Cliffs (1971)
40. Rowland, R.: L'informatica e il mestiere dello storico. In: Quaderni Storici, pp. 26–78 (1991)
41. Salton, G., Lesk, M.: Computer evaluation of indexing and text processing. Journal of the ACM (JACM) 15(1), 8–36 (1968)
42. SPARQL Query Language for RDF,
    http://www.w3.org/TR/rdf-sparql-query/
43. Wache, H., Vögele, T., Visser, U., Stuckenschmidt, H., Schuster, G., Neumann, H., Hübner, S.: Ontology-based integration of information - a survey of existing approaches. In: Proceedings of the Twelfth International Joint Conference on Artificial Intelligence (IJCAI 2001) Workshop on Ontologies and Information Sharing, pp. 108–117 (2001)
44. Wenger, E.: Communities of practice, learning, meaning and identity, Cambridge (1998)
45. W3C . OWL Web Ontology Language Overview – W3C Recommendation (February 10, 2004)
46. W3C . RDF Vocabulary Description Language 1.0: RDF Schema – W3C Recommendation (February 10, 2004)
47. W3C . RDF/XML Syntax Specification (Revised) – W3C Recommendation (February 10, 2004)
48. W3C . Extensible Markup Language (XML) 1.0 (Fifth Edition) – W3C Recommendation (November 26, 2008)
49. Wordnets in the world,
    http://www.globalwordnet.org/gwa/wordnet_table.htm

# Low-Level Document Image Analysis and Description: From Appearance to Structure*

Emanuele Salerno, Pasquale Savino, and Anna Tonazzini

**Abstract.** This chapter deals with the problem of processing and analyzing digital images of ancient or degraded documents to increase the possibilities of inferring their structures. Classification and recognition are needed to infer structure but, when dealing with degraded documents, they are particularly difficult to apply directly to unprocessed images. This is why an intermediate step is needed that extracts automatically the "perceptual components" of the documents from their appearance. By "appearance" of a document, we mean the "raw" data set, containing the "sensorial components" of the object under study. Ancient documents of historical importance pose specific problems that are now being solved with the help of information technology. As much information as possible should be drawn from the physical documents and should be structured so as to permit specialized searches to be performed in large databases. The tools we use to treat unstructured, low-level information are both mathematical and descriptive. Under a mathematical point of view, we model our appearance as a function of all the perceptual components, or *patterns* we want to identify. Once the model has been established, its parameters can be learned from the data available and from reasonable assumptions on both the model itself and the patterns. Our descriptive tools form a specialized metadata schema that can help both the storage and the indexing of all the digital objects produced to represent the original document, and provides a complete description of all the processing performed. Suitable links fully interconnect the various descriptions in order to relate the different representations of the physical object and to trace the history of all the processing performed. Inferring structure is much easier by analyzing the patterns and their mutual relationships than by analyzing the appearance.

Emanuele Salerno · Pasquale Savino · Anna Tonazzini
CNR - Istituto di Scienza e Tecnologie dell'Informazione, Via Moruzzi 1, 56124 Pisa, Italy
e-mail: {firstname.surname}@isti.cnr.it

* This work was partially supported by European funds through POR Calabria FESR 2007-2013, PIA 2008 project No. 1220000119. Partners: TEA sas di Elena Console & C. (CZ), Istituto di Scienza e Tecnologie dell'Informazione CNR (PI), Dipartimento di Meccanica Università dalla Calabria (CS).

# 1  Introduction

## 1.1  *Motivation, Objectives and Contributions*

Learning the structure of a document always requires an accurate classification of
the various patterns appearing on its support (paper, parchment, etc.). This task, in
turn, requires a reliable segmentation, which is very difficult when based on unprocessed images of ancient or heavily degraded documents. Our statistical approach
to document analysis [21] models the appearance of the document as a mixture of
various patterns each having its perceptual (and structural) significance. Disentan-
gling these patterns from the sensed mixtures can help image analysis in various
ways. First, virtual restoration [19, 25]: if some of the patterns contributing in the
appearance is unwanted (e.g., if it represents a distortion or an interference to the
meaningful part of the document), it can be removed, thus enabling a more easy and
accurate classification of the other parts[1]. Second, segmentation and classification
[15]: perceptually distinct parts of the document can embody structurally different
parts (such as main text, footnotes, titles, colophons, pictures, etc.). Being identified as separate patterns, these can be segmented, classified and then recognized
much more easily than from the raw appearance of the document. Third, extraction
of hidden features [20]: some patterns that are barely detectable from the appearance can be enhanced or even isolated, and can be used *per se* or as cues to infer
structure. This is the case with stamps, watermarks, margin notes, etc., which often
carry valuable information on the history and the authenticity of a document. At this
stage, our approach to document image analysis is thus a low-level one, since it is
only based on pixel information and does not exploit geometric or other high-level
features. A fundamental step to infer structure from the low-level output is to find a
standard procedure to describe precisely the features of all the maps available and
all the settings employed by the image processing algorithms as well. This is done
via a metadata schema that also allows us to effectively store all our images into a
large searchable database [2].

This chapter first describes the data models we use to account for the appearance
of a document (Section 2), and then some of the procedures we apply to extract
meaningful patterns from their spectral or spatial peculiarities (Section 3). Section
4, then, is an overview of the metadata schema we propose, and Section 5 shows
some of the results we obtain through our procedures.

## 1.2  *Background*

In the most intuitive and direct sense, image restoration techniques are applied to
document images to remove or mitigate the degradations originated by ageing of

---

[1] It is to note, however, that for some scholars all the patterns are meaningful, since they all
mark the history of the document and, as such, they must not get lost.

the originals (tears, stains, seethrough...) or by nonidealities of the image capture system (blur, sensor noise...). An immediate effect of this kind of restoration is the enhancement of the readability of the documents, either by a human reader or by an automatic procedure, such as optical character recognition (OCR).

Restoring the original appearance of a degraded document can also help to infer (automatically or not) its structure, since the distinctive features that mark the various structural components of the document are distinguishable more easily in the restored than in the degraded version. From an appearance-only representation, however, the meaningful patterns must only be identified by spatial features that are often difficult to define and detect.

Ancient and historically important documents are often heavily degraded and weakly structured, but their importance justifies the application of specialized acquisition procedures, such as diversity imaging. Depending on the type of diversity employed to capture the images, different patterns can be distinguished on the basis of their peculiarities in the different components of the data. As an example, if spectral diversity is used, spectrally distinct patterns can be isolated by exploiting the channel selectivity offered by the data.

Spectral diversity can be obtained simply by color cameras (which produce three-component, e.g. *RGB*, images) or by specialized multispectral or hyperspectral systems, which can produce even hundreds of narrowband channels from both the visible and nonvisible ranges. From this kind of data, we proposed to use linear blind source separation techniques to extract the patterns superimposed to each other in the multiple views available [21]. Although based on a largely approximated model, this approach proved to be useful for both virtual restoration and hidden feature extraction. In the particular case of double-sided documents, where the appearance can be degraded by interferences from one side to the other[2], the same blind source separation techniques associated with a special kind of viewpoint diversity, namely, *recto-verso* acquisition, make it possible to distinguish effectively between the patterns that are in the front page (*recto*) and in the back page (*verso*) of the document. This can be done by only assuming two patterns, corresponding to the "original" appearances of the recto and the verso sides, independently of the number of spectrally distinct patterns present in the images. This means that a separation of the two sides from the related seethrough interferences can even be done from grayscale images (that is, just from a two-component input [23]). Adding spectral diversity to *recto-verso* data also allows us to free the pages from seethrough effects and separate spectrally distinct patterns [24].

It is quite useful to archive the documents being processed in a digital library that also contains a description of their content and of the processing they were subject to. This enables an easy search of documents according to their content, to the processes applied on them, and to any possible component detected in them.

The documents in the digital library need a detailed description of their content and structure. Metadata are widely used to provide structured information about data and are useful for different purposes, ranging from content-based retrieval

---

[2] These types of degradations are collectively referred to as *seethrough*.

to document administration. In particular, *descriptive metadata* describe the document content with the purpose of supporting its retrieval. Descriptive metadata may provide information at different levels of detail for different types of data. To this purpose, several *de-facto* standards are available. For example, the Dublin Core Metadata (DCMI) [8] are used to provide simple and easily exchangeable metadata, the MPEG-7[14] standard is primarily used to describe multimedia information, MARC-21 is a standard defined by the Library of Congress and widely used by librarians, while IFLA/FRBR [9] is a general conceptual framework used to describe heterogeneous digital media resources. Specific applications adopt the metadata format that is more suitable to their needs. In many cases, the existing formats do not fit the application needs, and defining a new metadata format or extending an existing one becomes necessary. *Administrative metadata* provide a description of the documents that enables their management within the digital library. They are usually adopted to control the access to the documents and to their components, to support their protection against misuse, to preserve the digital resources, but can also be used to describe how the document content has been obtained, and which processing was applied to it.

## 2   Modeling the Appearance of an Ancient Document

Let us assume we have a diversity acquisition system that produces a 2D vector map $\mathbf{x}(i, j)$ with $N$ components, called *appearance*. The nature of the components of $\mathbf{x}$ (the *channels*) depends on the type of diversity used to capture the image[3]. Let us also assume that all the channels at pixel $(i, j)$ carry information from the same spatial location, that is, all the channels are spatially coregistered. In general this can be obtained through suitable image registration algorithms, whose details are not treated here (see [24] and references therein). The appearance can be seen as the superposition of a number $M$ of information layers (we call them *patterns*), represented collectively as a vector function $\mathbf{s}(i, j)$ whose components describe perceptually distinct configurations often carrying higher-level information. In the case of a document image, the patterns can represent the main text, pictures, titles, footnotes, the typical texture of the document support, and any other interfering pattern originated by the degradation of the physical document or nonidealities in the image capture system. Our data model considers each channel as a function of all the patterns.

$$\mathbf{x}(i, j) = \mathbf{f}(\mathbf{s}), \tag{1}$$

where the vector function $\mathbf{f}$ (the *mixing operator*) can be of any kind (linear or nonlinear, instantaneous or with memory, stationary or space-variant, etc.) Model (1)

---

[3] That is, spectral band, for color or multispectral acquisitions, viewpoint or acquisition time, for multiview or multiframe acquisitions, respectively, or diversity from any other feasible combination of illumination and capture modalities.

thus represents any kind of superposition among the different patterns. What this means in the case of a diversity document image is clear: the appearance results from a generalized superposition of the spatial-color patterns present on the page. Possible nonlinearities can result from the physics of the document or from a poor tuning of the capture system. Possible noninstantaneousness can emerge from the different optical features of the capture system in different channels or from the particular capture modalities of some patterns (transmission imaging, etc.) Possible nonstationarity can derive from nonuniformity in the document support. The image formation modalities in the different channels can be so complicated that function **f** should normally be considered unknown, or at least known with a coarse accuracy. This is the very start of our problem: the goal is to isolate the patterns **s** from the appearance **x**, taking into account that **f** is not perfectly known. Depending on the particular case, we could need to estimate **f** and reconstruct all the components of the pattern vector **s**, or just to extract a few components of **s**, the others being considered as nuisance quantities. Of course, the former is a very difficult task to be performed from scratch. Any particular application suggests a parametric form for the mixing operator and/or relevant properties of the individual patterns that can help the reconstruction. Hereafter, we show some of the forms that model (1) can assume.

First of all, function **f** could (normally do) contain a stochastic part, perhaps the most trivial case being the presence of an additive noise contribution $\mathbf{n}(i, j)$. Since this is always the case with sensor noise, we always assume the presence of such a contribution, although we do not introduce it explicitly into the data model unless it is used to derive the estimation algorithm. In fact, this is seldom the case in document image processing, since the image capture hardware and procedures are normally very well controlled, so that the noise component is well below the other error sources (oversimplified data model, unwanted patterns in the original, etc.)

Our first attempt to constrain the data model was to assume a linear and instantaneous data model, thus making the problem suitable to be solved via the standard independent component analysis (ICA) techniques (see, e.g., [11]). The application of ICA requires the patterns (or *sources*, in the ICA terminology) to be mutually independent, thus our attempt was motivated by the conjecture that the individual patterns (even if not strictly independent) are normally less correlated than the appearance channels. The mixing model in this case becomes

$$\mathbf{x}(i, j) = A\mathbf{s}(i, j) \tag{2}$$

where the unknown operator **f** has been parameterized as a constant *mixing* matrix $A$, and any individual pixel in the appearance only depends on the same pixel in the source vector. This model is very simple, and very fast ICA algorithms can be applied to estimate it. The quality of the results depends on the ability of the simple model to produce realistic data. Indeed, in some applications we obtained something similar to a true separation; in other applications, where the model was less justified, we obtained good results anyway, but just from the point of view of image enhancement. Two problems can hamper this approach: one is the already

mentioned inadequacy of the model, the other is a possible lack of mutual independence, which undermines the hypotheses of ICA. The former is more serious
than the latter, in our experience, since we have seen that good results can even be
obtained when mutual independence is not verified. Indeed, a simple decorrelation,
obtained by techniques such as principal component analysis (PCA), is often sufficient to free some of the patterns from the interfering material. In one case ICA and
PCA have proved to be equivalent [23]: the separation of main text and seethrough
from recto-verso grayscale data. In this case, model (2) is specialized by a $2 \times 2$
symmetric mixing matrix. This peculiarity was also exploited to perform the same
task from color or multispectral data [24], which also allow a reconstruction of the
original colors to be performed. Model (2) has also been used to remove seethrough
from single color scans of double-sided documents [21, 19], and to extract partially
hidden patterns or erased texts in palimpsests [20, 15].

A linear noninstantaneous model can partially account for both the optical performance of the capture system and the diverse modalities of light propagation. The
elements of the mixing matrix in a stationary noninstantaneous model are not constant anymore, but become convolutional kernels, so that the matrix-vector product
in Eq. (2) becomes a matrix convolution:

$$\mathbf{x}(i, j) = [A*\mathbf{s}](i, j) \tag{3}$$

In the cases where noninstantaneousness is due to the interaction of light with the
physical document, the kernels are normally unknown, and must be estimated along
with the patterns. This happens, for example, when some of the channels are transmission images and the light is scattered by the document support before reaching
the sensor, or when one or more patterns enter the appearance because they show
through the support[4]. Solving the blind source separation problem in this case is not
always easy, unless specific constraints on the mixing kernels are available. Conversely, when noninstantaneousness is due to the capture equipment, the convolution
kernels are normally known very well. A common case, for example, occurs with
multispectral imaging apparatuses, whose focusing point spread functions (PSF)
normally depend on the channel, but are known from the design and the calibration
of the acquisition system. In this case, Eq. (3) can be particularized as

$$\mathbf{x}(i, j) = [H * A\mathbf{s}](i, j) \tag{4}$$

where $A$ is again an $N \times M$ constant matrix, and $H$ is an $N \times N$ diagonal matrix
whose entries are the known PSFs of the $N$ channels. Of course, estimating the parameters and the sources when the problem can be modeled via (4) is normally much
easier than from the general model (3). We do not treat here the space varying problem, characterized by general, shift-variant kernels, although some nonstationarity
is always present for the nonuniformity of the document support. Some estimation
algorithms show a certain robustness against moderate nonstationarity effects despite these are not introduced explicitly in the data model.

---

[4] That is, by the definition we use here, when seethrough is present.

Finally, let us consider the case where the mixing is nonlinear. As already stated, a model for this phenomenon is given by Eq. (1), but such a general model does not help much to solve the problem, if no information is available on the nonlinear vector function **f**. Typical examples of nonlinear models emerge from an analysis of the showthrough phenomenon. Indeed, considering the physics of the interaction of the light with a document, it is apparent that the verso pattern contributes nonlinearly to the appearance of the recto page. Two nonlinear models for showthrough have been proposed in [17] and [12]. Both assume grayscale recto-verso data, and a common form for them can be

$$
\begin{cases}
x_{recto}(i, j) = s_{recto}(i, j) + f_{vr}(s_{verso}) \\
x_{verso}(i, j) = s_{verso}(i, j) + f_{rv}(s_{recto})
\end{cases}
\tag{5}
$$

where the meaning of the symbols is clear, but the quantities involved are optical densities rather that reflectances. In [17] the nonlinear functions $f_{vr}$ and $f_{rv}$ are derived from an approximated theoretical analysis of showthrough[5], and contain a convolutional part whose PSF is compensated via adaptive filtering. Approximating the showthrough patterns by the corresponding scanned versions (that is, the appearances of the two sides) enables a noniterative restoration algorithm to be derived. In [12], the type of nonlinearity is derived empirically, approximated and parametrized, and then estimated blindly through a recurrent network. A fixed showthrough PSF is then introduced to mitigate the effects of using instantaneous relationships to model a noninstantaneous phenomenon. In many cases, models of the type (5) allow the drawbacks of linear instantaneous models to be avoided.

## 3    Extracting Pattern Information from Low-Level Processing

With a generic model specified as in (1), we mean that each intensity map that composes the appearance vector is given by some mixture of a number of supposedly "original" intensity maps (the patterns), each representing some distinct feature of the original document. The goal of "separating" the individual patterns, however, is somewhat different from a traditional (hard) classification, since in the standard case each pixel in the appearance maps is labeled by one of a number of mutually exclusive classes (e.g., main text, background, picture...). If we solve one of our models, conversely, we extract a number of source functions that are defined over the whole image domain, and a single pixel can be labeled by more than one class. The value of each source function in a pixel denotes the "intensity" of a class in that pixel, namely, the degree of membership of that pixel to one class. Actually, once all the components of vector $s(i, j)$ have been reconstructed, we do not know which classes they are related to. Thus, this is a segmentation, where each pixel is given the degrees of membership to a number of "segments", with no immediate possibility

---

[5] Showthrough is the type of seethrough interference that is caused by the interaction between the document and the probing light produced by the scanning system.

to assign a class to anyone of them. The segments are not a partition of the image domain as in classical segmentation, but a superposition of continuous functions defined all over the image domain. This is the concept of "soft segmentation" [18], which is useful in many applications where the patterns can be superimposed to one another. Hereafter, we give some details about our experience with separation algorithms only based on linear models. We do not report results with nonlinear models since these are the subject of a research that is still in progress.

## 3.1   Linear Instantaneous Case

Model (2) is based on two fundamental assumptions: i) Since we consider images of documents containing homogeneous texts or drawings, we can reasonably assume that the color of each source is almost uniform, i.e., we have mean reflectance indices $a_{kl}$ for the $l$-th source at the $k$-th observation channel; ii) The source patterns $s_l(i, j)$, $l = 1, 2, ..., M$ denote the "quantity" of the $M$ patterns that concur to form the color at pixel $(i, j)$.

In the simplest single-side case, i.e., for RGB scans, vector $\mathbf{x}$ has dimension $N = 3$. Most documents can be seen as superpositions of only three ($M = 3$) different sources, namely, a "background", a "main text" and an "interfering texture". In general, however, the size of the "color" vector obtained from multispectral or hyperspectral sensors is larger than 3. Likewise, we can also have $M > 3$ if additional distinct patterns are present in the original document. In principle, depending on the capture system and the particular document analyzed, $N$ and $M$ can assume any value. All our approches, however, can manage to separate all the patterns if $N \geq M$. In some cases, a few pure patterns can even be extracted from an observation vector whose size is smaller than the total number of patterns (that is, if $N < M$), but all the remaining patterns are necessarily mixed in one or more output maps.

When the observations are two grayscale images of the recto and verso of a document page, affected by seethrough interferences, the linear instantaneous model derives from a simple particularization of Eq. (2)

$$\begin{cases} x_{recto}(i, j) = a_{11}s_{recto}(i, j) + a_{12}s_{verso}(i, j) \\ x_{verso}(i, j) = a_{21}s_{recto}(i, j) + a_{22}s_{verso}(i, j) \end{cases} \tag{6}$$

where either the recto or the verso maps are obtained by flipping horizontally the observed maps, in order to match the seethrough appearances to the original patterns from which they are generated. In this case, $a_{12}$ and $a_{21}$ represent the attenuations through which the main recto and verso patterns appear as seethrough in the opposite sides. Such attenuations depend on the features of the document support and other factors, such as ink fading. Since it is often reasonable to assume an isotropic behavior for the document support, we can often set $a_{12} = a_{21}$. Also, it is always reasonable to assume $a_{11} = a_{22}$. These properties can be enforced explicitly or implicitly to solve the seethrough cancellation problem.

In the general linear instantaneous model (2), the elements $a_{kl}$ of matrix $A$ are not known. This means that, when no additional assumption is made, the problem of extracting $\mathbf{s}$ from $\mathbf{x}$ is underdetermined. One condition under which this type of problems can have a unique solution (up to scaling and permutation, see [11]) is that the source functions are mutually independent. In this hypothesis, and if some additional assumptions are verified[6], both the sources and the mixing coefficients can be estimated from the data alone. In the multispectral, single side case, this means that the various patterns must be spectrally different. In the recto-verso case, some specific physical constraints ensure this condition to be verified and also allow us to relax the strict independence requirement, so that separation can be achieved by simply orthogonalizing the data images.

The recto-verso model can also be extended to the multispectral case:

$$\begin{cases} x_{recto}^k(i,j) = a_{11}^k s_{recto}^k(i,j) + a_{12}^k s_{verso}^k(i,j) \\ x_{verso}^k(i,j) = a_{21}^k s_{recto}^k(i,j) + a_{22}^k s_{verso}^k(i,j) \end{cases} \quad , k = 1,2,...N \qquad (7)$$

where the superscript $k$ denotes one of the $N$ spectral channels. The general $2N \times 2N$ linear system thus becomes either a set of $N$ linear $2 \times 2$ systems or, equivalently, a $2N \times 2N$ block diagonal system with $2 \times 2$ blocks. In the case (7), reasonable properties of the sources are that, for any possible pair of channels $(k,h)$, the recto and verso components $s_{recto}^k$ and $s_{verso}^h$ are almost uncorrelated, whereas the recto sources $s_{recto}^k$ and $s_{recto}^h$ are strongly correlated since they are just different frequency components of the same reflectance function. The same is true for any pair of verso sources. This means that, solving the $2N \times 2N$ system by ICA or channel decorrelation would be a mistake, since zero correlations would be forced between strongly correlated pairs. However, since the $N$ linear $2 \times 2$ systems associated to the individual channels are independent of each other, we can solve them separately. By this strategy, when the pages are captured by RGB sensors, the original document colors can be reconstructed by recomposing the restored color channels. This fact has been exploited to produce restored documents that, while cleansed of the unwanted interferences, maintain their original appearances as much as possible.

### 3.1.1   Solution through Independent Component Analysis

The ICA approach [11] enforces mutual independence on the estimated sources. This is equivalent to assume a factorized form for the joint prior distribution of $\mathbf{s}$:

$$P(\mathbf{s}) = \prod_{l=1}^{M} P_l(s_l) \qquad (8)$$

The estimated sources $\hat{s}_l$ must result from a linear combination of the channel maps, that is, $\hat{s}_l(i,j) = \mathbf{w}_l^T \mathbf{x}(i,j)$, where vector $\mathbf{w}_l$ contains the combination coefficients.

---

[6] Other necessary assumptions are that $A$ must be full-rank, with $N \geq M$, and all but at most one of the components $s_i$, seen as random processes, must be nongaussian.

The separation problem can be formulated as the maximization of the distribution in (8), subject to the constraint $\mathbf{x} = A\mathbf{s}$. This is equivalent to searching a matrix $W = (\mathbf{w}_1^T, \mathbf{w}_2^T, ..., \mathbf{w}_N^T)^T$ that, when applied to the data $\mathbf{x} = (x_1, x_2, ..., x_N)$, produces a vector $W\mathbf{x}$ whose components $\hat{s}_l$ are maximally independent. By taking the logarithm of Eq. (8), the problem solved by ICA in the square case ($M = N$) is then:

$$\widehat{W} = \arg\max_W \sum_{i,j} \sum_l \log P_l \left( \mathbf{w}_l^T \mathbf{x}(i,j) \right) + T \log |\det W| \qquad (9)$$

and matrix $\widehat{W}$ is an estimate of $A^{-1}$ up to arbitrary scale factors and permutations on the columns. Hence, each component $\hat{s}_l$ is one of the original source patterns up to a scale factor. A substantial difficulty arising from solving problem (9) is that normally the source priors $P_l$ are not known. As the ICA principle is effective for nongaussian sources, we only know that each $P_l$ must describe either a supergaussian or a subgaussian variable. Just requiring $P_l$ to behave either way has proved to give very good estimates for the mixing matrix and for the sources as well, no matter what the true source distributions are [4]. The FastICA algorithm [11] solves problem (9) by a fixed-point iteration scheme, giving a choice among a number of non-linearities to model various types of nongaussian distributions.

### 3.1.2 Solution through Channel Decorrelation

Although we obtained some promising result by ICA [21], there is no apparent physical reason why our sources should be mutually independent, so, even if the data model were correct, the ICA principle is not ensured to be able to separate the different patterns. However, it is intuitively and experimentally apparent that decorrelating the observed image channels can help maximizing the information content in each component of the output vector. This amounts to force the data cross-covariances to zero, i.e., to orthogonalize the data images. Some standard linear color space transformations have proved to approach this result. Indeed, the authors of [13] compared the performances of many such transformations, and found that the ones that obtain maximum-variance components are the most effective. The reference to evaluate the performances of any fixed transformation was the Karhunen-Loeve transform[7], which is known to give orthogonal output vectors, one of which has maximum variance. Rather than looking for fixed transformations with approximated results, we seek a "custom" transformation through an $M \times N$ matrix $W$

$$\mathbf{y}(i,j) = W\mathbf{x}(i,j) \qquad (10)$$

such that all the cross-covariances between the output components vanish or, in other words, the components of vector $\mathbf{y}$ are mutually orthogonal:

$$\langle y_m \cdot y_n \rangle = 0, \quad \forall m, n = 1, ..., M, \quad m \neq n \qquad (11)$$

---

[7] Also called *principal component analysis* (PCA), whose aim is to find the most informative among a number of random variables [6].

where the notation $\langle \cdot \rangle$ means expectation and, with no loss of generality, we assume zero-mean data vectors. It is clear that no unique matrix $W$ can be found, since, given an orthogonal vector set, any rigid rotation thereof is still orthogonal. Our $N \times N$ data covariance matrix is:

$$R_{\mathbf{xx}} = \left\langle \mathbf{x} \cdot \mathbf{x}^T \right\rangle \tag{12}$$

Since the distribution of vector $\mathbf{x}$ is not available, we can only compute the expectations needed from the data sample at our disposal, assumed as stationary. From (10) and (11), our problem is to find a matrix $W$ such that the covariance matrix of $\mathbf{y}$ is diagonal, that is,

$$R_{\mathbf{yy}} = \left\langle W\mathbf{x} \cdot \mathbf{x}^T W^T \right\rangle = W R_{\mathbf{xx}} W^T = D_M \tag{13}$$

where $D_M$ is any diagonal matrix of order $M$. Let us perform the eigenvalue decomposition of matrix $R_{\mathbf{xx}}$:

$$R_{\mathbf{xx}} = V_{\mathbf{x}} \Lambda_{\mathbf{x}} V_{\mathbf{x}}^T \tag{14}$$

where $\Lambda_{\mathbf{x}}$ is the diagonal matrix of the $M$ nonzero eigenvalues of $R_{\mathbf{xx}}$, and $V_{\mathbf{x}}$ is the $N \times M$ matrix whose columns are the related eigenvectors. In all practical cases, of course, all the eigenvalues are nonzero for the presence of noise. $N - M$ of them, however, have the same order of magnitude as the noise and are not significant to represent the data set. Equation (14) can still be written by only using the $M$ dominant eigenvalues to build matrix $\Lambda_{\mathbf{x}}$. It is easy to verify that all the following choices for $W$ satisfy Eq. (13):

$$W_o = V_{\mathbf{x}}^T \tag{15}$$

$$W_w = \Lambda_{\mathbf{x}}^{-\frac{1}{2}} V_{\mathbf{x}}^T \tag{16}$$

$$W_s = V_{\mathbf{x}} \Lambda_{\mathbf{x}}^{-\frac{1}{2}} V_{\mathbf{x}}^T \tag{17}$$

In particular, $W_o$ produces a PCA solution [6] with $D_M = \Lambda_{\mathbf{x}}$. Matrix $W_w$ yields a set of orthogonal components with unit norms ($D_M = I_M$). This is why using matrix $W_w$ is called *whitening*, or *sphering*[8]. Note that a whitening matrix can be multiplied from the left by any orthogonal matrix, and the result is still a whitening. In particular, if we use matrix $W_s$, we have a whitening matrix with the further property of being symmetric[9]. In [6], it is observed that, when $A$ is square and symmetric, left-multiplying $\mathbf{x}$ by $W_s$ is equivalent to perfoming ICA. This is the theoretical result we exploited to solve the recto-verso case. In the multispectral, single side case, we have been applying the above matrices to images of ancient documents, both manuscript and printed, with the aim at emphasizing possible hidden features. For each test image, the results are of course different for different orthogonalizing

---

[8] Left-multiplying a vector by matrix $W_w$ is also called *Mahalanobis transform*.

[9] Observe that Eq. (17) yields an $N \times N$ matrix, and the related transform is only valid when $M = N$.

matrices. In some cases, whitening can even isolate some of the pure patterns, which is part of the final aim of ICA processing.

## 3.2   Linear Convolutional Case

The original patterns that form the appearance of a document are often strongly blurred. This blur can be different from pattern to pattern and from observation to observation. As an example, in the recto of a double-sided document, the pattern interfering from the verso is always blurred for the scattering of light through the support. The same pattern is not, or much less, blurred in the verso side. Also, in many cases the noise affecting the scans cannot be neglected. In all these cases, the performances of the algorithms based on noiseless instantaneous models are normally poor, and we tried to adopt a linear stationary model that, still neglecting possible nonlinearities, includes unknown convolution kernels as in (3) plus additive noise components. In this formulation, the $k$-th channel data map is given by

$$x_k(i,j) = \sum_{l=1}^{M} (H_{kl} * s_l)(i,j) + n_k(i,j), \quad k = 1,2,...N, \quad l = 1,2,...M \quad (18)$$

where $H_{kl}$ is the convolution kernel acting on the $l$-th pattern in the $k$-th channel. In the recto-verso scenario, Eq. (18) provides an extension to Eq. (7):

$$\begin{cases} x_{recto}^k(i,j) = s_{recto}^k(i,j) + \left(H_{12}^k * s_{verso}^k\right)(i,j) + n_{recto}^k(i,j) \\ x_{verso}^k(i,j) = \left(H_{21}^k * s_{recto}^k\right)(i,j) + s_{verso}^k(i,j) + n_{verso}^k(i,j) \end{cases}, \quad k = 1,2,...N \quad (19)$$

where the symmetry properties of the linear instantaneous model still hold true. Through model (18), our problem becomes one of blind source separation from noisy convolutional mixtures or, in other words, of joint blind deconvolution and separation.

In [25], the problem of estimating the blur kernels, denoted collectively by $H$, and the deblurred sources has been approached in a Bayesian setting, and formulated as a joint maximum a posteriori probability estimation problem:

$$(\hat{s},\hat{H}) = \arg\max_{s,H} P(s,H|x) = \arg\max_{s,H} P(x|s,H)P(s)P(H) \quad (20)$$

where the sets of variables $s$ and $H$ are assumed independent of each other, and the quantities $P(s,H|x), P(x|s,H), P(s)$ and $P(H)$ are the joint posterior distribution, the likelihood, and the prior distributions, respectively. We approach this optimization problem through the following scheme, alternating componentwise maximization with respect to different groups of variables:

$$\hat{H} = \arg\max_{H} P(x|s(H),H)P(s(H))P(H) \quad (21)$$

where

$$\mathbf{s}(H) = \arg\max_{\mathbf{s}} P(\mathbf{x}|\mathbf{s},H)P(\mathbf{s}) \tag{22}$$

Our likelihood $P(\mathbf{x}|\mathbf{s},H)$ is derived by assuming the sensor noise to be white and Gaussian with zero mean, and the prior $P(\mathbf{s})$ to be the product of the priors on the individual sources. We adopted a local autocorrelation model for each source in the form of generic local smoothness Markov random fields, augmented to account for the regularity features of realistic edge maps. The Markov-random-field models are particularly suitable to describe both printed and handwritten text images [22]. In addition, accounting for a well behaved edge process is particularly useful for deblurring. The information accounted for by $P(H)$ regards positivity constraints, allowed magnitude ranges, and, in the recto-verso case, the symmetry properties mentioned above. To solve the problem, we proposed an iterative strategy that performs the maximization in (21) by simulated annealing and, at each iteration, computes the sources through Eq. (22) by a graduated-non-convexity algorithm [5, 3]. Our experimental validation, reported in [25], proves the effectiveness of the algorithm on documents featuring two or more overlapping patterns, such as texts, stamps, paper watermarks, and seethrough. In particular, accounting for the blur due to typical degradations such as smearing, diffusion and fading of ink has proved to be necessary to let the various patterns match in the data and, at the same time, to permit the patterns extracted to be refocused.

## 4 A Metadata Schema to Describe Data, Procedures, and Results

To infer the document structure, the results of low-level image analysis must be made available to successive phases, first of all to a module that performs the classification of the patterns extracted. Including the appearance and the patterns into a searchable data repository, along with a complete record of the processing performed, enhances greatly the usefulness of the low-level results and enables the higher-level tools to use all the information and data extracted during the low-level analysis. The searchable data repository stores the description of all the documents processed, their appearances (i.e., all the channels), the processing steps applied to the channels, and the patterns extracted from each group of channels. The repository also stores the representation of the maps corresponding to channels and patterns. All the relationships between different entities are recorded, so that it is possible to know how a pattern was obtained, and to perform further processing from intermediate results, if this is required by the classification strategies. The data repository also enables a content-based retrieval of the data and the processing results, to support the subsequent phases of structure recognition that make use of this content.

The creation of a searchable data repository entails adopting an effective metadata schema to identify precisely all the descriptive and technical features of all the digital objects to be stored and of the processing results. Indeed, unlike traditional archives, this data repository is not limited to support the retrieval of an object from

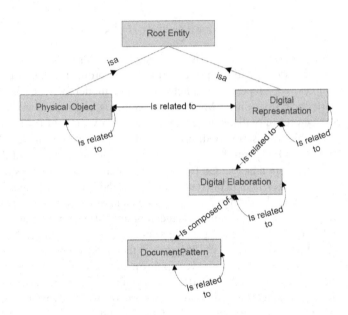

**Fig. 1** The structure of the metadata schema

its properties or its content. Since the digital objects are modified by many complex processes, the data repository also records all the transformations an object was subject to, and the processes that produced each specific output. The retrieval procedure uses metadata describing both the objects and the processes.

The metadata schema adopted describes the metadata of the physical object (the document) and the metadata of its digital representation. Where possible, the metadata element names directly match the DCMI [8] element names, considering its simplicity and the possibility to convert several schemas into DCMI terms. The main difference is in the introduction of more entities than just *Creation*, as well as in the qualification of the DC relation that is expected to be refined within the DC standard. Furthermore, there is a need to describe more than just the resources themselves. This entails the necessity of several entities (each with its own metadata format), and a data model to point out the relationships between them (e.g., "Is A", "Is Part Of", "Is Depicted In", "Is Related To"). The proposed metadata model derives from the models used to describe structured and highly interconnected information, such as Audio-Video material [10, 16, 7] and heterogeneous multimedia cultural heritage material [2], and consists of four entities, as sketched in Figure 1.

The *PhysicalObject* entity contains metadata elements that describe the physical document being processed. They include elements such as the *Title* of the object, its *Description*, its *Location*, *Type*, etc. A physical object may have a relation to other physical objects, for example to the ones belonging to the same collection or produced by the same author, and has a relation to one or more digital representations. The features of each digital representation are described by the *DigitalRepresentation* entity, which describes a visual surrogate of a physical object. By the terminology adopted in this chapter, these are the channels coming from the digital acquisition. The metadata elements of *DigitalRepresentation* include the *Format* used to represent the images, the *Acquisition modality* used, and a reference to the *Digital images* containing the acquisitions. The digital representations of a physical object are processed by several algorithms to improve their quality, to separate the patterns, etc. The result is described by the *DigitalElaboration* entity. Each *DigitalRepresentation* may produce one or more digital elaborations, and any digital elaboration may be obtained as the result of another elaboration. The *DigitalElaboration* entity contains elements such as the *Format* used to represent the images corresponding to the different spectral bands, the *ElaborationProcess* performed on the digital representation (or digital elaboration), when it was done, in order to obtain the current digital elaboration, and a reference to the *Digital images* containing the elaborations. Finally, a *DigitalElaboration* is composed of *DocumentPatterns*, which are the document parts extracted during the elaboration. It must be stressed that, at this stage, these parts are not recognized as belonging to specific classes of document elements. Classification is expected to be performed as part of a higher-level document analysis.

Metadata describing a document and its elaborations are represented in XML format [26] and stored in a digital library supporting searches based on document content as well as on the relations existing between document representations [1].

The rich information stored in the metadata archive at the completion of the low-level document analysis can be used by high-level document analysis, which may take advantage of the document elements extracted, their relationships with other elements extracted from the same document, the structural information about them, and the possible relationships with elements belonging to other documents. Indeed, classification may use results of classifications performed on other documents, which can contain elements similar to the ones being processed.

## 5 Experimental Evaluation

This section shows examples where the strategies presented in Sections 2 and 3 are used to either clean the document appearance (mitigation of interferences) or extract partially hidden or entangled patterns, such as stamps, watermarks, and erased strokes. The examples include a description of how the metadata are associated to the raw images and to the processing history and results.

The first example is the extraction of a barely visible watermark from a data set of two infrared images, one in both reflection and transmission, with front and back illumination, and the other in transmission, with back illumination (Fig. 2).

One of the output channels of a $2 \times 2$ ICA on these data contains a pattern reproducing the watermark. Figure 3 shows the ICA output as is. Simple postprocessing can be applied to further enhance the pattern extracted.

Another example is related to an RGB data set captured from an illuminated manuscript, with a colored dropcap in it (Fig. 4). Two outputs from a $3 \times 3$ ICA

**Fig. 2** Reflection-transmission (left, front and back illumination) and transmission (right, back illumination) infrared data images of a document printed on watermarked paper

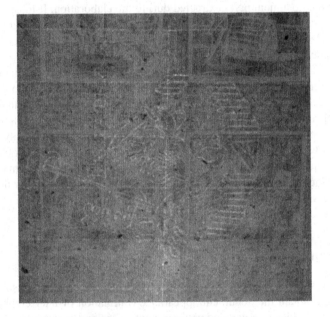

**Fig. 3** Watermark pattern extracted from the data images in Fig. 2

feature the isolated dropcap pattern (whose color can also be recovered) and the main text pattern.

To appreciate the difference between linear instantaneous and convolutive processing, we use the grayscale recto-verso pair shown in Fig. 5. In Fig. 6, we show the result of symmetric whitening as described at page 353. Observe that the partially cancelled seethrough strokes are surrounded by a halo due to the convolution effect that has been neglected and, perhaps, to small registration errors. Since the

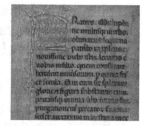

**Fig. 4** Left, RGB scan from an illuminated manuscript (grayscale version). Dropcap (center) and main text (right) patterns extracted by linear instantaneous ICA.

ra puesta a derribar la má
ros, aborrecidos de tantos
seis, no habríais alcanzad

chando lo que mi amigo n
sus razones, que sin pon
mismas quise hacer este
eción de mi amigo, la bue
lo tal consejero, y el alivi
historia del famoso Don (
or todos los habitadores d
amorado, y el más valien
ío en aquellos contornos

**Fig. 5** Grayscale double-sided acquisition of a document affected by seethrough. Left: recto side; right: verso side (horizontally flipped and coregistered with the recto).

level of this halo is higher than the one of the residual seethrough, it also produces an error in the estimation of the background color. Clipping the highest region in the image histogram would partially correct this effect. The result obtained by the convolutional procedure (21)-(22) using the data model (19) with $N = 1$ is shown in Fig. 7. It is apparent that in this case the seethrough cancellation has been much more effective than above, and that no halo is present. Also, the intrinsic nonstationarity of the seethrough effect has been partially recovered by virtue of the local smoothness enforced by the Markov random field priors. The inclusion of seethrough convolution kernels has thus solved one of the problems related to the noninstantaneousness of the data model. A residual problem, however, easily visible in Fig. 7, is an erosion of the printed characters where the seethrough has been subtracted. This is specifically due to linearity, and can only be avoided by resorting to a nonlinear model.

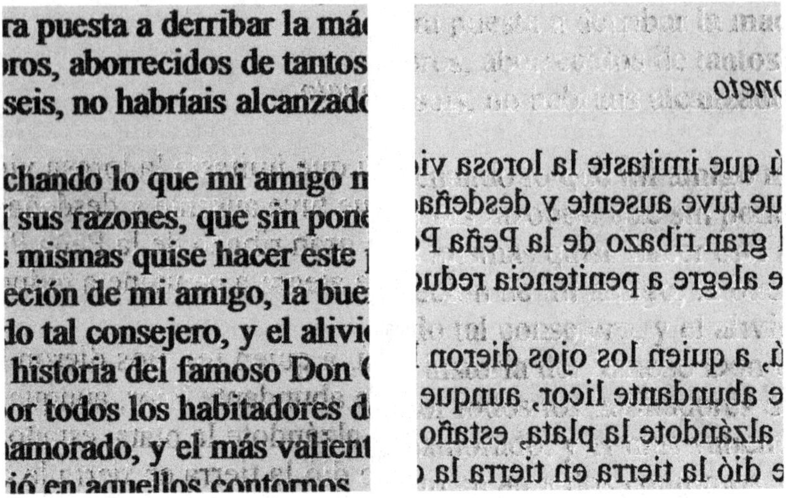

ra puesta a derribar la má(
)ros, aborrecidos de tantos
seis, no habríais alcanzad(

chando lo que mi amigo n
í sus razones, que sin pone
; mismas quise hacer este ¡
eción de mi amigo, la bue
lo tal consejero, y el alivi(
historia del famoso Don (
or todos los habitadores d
lamorado, y el más valient
ió en aquellos contornos

**Fig. 6** Seethrough cancellation on the data in Fig. 5 by symmetric whitening

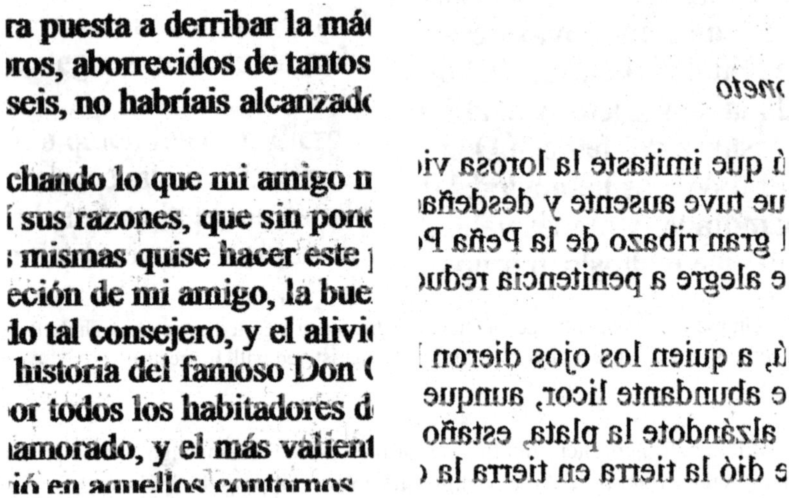

ra puesta a derribar la má(
)ros, aborrecidos de tantos
seis, no habríais alcanzad(

chando lo que mi amigo n
í sus razones, que sin pone
; mismas quise hacer este ¡
eción de mi amigo, la bue
lo tal consejero, y el alivi(
historia del famoso Don (
or todos los habitadores d
lamorado, y el más valient
ió en aquellos contornos

**Fig. 7** Seethrough cancellation on the data in Fig. 5 by algorithm (21)-(22)

A further, and concluding, example demonstrates an application of seethrough cancellation and feature extraction from resto-verso RGB scans, and is described in detail to permit the resulting metadata structure to be introduced. Figure 8 shows the recto from a double-sided RGB scan of a heavily distorted manuscript. The verso is not shown.

**Fig. 8** Recto side from an RGB double-sided scan heavily affected by seethrough (grayscale version)

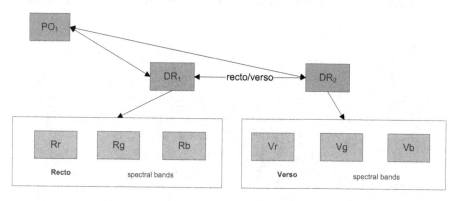

**Fig. 9** Example of the schema of the Physical Object, $DR_1$ and $DR_2$

Figure 9 shows the metadata associated to the manuscript (the Physical Object $PO_1$) and to the Digital Representation of the recto and of the verso scans.

The scans of the recto and verso sides are represented by two Digital Representations ($DR_1$ and $DR_2$). $DR_1$ (the recto side) is constituted of 3 grayscale images, $Rr$, $Rg$, and $Rb$, corresponding to the red, green and blue channels of the physical object, respectively. Analogously, the Digital Representation $DR_2$ of the verso consists of 3 maps, $Vr$, $Vg$, and $Vb$, corresponding to the red, green and blue channels, respectively. $DR_1$ and $DR_2$ are connected through a recto/verso link. After flipping the verso horizontally, the first procedure we must apply is spatial coregistering: one of the six channel maps available is taken as the reference, and all the remaining maps

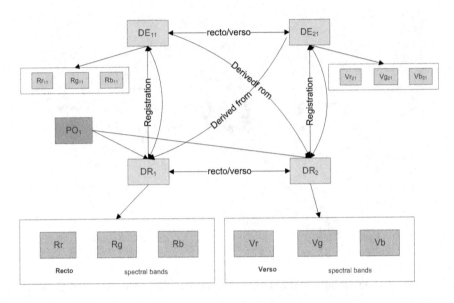

**Fig. 10** Example of the schema of the Physical Object, digital scans and image registration

**Fig. 11** Restored recto from an RGB double-sided scan obtained through channel-by-channel symmetric whitening (grayscale version)

**Fig. 12** Stamp pattern extracted by instantaneous linear ICA from the RGB data in Fig. 11

are transformed so as to locate all their pixels precisely on the reference map. This produces two Digital Elaborations, $DE_{11}$ and $DE_{21}$, still having 3 channels each, and connected again through a recto-verso link. Figure 10 shows the metadata entities created so far, together with the links among them. For example, the links from both $DR_1$ and $DR_2$ to $DE_{11}$ specify that the registration involved $DR_1$ as well as $DR_2$. The next process is seethrough reduction by the model in (7) with $N = 3$ and the orthogonalization procedure (17) applied separately to the three $2 \times 2$ subsystems related to the red, green and blue components of the data. The six output images can then be recomposed to form the restored recto and verso color images. This produces the Digital Elaborations $DE_{12}$ and $DE_{22}$, of 3 channels each, again connected through a recto-verso link.

The result for the recto is shown in Fig. 11, where the reduction of seethrough and the good reconstruction of the original colors are apparent. At this point, further processing is possible on the processed data. For example, one could want to extract the stamp pattern from the RGB image in Fig. 11 by a $3 \times 3$ linear instantaneous ICA. This further elaboration of $DE_{12}$ results in the Digital Patterns $DP_1$, $DP_2$ and $DP_3$, constituted of 3 grayscale images. The output channel containing the stamp pattern ($DP_3$) is shown in Fig. 12. The representation (through the proposed metadata schema) of all elaborations performed on the manuscript is shown in Figure 13. The figure only illustrates the instances of the different entities and their relationships. The values of the metadata elements are not shown.

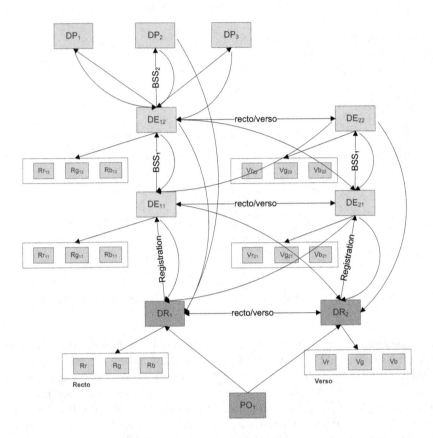

**Fig. 13** Representation of the Physical Object, of its Digital Representation, Digital Elaborations and Document Patterns

## 6   Conclusions and Future Work

Ancient documents are often weakly or nonuniformly structured, so inferring automatically their structures can be very far from being easy. For various reasons, they can also be heavily distorted, so that even the very basic classification procedures are likely to fail when applied to raw data. These difficulties can always arise with any nonstandard document, even with modern ones. On the other hand, the importance of today's effort to digitize the rich heritage conserved in our libraries and archives will increase as much as accessing the digital libraries will be made easy and effective. A great help in this sense will come from the availability of structural and content information about the digital objects and, of course, this will become affordable for most stakeholders inasmuch as automatic procedures for structure learning are available.

We argue that low-level processing can be very important to provide clean data for the subsequent tasks of segmentation and classification, which are the first steps to learn structural information. With reference to ancient documents seen as particularly difficult cases, but with no loss of generality, we report here some low-level processing strategies we experimented in recent years, along with some results obtained by these strategies applied to digital images of real documents. The class of algorithms we present here is based on generative models that require diversity data, that is, in any possible sense, vector data images. On the one hand, this means that our procedures can be exploited fully on data coming from specialized imaging systems (multispectral, etc.), whose use is only justified on particularly precious or historically important documents. On the other hand, however, and this is demonstrated by the experimental results shown here, impressive results can also be obtained from simple RGB images or pairs of grayscale scans. This opens a wide range of opportunities on already existing data, captured without considering the possibility of extracting information automatically.

By extending existing metadata representations, the metadata schema we are proposing describes the semantic content of the documents in its whole, and the semantic content of the results obtained after processing. Suitable links interconnect the various descriptions to relate the different representations of a physical document, and to trace the processes related to all elaborations, from digital acquisitions, to the enhanced versions, to the extracted features. This information can be used to assist the classification of new image components according to similar images previously classified, bringing to a significant improvement of the overall document structure recognition process.

All the processing methods shown here are not very expensive computationally if based on an instantaneous model, including the iterative methods used for ICA. The decorrelation techniques, in turn, are just standard linear algebra. This opens future opportunities to implement an increasing amount of sophisticated computation by embedded systems, so that integrating some low-level processing in specialized capture devices could become both feasible and convenient. Using linear convolutional models avoids some of the drawbacks of instantaneous processing, but leads to more expensive algorithms. One of our future directions of research will be to make these algorithms more efficient. A rich schema of descriptive metadata is essential to include semantics into the digital descriptions that come out from document image capture and from both the low- and high-level processing and analysis stages. Devising largely automatic tools to do this job would be decisive to solve the problems related to the conservation of human knowledge. As a first step, we are planning to link our metadata editor and our capture and processing codes.

Besides implementation aspects, there are a number of problems that are still open to innovative solutions. Surpassing the simple linear and instantaneous models is extremely important to develop the most specialized and accurate tasks. This invariably drive us to complicated methods, which are less tractable, both analytically and numerically, and normally result in expensive algorithms. Moreover, increasing the accuracy of a data model may lead to decreasing its general applicability: as much a method is based on a physically accurate model as narrow is its range of application. A wide area is thus left to research devoted to modelization and development of new algorithms. At present, besides studying prior pattern probabilities to include into instantaneous and convolutional linear models, we are exploring the capabilities of *ad hoc* nonlinear models for restoration. Assessing the performances of these new methods in as many cases as possible, and comparing the results with the ones obtained through the methods described here, will enable us to identify the applications where using a complicated procedure offers advantages that make it preferable to simple but approximated strategies.

## List of Acronyms

DCMI   Dublin Core Metadata Initiative
FRBR   Functional Requirements for Bibliographic Records
ICA   Independent Component Analysis
IFLA   International Federation of Library Associations and institutions
OCR   Optical Character Recognition
PCA   Principal Component Analysis
PSF   Point Spread Function
RGB   Red, Green, Blue color space
XML   eXtended Markup Language.

## References

1. Amato, G., Gennaro, C., Rabitti, F., Savino, P., Milos, A.: A multimedia content management system for digital library applications. In: ECDL 2004, pp. 14–25 (September 12-17, 2004)
2. Amato, G., Cigarrán, J.M., Gonzalo, J., Peters, C., Savino, P.: MultiMatch – multilingual/Multimedia access to cultural heritage. In: Kovács, L., Fuhr, N., Meghini, C. (eds.) ECDL 2007. LNCS, vol. 4675, pp. 505–508. Springer, Heidelberg (2007)
3. Bedini, L., Gerace, I., Tonazzini, A.: A deterministic algorithm for reconstructing images with interacting discontinuities. CVGIP: Graph. Models Image Process. 56(2), 109
4. Bell, A.J., Sejnowski, T.J.: An Information-Maximization Approach to Blind Separation and Blind Deconvolution. Neural Comp. 7, 1129
5. Blake, A., Zissermann, A.: Visual Reconstruction. MIT Press, Cambridge (1987)
6. Cichocki, A., Amari, S.-I.: Adaptive Blind Signal and Image Processing. Wiley, New York (2002)
7. Debole, F., Savino, P., Eckes, G.: Searching and browsing film archives: The European film gateway approach. In: Proc. 4th Int. Congr. Sci. & Technol. Safeguard Cultural Heritage in the Mediterranean Basin, Cairo, Egypt, December 6-8, vol. I, pp. 359–364 (2009)
8. The Dublin Core Metadata Initiative, http://dublincore.org/
9. Functional Requirements for Bibliographic Records, http://www.ifla.org/en/publications/functional-requirements-for-bibliographic-records
10. Gennaro, C., Rabitti, F., Savino, P.: The Use of XML in a video digital library. Intelligent Search on XML Data, 19–38 (2003)
11. Hyvärinen, A., Karhunen, J., Oja, E.: Independent component analysis. Wiley, New York (2001)

12. Merrikh-Bayat, F., Babaie-Zadeh, M., Jutten, C.: A Nonlinear blind source separation solution for removing the show-through effect in the scanned documents. In: Proc. Eur. Signal Processing Conf, EUSIPCO 2008, Lausanne, Switzerland, August 25-29 (2008)
13. Ohta, Y., Kanade, T., Sakai, T.: Color information for region segmentation. Computer Graphics and Image Processing 13(3), 222
14. Salembier, P., Sikora, T., Manjunath, B.: Introduction to MPEG-7: Multimedia content description interface. Wiley, New York (2002)
15. Salerno, E., Tonazzini, A., Bedini, L.: Digital image analysis to enhance underwritten text in the Archimedes palimpsest. Int. J. Doc. Anal. Rec. 9(2-4), 79
16. Savino, P., Peters, C.: ECHO: a digital library for historical film archives, Int. J. Int. J. on Digital Libraries 4(3-7), 1
17. Sharma, G.: Show-through cancellation in scans of duplex printed documents. IEEE Trans. Image Processing 10(5), 736
18. Tai, Y.W., Jia, J., Tang, C.K.: Soft color segmentation and its applications. IEEE Trans. Patt. Anal. Mach. Intell. 29, 1520
19. Tonazzini, A., Salerno, E., Mochi, M., Bedini, L.: Bleed-through removal from degraded documents using a color decorrelation method. In: Marinai, S., Dengel, A.R. (eds.) DAS 2004. LNCS, vol. 3163, pp. 229–240. Springer, Heidelberg (2004)
20. Tonazzini, A., Salerno, E., Mochi, M., Bedini, L.: Blind source separation techniques for detecting hidden texts and textures in document images. In: Campilho, A.C., Kamel, M.S. (eds.) ICIAR 2004. LNCS, vol. 3212, pp. 241–248. Springer, Heidelberg (2004)
21. Tonazzini, A., Bedini, L., Salerno, E.: Independent component analysis for document restoration. Int. J. Doc. Anal. Rec. 7(1), 17
22. Tonazzini, A., Bedini, L., Salerno, E.: A Markov model for blind image separation by a mean-field EM algorithm. IEEE Trans. Image Processing 15(2), 473
23. Tonazzini, A., Salerno, E., Bedini, L.: Fast correction of bleed-through distortion in grayscale documents by a blind source separation technique. Int. J. Doc. Anal. Rec. 10(1), 17
24. Tonazzini, A., Bianco, G., Salerno, E.: Registration and enhancement of double-sided degraded manuscripts acquired in multispectral modality. In: Ramos, O., Karatzas, D. (eds.) Int. Conf. Doc. Anal. Rec., ICDAR 2009, pp. 546–550. IEEE Computer Society Press, Los Alamitos (2009)
25. Tonazzini, A., Gerace, I., Martinelli, F.: Multichannel blind separation and deconvolution of images for document analysis. IEEE Trans. Image Processing 19(4), 912
26. Extensible Markup Language, http://www.w3.org/XML/

# Model Learning from Published Aggregated Data

Janusz Wojtusiak and Ancha Baranova

**Abstract.** In many application domains, particularly in healthcare, an access for individual datapoints is limited, while data aggregated in form of means and standard deviations are widely available. This limitation is a result of many factors, including privacy laws that prevent clinicians and scientists from freely sharing individual patient data, inability to share proprietary business data, and inadequate data collection methods. Consequently, it prevents the use of the traditional machine learning methods for model construction. The problem is especially important if a study involves comparisons of multiple datasets, where each is derived from different open-access publications where data are represented in an aggregated form. This chapter describes the problem of machine learning of models from aggregated data as compared to traditional learning from individual examples. It presents a method of rule induction from such data as well as an application of this method to constructing of the predictive models for diagnosing liver complications of the metabolic syndrome – one of the most common chronic diseases in humans. Other possible applications of the method are also discussed.

## 1 Introduction

Open – access publications are one of the most important sources of scientific data vital for healthcare research and industry. These publications can be automatically searched and retrieved from the internet and in many instances they are used to build foundation for further studies. Unfortunately, it is often not possible to obtain

Janusz Wojtusiak
Department of Health Administration and Policy,
George Mason University Northeast Module,
Room 108 4400 University Drive, MSN 1J3
Fairfax, VA 22030, USA
e-mail: jwojt@mli.gmu.edu

Ancha Baranova
The Center for Biomedical Genomics
Room 182 Discovery Hall, MSN 4D7
10900 University Blvd
Manassas VA, 20110
e-mail: abaranov@gmu.edu

M. Biba and F. Xhafa (Eds.): Learning Structure and Schemas from Documents, SCI 375, pp. 369–384.
springerlink.com      © Springer-Verlag Berlin Heidelberg 2011

the original datasets on which the published studies were performed. This is because more often than not, scientists are reluctant or not allowed to share their original data. This is particularly the case in medical, behavioral, and social studies in which data are protected by patients' privacy laws. Many studies also use confidential financial, management or security datasets that cannot be shared outside an organization. The most common reasons for which data are not available are:

- **Patient privacy.** Privacy laws preventing sharing and using individual patient data are enforced in most countries. While particulars may differ, the privacy laws often require informed consent of each patient for his or her personal data to be used for a specific study. Examples of such laws are the U.S. Health Portability and Accountability Act of 1996 [1], and the Directive 95/46/EC of the European Parliament and the Council of 24 October 1995 that cover the protection of individuals with regard to the processing of personal data and the movement of such data [29].

- **Confidentiality.** In many cases collected datasets remain confidential as they include information that cannot be shared due to business or other reasons. This is often the case with financial information (i.e. hospital billing datasets) or existing patients' records (i.e., electronic medical records) that a company keeps protected from competitors. To attain access to such data, special agreements are required, and these are often impossible to arrange.

- **Keep data within a research group for further use.** All aspects of collecting data take enormous efforts. Those who have access to reliable datasets have better chances of publishing their research results and, therefore, better performance reviews and possibility of funding. However, while most researchers do agree that all research data should be shared, not many are actually willing to share their own datasets.

- **Lack of public trust in sharing data.** People are concerned about storage and sharing of their personal information as data by public sector and private organizations. This includes but is not limited to data like DNA fingerprints and electronic medical records.

- **No individual data are recorded.** In many cases no individual data are collected, recorded, or reported in the final dataset(s). Some examples of this type of datasets include public health data that combine reports from multiple local governments or organizations that communicate to public only summaries for each area of assessment, and results of large scale experiments in which datasets are simply too large to store and, therefore, need to be immediately processed and aggregated before storing. Such datasets are found, for instance, in astronomy and physics.

Recently, some research funding agencies, including the National Institutes of Health, implemented a policy requiring that data collected in supported studies, mainly clinical trials, be available to others for research purposes. This policy, however, covers only a small fraction of studies performed worldwide.

Most publications, for many of the above reasons, will include only summaries of data in aggregated forms. This is partially due to public health surveillances and data collection systems, which rely on aggregated data collected by distributed institutions [7]. Traditional machine learning methods are not suited for such datasets, as they are designed to work with individual data points. This chapter focuses on the use of aggregated data extracted from medical publications. However, the methodology is translatable and can be applied in other domains.

## 2 Mission, Objectives, and Contributions

Analysis of published aggregated data requires a new class of machine learning methods. Aggregated data are most often presented as means and standard deviations for several parameters measured over a group of observations (i.e., in certain cohort of patients). This chapter introduces a concept of a learning process based on aggregated data, and describes a rule-based approach to creating predictive models from such datasets. Specifically, the approach employs an AQ-based learning process that uses aggregated data to derive attributional rules that are more expressive than standard IF ... THEN rules. This knowledge representation is briefly described in Section 5, and the learning algorithm in Section 6.

There are several requirements for successful application of machine learning to aggregated data. Many of these criteria are also applicable to traditional machine learning from individual examples.

-   **Accuracy.** Models have to provide reliable predictions, which is in most cases their main function. Although models are learned from aggregated data, their accuracy is measured using individual datapoints within a traditional type of validation datasets. Multiple measures of accuracy are available, all of which perform some form of accounting of correct and incorrect predictions and combinations thereof. The most commonly used measures of accuracy include precision, recall, sensitivity, specificity, F-score, and others. When only aggregated data are available, these measures can be estimated as described in Section 6.3.

-   **Transparency.** Medical and healthcare studies require models to be understood easily by people not trained in machine learning, statistics, and other advanced data analysis methods. In this sense, providing just the reliable predictions is not sufficient, as models should also "explain" why a specific prediction is made and what the model actually does. The concept of understandability and interpretability is very well known in expert systems and early work on artificial intelligence, but has been largely ignored by many modern machine learning methods.

- **Acceptability.** Models need to be accepted by potential users. While partially related to transparency, acceptability requires that the models don't contradict the knowledge of existing experts or are otherwise "reasonable."

- **Efficiency.** Both model induction and model application algorithms need to be efficient. Although machine learning from aggregated data in many cases does not involve very large datasets, data that is derived from relevant publications can represent between tens and thousands of cohorts. Although much smaller than considered for data mining algorithms, aggregated datasets should not be subjected to analysis by inefficient algorithms with very large computational complexity.

- **Exportability.** Results of machine learning should be directly transferable to decision support systems. It is not unusual that these learned models will work along with other existing models and need to be compatible. For example, learned models can be translated or directly learned in the form of rules in Arden Syntax [14], a popular representation language in clinical decision support systems.

## 3 Related Work

The problem of analyzing results of published studies is well known. Systematic reviews are used to gather, process, and analyze findings within a collection of closely related studies. Their goal is to arrive at conclusions supported by many other studies. Meta-analysis methods, often used in systematic reviews, are used to calculate statistical descriptions that characterize data used in multiple studies. Systematic reviews and meta-analysis of published studies are important research tools, and are particularly popular in healthcare, policy, social sciences, and law. By combining the results of multiple studies, meta-analysis is able to increase the confidence in study conclusions and cross-validate the results of the particular study. Extensive theory has been built on how to aggregate results from multiple studies and derive statistically valid conclusions [15]. These methodologies are used in preparing systematic reviews such as those by the Cochrane Collaboration [13][4] in healthcare, and the Campbell Collaboration [5][9]. in public policy and law.

In addition to other disciplines concerned with identifying knowledge in published studies, although other forms exist including rarely use the same sets of attributes. Literature-based discovery (LBD) seeks to identify unknown relationships in data drawn from published results [12][30]. By bringing together results published in several papers, new relationships that were not considered in the original studies can be found. Several methods have been created to support LBD [3]. While the general framework of LBD is somewhat similar to the described method, its goal is to discover relationships, rather than build models.

Significant work has been done to develop methods for finding and classifying publications to be included in systematic reviews [18]. These include both research and commercial systems working with publication databases.

Surprisingly, despite the extremely fast growth of machine learning, a discipline that developed powerful data analysis and knowledge discovery tools, little work has been done to use advanced learning methods to support systematic reviews and meta-analysis. Two machine learning areas that are closely related to the described method are statistical relational learning [11] [6] and inductive logic programming [16]. Both areas deal with the more general problem of learning from datasets with complicated structures, rather than the specific problem of learning from published results.

## 4 Aggregated Data

In this chapter we assume a usual situation in which each patient is an individual datapoint described using a set of attributes $A_1 \ldots A_k$. Typical machine learning programs use such individual datapoints in the form of attribute-value examples (1) where $v_1, v_2, \ldots, v_k$ are values of attributes $A_1, \ldots A_k$.

$$(v_1, v_2, \ldots, v_k) \tag{1}$$

Typically each example is described using the same attributes, thus the input dataset used for learning is in the form of a flat attribute-value table. In the case when some attributes are not present in a description of a specific example, meta-values (a.k.a. missing values) can be used. This form of data is, however, almost never included in published manuscripts for reasons outlined in the introduction.

The most typical form of data available in publications is "aggregated tables," although other forms including correlations coefficients, regression models, and others. An aggregated table includes a summary of data aggregated for one or more group of examples (i.e., patient cohorts), usually given as means and standard deviations or frequencies of attributes' values in that groups. Some papers report standard errors, variances, or confidence intervals that can be usually converted into standard deviations. In this chapter we assume that $G_1, \ldots, G_n$ are groups of patients described in publications $P_1, \ldots P_p$. One publication often includes more than one group of patients (i.e., disease and healthy controls, or before and after treatment). Usually, all patients within one group are described using the same set of attributes, although patient groups described in different publications rarely use the same sets of attributes.

Table 1 illustrates example data derived from multiple publications related to metabolic syndrome. It includes means and standard deviations of several attributes in two groups (NAFLD and controls) derived from studies about non-alcoholic fatty liver disease (NAFLD).

**Table 1** Example aggregated data derived from multiple publications. It is a subset of datasets used to induce rules described in Section 7.

|         | NAFLD          | NAFLD       | SS             | NASH            | SS            | C_NASH     |
|---------|----------------|-------------|----------------|-----------------|---------------|------------|
| M/F     | 15–2           | 155–19      | ?              | ?               | 11–10         | 10–9       |
| Age     | 44+/-3         | 41+/-11     | ?              | ?               | 41+/-13       | 43+/-14    |
| Weight  | 86.2+/-3.5     | ?           | ?              | ?               | ?             | ?          |
| BMI     | 27.4+/-0.8     | 27.3+/-3.2  | ?              | ?               | 33+/-45       | 31+/-4     |
| Height  | 1.77+/-0.02    | ?           | ?              | ?               | ?             | ?          |
| FG      | 107.1+/-7.56   | 98.2+/-26.0 | 108.90+/-9.09  | 108.18+/-8.72   | 93+/-13       | 91+/-11    |
| FI      | 13.4+/-1.5     | 15.1+/-7.9  | 13.9+/-2.0     | 12.7+/-2.2      | 23.98+/-16.78 | 12.94+/-9.6 |
| TC      | 208.07+/-11.92 | ?           | 211.53+/-19.23 | 199.61+/-7.69   | 207+/-36      | 280+/-50   |
| HDL     | 48.36+/-3.9    | 47+/-11     | 44.85+/-7.8    | 50.31+/-10.92   | 36+/-6        | 47+/-14    |
| LDL     | 126.36+/-11.15 | ?           | 121.68+/-11.7  | 128.31+/-12.09  | 135+/-26      | 131+/-41   |
| T       | 181.56+/-31.15 | 138+/-93    | 191.35+/-40.05 | 178+/-26.7      | 230+/-85      | 165+/-75   |
| HOMA    | 3.61+/-0.55    | ?           | 3.30+/-0.40    | 3.75+/-0.60     | 7.0+/-5.4     | 3.2+/-3.0  |

FG = Fasting Glucose (mg/dl), FI = Fasting Insulin (mUI/l) , TC = Total Cholesterol, T = Triglycerides

Aggregated values of attributes are given in the form of pairs $(\mu_A, \sigma_A)$. Where $A$ is a measured attribute, and $\mu_A$ and $\sigma_A$ denote its mean and standard deviation measured over a group, for which the aggregation was done. Given that means and standard deviations for several parameters are available, each group can be described by an *aggregated example* given as (2). It can be simplified into (3) when order of attributes is defined.

$$(A_1 = (\mu_{A_1}, \sigma_{A_1}), A_2 = (\mu_{A_2}, \sigma_{A_2}), \ldots, A_k = (\mu_{A_k}, \sigma_{A_k})) \qquad (2)$$

$$((\mu_{A_1}, \sigma_{A_1}), (\mu_{A_2}, \sigma_{A_2}), \ldots, (\mu_{A_k}, \sigma_{A_k})) \qquad (3)$$

For non-numerical attributes, a typically used aggregated form lists frequencies of values in a group, explicitly showing distribution of examples. For example, a group of patients may include 20% smokers and 80% non-smokers.

Another type of data describes entire groups. In medical or social publications these can be related to inclusion criteria for a specific study and additional facts about participants. Although describing groups of data, these attributes refer directly to individual examples. For, example if a study is performed among white males, then each individual subject in the data has precisely this value for attributes describing ethnicity and gender.

Sample sizes (numbers of examples in groups) are always provided. Although they do not provide any information about the subjects themselves, but rather about groups, sample sizes constitute important information crucial during model induction and its coverage estimation. Given both aggregated and not aggregated data, examples take the form (4).

$$(size, (\mu_{A_1}, \sigma_{A_1}), (\mu_{A_2}, \sigma_{A_2}), \ldots, (\mu_{A_k}, \sigma_{A_k})), v_k + 1, \ldots v_l) \qquad (4)$$

Note that in one study an attribute can be in the aggregated form, in the second study the same attribute can be in non-aggregated form (i.e. used as inclusion criteria), and completely not available in the third study.

The forms of data outlined above differ from those typically used in machine learning from examples, dominated by learning from individual attribute-value examples in the form (1). Although handling qualitative statements and background knowledge, which is a part of structured machine learning [8], and has been well studied in inductive logic programming [16] [25] and statistical relational learning [11], no special methods for learning from published results that stress aggregated data are available. Furthermore, although relational learning assumes using aggregates [26] [28], they deal only with individual examples with additional characteristics that are being aggregated. This is in contrast to the presented method in which an aggregated example represents a group.

# 5 Attributional Rules as Knowledge Representation

An earlier part of this chapter discussed requirements for models induced from aggregated data. Rule-based knowledge representation is known to satisfy several of the criteria. However, standard IF...THEN rules are using only conjunctions of simple statements and have limited expression power. Therefore, more expressive forms of rules are used in the presented work.

The main representation of knowledge used in the described method is *attributional rules* [21] whose one form is given by (5). Both *CONSEQUENT* and *PREMISE* are conjunctions of attributional conditions (6). The symbols <=, and |_ denote implication and exception operators, respectively. *EXCEPTION* is either an exception clause in the form of a conjunction of attributional conditions or an explicit list of examples constituting exceptions to the rule. *ANNOTATION* is an additional statistical description, including, for example, the rule's coverage.

$$CONSEQUENT <= PREMISE \mid\_ EXCEPTION : ANNOTATION \qquad (5)$$

$$[L\ REL\ R : A] \qquad (6)$$

An attributional condition corresponds to a simple natural language statement. Its general form is (6), in which $L$ is an attribute, a counting attribute (derived from other attributes), or a simple arithmetical expression over numerical attributes; $R$ is an attribute value, internal disjunction or conjunction of values, a range, or an attribute; $REL$ is a relation applicable to $L$ and $R$; and $A$ is an optional annotation that provides statistical information characterizing the condition. The annotation includes numbers of cases satisfied by the condition and its consistency. When $L$ is a binary attribute $REL$ and $R$ may be omitted. Several other forms of attributional rules are available, all of which resemble statements in natural language, and thus are interpretable by people not trained in machine learning [21].

The above choice of rule-based knowledge representation is based on the fact that it satisfies transparency and exportability criteria for models stated in Section 2. It can also provide accuracy comparable with other representations, without the need to employ special procedures that convert black-box representations to human-oriented explanations [2].

The following section describes an algorithm for inducing attributional rules from data, and its extension needed to handle aggregated data.

# 6 Rule Induction

## 6.1 AQ Algorithm

Many algorithms are available for inducing rules from data. Despite their differences, the algorithms have two common elements: rule construction, and rule evaluation. Although the described method for learning from aggregated data can be adapted to most rule learning systems, in this chapter the focus is on the AQ approach to rule learning [19] [20] [32]. This focus is important because the method has several advantages that make it suitable for learning from aggregated data. AQ generates attributional rules described above, deals with multiple data types [23] and meta-values [22], includes different generalization and reasoning methods, and is fairly flexible due to the large number of parameters that control the learning process. The method follows the popular separate-and-conquer approach to rule learning that is summarized by [10], and is capable of using powerful statistical measures of rules quality that incorporate aggregated data found in published results.

The AQ learning works in two main stages: rule construction and rule optimization. At the core of the first stage is a star generation algorithm, which creates multiple generalizations (in the form of attributional rules), called *stars*, of a selected positive example that do not cover negative examples. A combination of rules selected from one or more stars is used as a generated hypothesis. Within the star generation, AQ applies an *extension-against operator[20]* whose goal is to find all possible rules that distinguish a given positive example, called *seed*, from a given negative example. In the original method, the extension-against is a purely logical operation and it is denoted by the --⎯ symbol. In its simplified form for non-aggregated data, a seed $s = (a_1,...a_k)$ extended against a negative example $n = (b_1,...b_k)$ is a set of one-condition rules shown in (7) for all attributes for which values in the seed and the negative example are different.

$$s \, \text{--} \vert \, n = \bigvee [A_i \neq b_i] \text{ for all } i \text{ such that } a_i \neq b_i \tag{7}$$

For example, $(7, 5, 3) \, \text{--} \vert \, (2, 5, 4) = [A_1 \neq 2] \vee [A_3 \neq 4]$. Here, the attribute $A_2$ is not used because it takes the same value in the seed and the negative example. The operator works the same way for symbolic (nominal, structured, etc.) and numeric (interval, ratio, etc.) attributes.

Multiple applications of the operator allow for the creation of rules that cover the seed and exclude negative examples. Intersection of all such rules covers the seed and rule out any negative example.

At this stage of rule construction, AQ applies a beam search to filter out potentially large number of generated rules. The method allows for multiple criteria of rule evaluation, most of which are based on statistical evaluation and complexity of rules. In the second stage, rules/hypotheses are optimized to maximize their predictive accuracy while maintaining simplicity. This process is somewhat similar to pruning, which is frequently done by learning programs. In AQ, rules are not only pruned, but can also be extended through a set of optimization operators working on attribute, condition, rule, and hypothesis levels.

When learning from aggregated data, information about distributions of values in aggregated examples is used. Consequently, the *extension-against operator* is no longer purely logical.

## 6.2 Rule Induction from Aggregated Data

A general schema of learning from published results is presented in Figure 1. Aggregated and individual data are used for rule generation and evaluation, while qualitative/quantitative results and background knowledge are used to constrain the generation of models. Each rule is evaluated not only for coverage-based quality (statistical measures), but also by its simplicity and given constraints. There are several possible approaches to the problem of rule induction from aggregated examples. This section briefly overviews these approaches, with the focus primarily on the third method that directly uses aggregated information within the AQ rule induction algorithm.

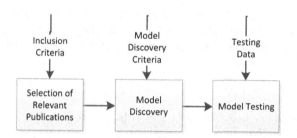

**Fig. 1** A flowchart describing the process of model development based on published aggregated data.

**Sampling.** One simple approach for learning from published results is to approximate the original datasets by sampling. An initial study [24], in which aggregated examples were sampled, indicated, however, that the method does not work well due to deficient information of interrelationships between attributes. An important advantage of the method is that any machine learning method from examples can be applied, because individual examples are created during the sampling process.

**Weighting.** Another simple method is to use aggregated examples in the form (8) which includes only mean values. Standard deviations are used to weight examples [33] [34].

$$(\mu A1, \mu A2,...,\mu Ak) \tag{8}$$

By doing so, there is no need to significantly modify rule learning algorithms. However, the method ignores important information of "overlaping" groups.

It is important to note that when using this method the rule induction algorithm treats aggregated examples as individual subjects and de facto learns rules that describe weighted groups. Despite its simplicity, the method gave good results in an initial study presented in Section 7.

Limitations of the above methods show the importance of using an algorithm that directly induces rules for classifying individual examples based on aggregated examples. Such algorithm needs to effectively use information about distributions and "recognize" the fact that it is dealing with aggregated examples representing groups not with individual subjects.

**Extension-against.** The AQ rule learning algorithm can use directly the form (4) of aggregated examples, and effectively incorporate the information about standard deviations when comparing aggregated examples. Assuming normal distribution, $N(\mu_{A_i}, \sigma^2_{A_i})$, over 95% of data described by the aggregated examples lay in the range given by condition (9).

$$[\mu_{A_i} - 2\sigma_{A_i}, \mu_{A_i} + 2\sigma_{A_i}] \tag{9}$$

Thus, when comparing two aggregated examples, a reasonable assumption is that two aggregated values are indistinguishable if the ranges (9) in the examples are intersecting. The modified extension against operator is defined by the formula (10) for numeric attributes.

$$s --| n = V [x_i \neq (\mu_{A_i^n} - 2\sigma_{A_i^n}, \mu_{A_i^n} + 2\sigma_{A_i^n})] \tag{10}$$

for all $i=1..k$ such that $[\mu_{A_i^s} - 2\sigma_{A_i^s}, \mu_{A_i^s} + 2\sigma_{A_i^s}] \cap [\mu_{A_i^s} - 2\sigma_{A_i^s}, \mu_{A_i^s} + 2\sigma_{A_i^s}]$.

For aggregated discrete attributes the extension-against operator is defined by the formula (11).

$$s --| n = V [A_i \neq v_1..v_k], i = 1..n \tag{11}$$

$$fs(A_i, v_j) < fn(A_i, v_j) + \varepsilon$$

Here, $fs$ denotes distribution of values in $s$ and $fn$ denotes distribution of examples in $n$. For example, if $D(A_1)=\{a,b,c,d\}$. $D(A_2) =\{r,g\}$, $s=(A_1=(0.3,0.05,0,0.65),$ $A_2=(0.2,0.8))$ and $n=(A_1=(0.5,0,0.3,0.2), A_2=(0.27,0.73))$, and $\varepsilon=0.1$, then $s --| n = [A_1 \neq a, c]$. The attribute $A_2$ is not used at all, because the difference between distributions for that attribute is within the margin $\varepsilon$.

For non-aggregated attributes, i.e. attributes that describe entire groups, the extension-against operator is not modified.

## 6.3 Calculating Coverage

A key component to any rule induction algorithm is the calculation of rules' coverage. Positive and negative coverage needs to be estimated for individual examples using aggregated examples representing groups. This is needed during both rule creation and rule optimization. In order to estimate numbers of examples from a group satisfying a condition [A=a...b] learned from aggregated data (A is a continuous attribute), the probability (12) of an individual example satisfying the condition can be multiplied by the number of examples in the group. $\Phi_{\mu\sigma^2}(A)$ is the cumulative distribution function, $\mu$ is mean value of A in the group, and $\sigma$ is standard deviation of A in the group.

$$p(A = a...b) = \Phi_{\mu,\sigma^2}(b) - \Phi_{\mu,\sigma^2}(a) \tag{12}$$

In order to estimate the numbers of examples satisfying a rule an independence of attributes is assumed. The rationale behind the assumption is that if two attributes were dependent, the rule learning program would not need to use both of them in the rule (as the value of one implies the value of another). Thus, probabilities (12) for all conditions in the PREMISE can be multiplied. The resulting joint probability is then multiplied by one minus the joint probability of the EXCEPTION, which gives the probability of an example in the group satisfying the rule. Finally, the estimated number of examples from the group satisfying the rule is calculated by multiplying the joint probability by the number of examples in that group. The operation is repeated for all groups for which aggregated data are available.

In the presence of additional information such as covariance between attributes, it is possible to calculate a better estimation of the joint probability than when assuming independence of conditions.

Using learned rules to classify new examples is straightforward, because they are intended to classify individual examples, not aggregated examples representing groups. Rules for classifying individual examples are learned from aggregated examples.

## 7 Evaluation

The initial methodology [33] [34] for learning from aggregated data has been applied to a small database derived from clinical research publications in a number of well-known peer-reviewed journals. Part of the database was presented in Table 1. The application was concerned with creating predictive models for diagnosing liver complications in metabolic syndrome (MS). Metabolic syndrome and its secondary complications pose a significant challenge for practicing diagnosticians. Abdominal obesity appears to be its predominant underlying risk factor. Metabolic abnormalities associated with MS, particularly a resistance to the insulin, predispose people to non-alcoholic fatty liver disease (NAFLD) and its more severe manifestation, nonalcoholic steatohepatitis (NASH). The health-related costs associated with these complications are substantial, thus early prediction and prevention of these conditions are of significant importance. Currently, it is not possible to make an accurate diagnosis of NAFLD and/or NASH without a liver biopsy. It is an invasive and costly procedure that is prone to complications, some minor, such as pain, and some more severe, including possibility of death as a result of bleeding or infection [27].

An attractive alternative, pursued in the research, was to use panels of the serum markers, because blood samples could be collected in a minimally invasive way. However, the predictions made in prior studies using currently available prediction algorithms lack consistency. Typically, clinical studies of MS and its complications are performed on single groups of patients collected in one hospital, and use only simple statistical measures for group comparisons and correlation plotting. No large datasets concerning metabolic syndrome are available and data describing measurables in each patients are not available, either. Thus, only methods that deal with results published in papers are applicable.

The data used for this study were in the aggregated form (3). They were collected from articles published in peer-reviewed journals including Hepatology, Obesity Research, International Journal of Obesity, and some others. For the pilot study, we retrieved aggregated clinical data from 20 separate hospital cohorts that included 12 groups of patients with present liver disease symptoms and 8 control groups of healthy subjects. Every single group of patients was described in terms of the mean of attributes measured for this group of patients. The total number of different attributes retrieved from papers was 152. In each study however, different attributes were measured, which added additional complexity to the problem. In fact, none of the attributes were present in all studies, even though these very similar studies were dealing with exactly the same clinical problem.

The goal was to construct a set of rules for predicting non-alcoholic fatty liver disease (NAFLD), simple steatosis (SS), and Nonalcoholic Steatohepatitis (NASH). Data also included a number of healthy cohorts, represented as control groups serving as a contrast set for learning. It should be noted that NAFLD is the most general condition that comprises both SS and NASH cases. Therefore, we first sought rules that differentiate NAFLD from healthy cases and then rules characterizing NASH, the most severe form of NAFLD.

Below we present two example rules derived by the method. The first rule states that there is presence of non-alcoholic fatty liver disease or its subtypes, if body-mass index is greater or equal 26.85, except for when aspartate aminotransferase level is at most 27.2 units/L and adiponectin level is at least 7.25 mg/ml. The rule's condition is satisfied by eight groups of patients belonging to NAFLD or its subtypes, and two control groups. The exception part of the rule which consists of two conditions filters out both control groups. The entire rule is satisfied by eight groups of patients belonging to NAFLD or its subtypes and non-control groups. The rule's quality is 0.816, and its complexity is 25. The second rule can be interpreted in an analogous way.

[Class=NAFLD]
    <== [BMI>=26.85: 8,2]
     |_ [AST<=27.2] & [Adiponectin>=7.25]
     : p=8,n=0,Q(w)=0.816,cx=25

[Class=NAFLD]
    <== [Adiponectin<=6.18: 8,1]
     : p=8,$n_{min}$=0,$n_{max}$=1,Q(w)=0.695,cx=5

Similar rules have been obtained for predicting simple setatosis and nonalcoholic steatohepatitis [34]. The rules are easy to interpret and are consistent with experts' existing knowledge. An explanation of the parameters is in the AQ21 User's Guide [31], the system that was used to implement the initial methodology.

Validation of these rules for predicting NAFLD resulted in a positive predictive value (PPV) of 85-87%, reflecting relatively high "rule-in" characteristic of the algorithm. The best rule for the prediction of NASH relied on combination of fasting insulin, HOMA and adiponectin values with an accuracy of 78%, with PPV of 71% and negative predictive value (NPV) of 37%.

The models generated by AQ21 are presented in the form of attributional rules, a highly transparent representation which is easy to understand by people not trained in advanced statistics, machine learning, and other computational technologies. Additionally, these kinds of models could be readily imported into existing clinical decision support systems and useful in the settings of point-of-care (POC) initial health assessment. Simplicity of the developed models allows for the use of them on "the back of envelope" in settings where advanced diagnostics are not available (i.e. in developing countries).

## 8 Discussion

The presented methodology for learning from aggregated data has been developed within the well known AQ rule induction algorithm. The algorithm is able to induce from aggregated data attributional rules for classifying individual examples. The process is depicted in Figure 2.

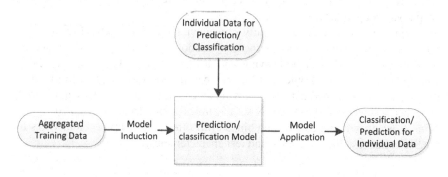

**Fig. 2** Induction of models from aggregated data and application to individual data.

Traditional machine learning from examples methods are able to deal with aggregated data when using the sampling method described in Section 6. It is also possible to extend some the methods to deal directly with aggregated data, however, such extension depends on how the specific algorithm works and how it treats individual examples.

Data used in the presented study were manually retrieved from selected publications. It's been recommended that this very labor-intensive process be performed by at least two independently working people and then the results compared and discrepancies discussed in a panel. The selection process of publications to be included in the analysis should also be done by the panel [17]. With the use of currently available technology the process cannot be fully automated, because it depends on the understanding of the publications. However, it is possible to aid personnel in performing this time consuming task. Relevant publications can be pre-selected using advanced search tools available for databases such as PubMed. Data tables can be automatically identified in publications and derived in a tabular form. Ontologies and dictionaries can be used to discover discrepancies in terminology and units. Finally, text mining methods can be used to identify and retrieve data not present in tabular forms (i.e. inclusion criteria).

## 9 Conclusion

New methods are needed to create accurate and transparent predictive models from de-individualized published clinical data in aggregated forms. Machine learning of attributional rules from published data, including aggregated clinical parameters, inclusion criteria, demographic information and target diagnoses, is able to derive such models.

This chapter described a methodology for machine learning or attributional rules from aggregated published data. The described methodology may complement current systematic reviews and meta-analyses such as Cochrane Reviews. With rapidly changing clinical knowledge, such an automated method with the ability to incrementally update knowledge can prove to be the needed method to keep reviews up to date.

Preliminary application of an early implementation of the method resulted in a set of attributional rules for predicting non-alcoholic fatty liver disease and its subtypes in patients with metabolic syndrome. It illustrated validity of the method on a real-world important problem. Machine learning software applied to the meta-analysis of the published data may provide an easy, non-invasive way to diagnose most patients with NAFLD and NASH. Clinical parameters highlighted by machine learning process can be combined with other non-invasive biomarkers for NASH to increase their accuracy and test characteristics.

# References

[1] Annas, G.J.: HIPAA Regulations — A New Era of Medical-Record Privacy? New England Journal of Medicine 348, 1486–1490 (2003)

[2] Baehrens, D., Schroeter, T., Harmeling, S., Kawanabe, M., Hansen, K., Müller, K.: How to Explain Individual Classification Decisions. Journal of Machine Learning Research 11, 1803–1831 (2010)

[3] Burza, P., Weeber, M.: Literature-based Discovery. Springer, Heidelberg (2008)

[4] The Cochrane Collaboration, The Cochrane Manual 4 (2008) (updated August 14, 2008)

[5] Davies, F., Boruch, R.: The Campbell Collaboration Does for Public Policy what Cochrane Does for Health. BMJ 323, 294–295 (2001)

[6] De Raedt, L.: Logical and Relational Learning. Springer, Heidelberg (2008)

[7] Diamond, C.C., Mostashari, F., Shirky, C.: Collecting And Sharing Data For Population Health: A New Paradigm. Health Affairs 28(2) (2009)

[8] Dietterich, T.G., Domingos, P., Getoor, L., Muggleton, S., Tadepalli, P.: Structured machine learning: the next ten years. Machine Learning 73(1), 3–23 (2008)

[9] Farrington, D.P., Petrosino, A.: The Campbell Collaboration Crime and Justice Group. Annals of the American Academy of Political and Social Science 578, 35–49 (2001)

[10] Fürnkranz, J.: Separate-and-conquer rule learning. Artificial Intelligence Review 13, 3–54 (1999)

[11] Getoor, L., Taskar, B. (eds.): Introduction to statistical relational learning. MIT Press, Cambridge (2007)

[12] Gordon, M., Lindsay, R.K., Fan, W.: Literature-Based Discovery on the World Wide Web. ACM Transactions on Internet Technology 2(4), 261–275 (2002)

[13] Higgins, J.P.T., Green, S. (eds.): Cochrane Handbook for Systematic Reviews of Interventions (2008), http://www.cochrane-handbook.org Version 5.0.0 (updated February 2008)

[14] Hripcsak, G.: Writing Arden Syntax medical logic modules. Computers in Biol-ogy and Medicine 24(5), 331–363 (1994)

[15] Hunter, J.E., Schmidt, F.L.: Methods of Meta-Analysis, Correcting Error and Bias in Research Findings, 2nd edn. Sage Publications Inc., Thousand Oaks (2004)

[16] Lavrac, N., Dzeroski, S.: Inductive Logic Programming: Techniques and Applications. Ellis Horwood, New York (1994)

[17] Lipsey, M.W., Wilson, D.: Practical Meta-Analysis. Sage Publications, Thousand Oaks (2000)

[18] Matwin, S., Kouznetsov, A., Inkpen, D., Frunza, O., O'Blenis, P.: A new algorithm for reducing the workload of experts in performing systematic re-views. Journal of the American Medical Informatics Association 17(4), 446–453 (2010)

[19] Michalski, R.S.: On the Quasi-Minimal Solution of the General Covering Prob-lem. In: Bled, Y. (ed.) Proceedings of the V International Symposium on Information Processing (FCIP 1969), vol. 3, pp. 125–128 (1969)

[20] Michalski, R.S.: A Theory and Methodology of Inductive Learning. In: Michalski, R.S., Carbonell, T.J., Mitchell, T.M. (eds.) Machine Learning: An Artificial Intelligence Approach, pp. 83–134. TIOGA Publishing Co, Palo Alto (1983)

[21] Michalski, R.S.: ATTRIBUTIONAL CALCULUS: A Logic and Representation Language for Natural Induction, Reports of the Machine Learning and Inference Laboratory, MLI 04-2, George Mason University. Fairfax, VA (2004)

[22] Michalski, R.S., Wojtusiak, J.: Reasoning with Missing, Not-applicable and Irrelevant Meta-values in Concept Learning and Pattern Discovery, Technical Report 2005-02, Collaborative Research Center 637, University of Bremen, Germany (2005)

[23] Michalski, R.S., Wojtusiak, J.: Semantic and Syntactic Attribute Types in AQ Learning, Reports of the Machine Learning and Inference Laboratory, MLI 07-1, George Mason University. Fairfax, VA (2007)

[24] Michalski, R.S., Wojtusiak, J.: The Distribution Approximation Approach to Learning from Aggregated Data, Reports of the Machine Learning and Inference Laboratory, MLI 08-2, George Mason University. Fairfax, VA (2008)

[25] Muggleton, S.H., De Raedt, L.: Inductive logic programming: Theory and me-thods. Journal of Logic Programming 19(20), 629–679 (1994)

[26] Perlich, C., Provost, F.: Distribution-based aggregation for relational learning with identifier attributes. Machine Learning 62, 65–105 (2006)

[27] Poynard, T., Ratziu, V., Charlotte, F., Messous, D., Munteanu, M., Imbert-Bismut, F., Massard, J., Bonyhay, L., Tahiri, M., Thabut, D., Cadranel, J.F., Le Bail, B., de Ledinghen, V.: LIDO Study Group, CYTOL study group, Diagnostic value of biochemical markers (NashTest) for the prediction of non alcoholo steato hepatitis in patients with non-alcoholic fatty liver disease. BMC Gastroenterology 6(34) (2006)

[28] Vens, C.: Complex aggregates in relational learning. AI Communications 21, 219–220 (2008)

[29] Verschuuren, M., Badeyan, G., Carnicero, J., Gissler, M., Asciak, R.P., Sakkeus, L., Stenbeck, M., Devillé, W.: and For The Work Group on Confidentiality and Data Protection of the Network of Competent Authorities of the Health Information and Knowledge Strand of the EU Public Health Programme (August 2003) ; The European data protection legislation and its consequences for public health monitoring: a plea for action. European Journal of Public Health 18(6), 550–551 (2008) doi:10.1093/eurpub/ckn014

[30] Weeber, M., Kors, J.A., Mons, B.: Online tools to support literature-based discov-ery in the life sciences. Briefings in Bioinformatics 6(3), 277–286 (2005)

[31] Wojtusiak, J.: AQ21 User's Guide, Reports of the Machine Learning and Infe-rence Laboratory, MLI 04-3, George Mason University. Fairfax, VA (2004)

[32] Wojtusiak, J., Michalski, R.S., Kaufman, K., Pietrzykowski, J.: The AQ21 Natural Induction Program for Pattern Discovery: Initial Version and its Novel Features. In: Proceedings of The 18th IEEE International Conference on Tools with Artificial Intelligence, Washington D.C (2006)

[33] Wojtusiak, J., Michalski, R.S., Simanivanh, T., Baranova, A.V.: The Natural Induction System AQ21 and Its Application to Data Describing Patients with Metabolic Syndrome: Initial Results. In: Proceedings of the International Conference on Machine Learning and Applications, Cincinnati, OH (2007)

[34] Wojtusiak, J., Michalski, R.S., Simanivanh, T., Baranova, A.V.: Towards application of rule learning to the meta-analysis of clinical data: An example of the metabolic syndrome. International Journal of Medical In-formatics 78(12), e104–e111(2009)

# Data De-duplication: A Review

Gianni Costa, Alfredo Cuzzocrea, Giuseppe Manco, and Riccardo Ortale

**Abstract.** The wide exploitation of new techniques and systems for generating, collecting and storing data has made available growing volumes of information. Large quantities of such information are stored as free texts. The lack of explicit structure in free text is a major issue in the categorization of such kind of data for more effective and efficient information retrieval, search and filtering. The abundance of structured data is problematic too. Several databases are available, that contain data of the same type. Unfortunately, they often conform to different schemas, which avoids the unified management of even structured information. The *Entity Resolution* process plays a fundamental role in the context of information integration and management, aimed to infer a uniform and common structure from various large-scale data collections, with which to suitably organize, match and consolidate the information of the individual repositories into one data set. *De-duplication* is a key step of the Entity Resolution process, whose goal is discovering duplicates within the integrated data, i.e., different tuples that, as a matter of facts, refer to the same real-world entity. This attenuates the redundancy of the integrated data and, also, enables more effective information handling and knowledge extraction through a unified access to reconciled and de-duplicated data. Duplicate detection is an active research area that benefits from contributions from diverse research fields, such as, machine learning, data mining and knowledge discovery, databases as well as information retrieval and extraction. This chapter presents an overview of research on data de-duplication, with the goal of providing a general understanding and

Gianni Costa · Alfredo Cuzzocrea · Giuseppe Manco · Riccardo Ortale
ICAR-CNR, Via P. Bucci, 41C, 87036 Rende (CS) - Italy
e-mail: {costa,cuzzocrea,manco,ortale}@icar.cnr.it

M. Biba and F. Xhafa (Eds.): Learning Structure and Schemas from Documents, SCI 375, pp. 385–412.
springerlink.com                                                    © Springer-Verlag Berlin Heidelberg 2011

useful references to fundamental concepts concerning the recognition of similarities in very large data collections. For this purpose, a variety of state-of-the-art approaches to de-duplication is reviewed. The discussion of the state-of-the-art conforms to a taxonomy that, at the highest level, divides the existing approaches into two broad classes, i.e., unsupervised and supervised approaches. Both classes are further divided into sub-classes according to the common peculiarities of the involved approaches. The strengths and weaknesses of each group of approaches are presented. Meaningful research developments to further advance the current state-of-the-art are covered as well.

# 1 Introduction

Recognizing similarities in large collections of data is a major issue in the context of information integration. The wide exploitation of new techniques and systems for generating, collecting and storing data has made available very large collections of data, such as personal demographic data, bibliographic information, phone and mailing lists. Often, the integration of such data is a problematic process, that involves dealing with two major issues, namely *structural* and *syntactic* heterogeneity.

Structural heterogeneity is essentially due to the lack of a common (explicit or latent) structure in the available data, that avoids a proper organization of such data for a more effective and efficient information retrieval, search and filtering. Structural heterogeneity depends on the nature of the available data. In the case of structured data, the individual repositories explicitly exhibit their own schemas and, thus, dealing with the arising structural heterogeneity involves finding a common schema for the representation of the integrated data. Instead, when the available data collections are unstructured, such as in the case of free text, it is likely that they are characterized by a latent segmentation into specific semantical entities (e.g., personal demographic information typically comprises names, addresses, zip codes and place names, which indicate a convenient organization for the these kind of data). Such a segmentation is not a-priori known and, hence, handling with structural heterogeneity of textual data means finding an explicit common schema from the various latent ones (if any). In principle, such a common schema would allow to fit the integrated textual data into some field structure, so that to exploit the mature relational technology for more effective information management. Structural heterogeneity is addressed through the exploitation of suitable methods and techniques that reconcile distinct collections of (structured or unstructured) data, by integrating them into one data set with a uniform schema.

However, once reconciled, the integrated data can be affected by syntactic heterogeneity. This is a fundamental issue in the context of information integration systems, that consists in discovering duplicates within the integrated data, i.e., syntactically different records that, as a matter of facts, refer to the same real-world entity. Duplicate detection is necessary to avoid data redundancy as well as inaccuracies in query processing and knowledge extraction [13]. A typical example is the reconciliation of demographic data sources into a data warehousing environment. Consider, e.g., a banking scenario, where the main interest is to rank the credit risk of a customer, by looking at the past insolvency history. In this setting, useful information about payments may come from different sources, each of which likely conforming to a different encoding of the data, so that names and addresses may be stored in rather different formats. The reconciliation of the various data sources is a first step towards the design of a decision support system. However, it is the de-duplication of the reconciled data that allows to correctly analyze the attitude of insolvency of the individual customers, thus enabling an effective decision making.

More generally, besides the post-processing of reconciled data, de-duplication techniques are also particularly useful in all those applicative settings where large collections of data are available. In such domains, such techniques can be exploited in a preliminary exploratory phase, for the purpose of reducing the number of duplicated data, which ultimately improves the quality of the underlying data.

Four challenging requirements of duplicate detection are: (*i*) the capability to handle huge volumes of data; (*ii*) efficiency; (*iii*) scalability; (*iv*) the availability of incremental algorithms.

In particular, the requirement to process large bodies of data generally imposes severe restrictions on the design of data structures and algorithms for de-duplication. Such restrictions are necessary to ultimately limit both the computational complexity of the de-duplication scheme (a required time, that is quadratic in the size of the underlying database, is prohibitively high) and its I/O operations on disk (several random accesses to secondary storage imply continuous paging activities).

Efficiency and scalability issues do play a predominant role in many applicative contexts, where large data volumes are involved, especially when the object-identification task is part of an interactive application, calling for short response times. For instance, the typical volume of data collected on a daily basis in a banking context amounts on average to $500,000$ instances, representing credit transactions performed by customers throughout the various agencies. In such a case, the naive solution of comparing such instances in a pairwise manner, according to some given similarity measure, is infeasible. As an example, for a set of $30,000,000$ tuples (i.e., data collected in a 2 months-monitoring), the naive strategy would require $O(10^{14})$ tuple comparisons, which is clearly prohibitive.

Also, the detection of duplicates should preferably be performed in an incremental manner, so that to properly account for the possibly streaming nature of the data. In such cases, the available data collection is incrementally augmented through the progressive integration of newly arrived data, whose prior de-duplication raises a stringent online requirement, i.e., that the redundancy of such data is promptly recognized. Practically speaking, the cost of incremental de-duplication should be (almost) independent of the size of the available data.

This chapter surveys seminal research in the field of duplicate detection and discusses consolidated results in the light of the aforementioned requirements. The discussion proceeds as follows. Section 2 introduces the process of duplicate detection and identifies two categories of approaches, namely, supervised and unsupervised techniques. Section 3 deals with the supervised approaches. Section 4 is devoted to the unsupervised approaches. In both sections 3 and 4, the strengths and the weaknesses of each encountered family of approaches are discussed. Finally, Section 5 concludes by highlighting directions of further research, that can advance the current state-of-the-art in duplicate detection.

## 2 Problem Description

Duplicate detection is a step of a more complex process, referred to as *Entity Resolution* [11, 13, 26, 41, 47, 73, 98], that plays a fundamental role in the context of information integration and management. The typical scenario in the design of information systems is the availability of multiple data repositories, with different schemas and assumptions on the underlying canonical data representation. Schema differences imply a segmentation of data tuples[1] into sequences of strings, that correspond to specific semantical entities. However, such a segmentation is not known in advance and this is a challenging issue for duplicate detection. Moreover, the adoption of various canonical data representations (such as the presence of distinct data separators and/or various forms of abbreviations) coupled with erroneous data-entry, misspelled strings, transposition oversights and inconsistent data collection further exacerbates the foresaid difficulties behind the recognition of duplicates. The goal of Entity Resolution is to suitably reconcile, match and consolidate the information within the different repositories [14], so that all data exhibits a uniform representation and duplicates are properly identified. The process of Entity Resolution consists of the following three steps.

- *Schema reconciliation*, which consists in the identification of a common representation for the information in the available data [1, 18, 21, 66, 71, 77].
- *Data reconciliation*, that is the act of discovering synonymies in the data, i.e. apparently different records that, as a matter of fact, refer to the same real-world entity.
- *Identity definition*, aimed to find groups of duplicate tuples and to extract one representative tuple for each discovered group. Representative tuples allow better information processing as well as a meaningful compression of the size of the original data.

The focus of this chapter in on the techniques for the detection of duplicated data, typically employed in the second step of the Entity Resolution process. Duplicate detection has given rise to a large body of works in several research communities, where it is referred to with as many umbrella names, such as, e.g., *Merge/Purge* [53], *Record Linkage* [39, 97], *De-duplication* [85], *Entity-Name Matching* [30], *Object Identification* [79].

---

[1] The term *tuple* is abstractedly used throughout the chapter to denote a reconciled fragment of data, that can be either a textual sequence of strings or a structured record.

In most of these approaches, a major issue is represented by the definition of a method for comparing tuples, that typically consist of many components. Recognizing similar tuples involves matching their constituting components. These can represent both numbers and strings. Depending on the structured or unstructured nature of the underlying data, such components can be either values of the fields of some database schema or tokens from textual data. In both cases, approaches to data de-duplication require dealing with mismatches between tuple components. While research on the identification of mismatches between numbers is not yet mature, a variety of schemes have been developed for dealing with various kinds of heterogeneity and mismatches between strings across different information sources [58].

A categorization of the most commonly adopted schemes for matching string components in the context of duplicate detection is provided in [58]. Such a categorization includes two major classes of string (dis)similarity functions [50], i.e., *character-based similarity metrics*, that are meant for dealing with differences in the individual characters of string components, and *token-based similarity metrics*, that instead aim to capture a degree of similarity between two tuples, even if their components are rearranged.

A detailed analysis of these metrics is beyond the scope of this chapter. For our purposes, it suffices to know that the availability of such schemes for matching individual component allows the design of suitable approaches to the detection of duplicates in the domains of both structured and textual data [15, 29, 75, 76, 85], in which individual tuples consist of multiple components. These approaches can be divided into two broad and widespread families, i.e., supervised and unsupervised techniques, which are respectively covered in section 3 and section 4.

## 3 Supervised Approaches to De-duplication

The common idea behind such category of approaches is to learn from the training data suitable models for tuple matching [6, 81, 92, 36]. This involves learning probabilistic models [39, 82] or deterministic classifiers [15, 30] characterizing pairs of duplicates from training data, consisting of known duplicates. Such methods assume that training data contain the wide variety of possible errors in practice. However, such a comprehensive collection of training data very rarely exists in practice. Partially, this issue was addressed by approaches based on active learning [85, 91]. These require an interactive manual guidance. In many scenarios, it is not plain to obtain satisfying training data or interactive user guidance.

For what specifically regards supervised approaches to de-duplication, we argue that, being state-of-the-art proposals of the literature very heterogenous among them, the most convenient classification to be proposed here is the one based on the *kind of data* to be de-duplicated, and can be reasonably devised as follows: (*i*) *supervised approaches for de-duplicating relational data*; (*ii*) *supervised approaches for de-duplicating multidimensional data*; (*iii*) *supervised approaches for de-duplicating Data-Mining data/results*; (*iv*) *supervised approaches for de-duplicating linked and XML data*; (*v*) *supervised approaches for de-duplicating streaming data.* The remaining part of this Section is organized in sub-sections that strictly follow the proposed classification for supervised approaches to de-duplication.

## 3.1  Relational Data De-duplication Approaches

In the context of relational data, [75] proposes an efficient algorithm for recognizing *clusters of approximatively duplicate records* in large databases. The main goal of the proposed algorithm consists in overcoming limitations exposed by a previous relevant approach, the *Smith-Waterman algorithm* [89], which has been early proposed for identifying common molecular subsequences in the context of molecular biology research, by means of three innovative steps. First, an optimized version of the Smith-Waterman algorithm is introduced in order to compute minimum edit distances among candidate duplicate records according to a *domain-independent approach*. Second, a meaningful union algorithm for keeping track of duplicate clusters as long as new duplicates are discovered is exploited. Third, a novel priority-queue-based method is used to finally retrieve clusters of duplicate records depending on the size and the homogeneity of the underlying database. Experimental results demonstrate the benefits of algorithm [75] over the state-of-the-art Smith-Waterman algorithm.

[35] moves the attention on the problem of improving *spatial join algorithms over spatial databases* via detecting duplicates that may occur in the set of candidate spatial database objets involved by (spatial join) queries. The most remarkable contribution of this research consists of a significant improvement of performance of two state-of-the-art spatial join algorithms, namely *PBMS* [83] and $S^3J$ [67], as confirmed by experiential results shown in [35]. [35] clearly demonstrates how duplicate record/object detection not only is useful for Data Mining and Data Warehousing, but even for Database query processing issues.

*Blocking techniques* [61] represent a traditional indexing approach for reducing the number of comparisons due to data de-duplication, at the cost of potentially missing some true matches. These techniques typically divide the target database into *blocks* and compare only the records that fall into the same block. One traditional method is to scan the database and compute a value of an appropriate *hash function* for each record. The value of the hash function defines the block to which this record is assigned. The limit of this approach is that conventional hashing techniques cannot be used for obtaining approximate duplicate detections, since the hash value of two similar records could not be the same (due to unexpected collisions).

[85] focuses the attention on the presence of duplicates in *data integration systems*, hence it elects de-duplication as one of the most important research challenge to be faced-off in such systems. Indeed, integrating data from multiple and heterogeneous sources easily exposes to the presence of duplicates, both data and concept duplicates. Starting from limitations of actual approaches, which are mainly

hand-coded in nature, [85] proposes an innovative *learning-based de-duplication method* whose main idea consists in interactively discovering critical training pairs, i.e. those training pairs that may effectively provide a benefit for the ongoing de-duplication process, via so-called *active learning* [31]. This method is then embedded into the core layer of a complete interactive de-duplication system, called ALIAS, for which reference architecture and main functionalities are provided. Overall, [85] indeed proposes a sort of *learning-based classifier*, whose performance is assessed via a comprehensive set of experiments on both synthetic and real-life data sets.

Christen [16] provides us with an extremely useful research experience on the performance and scalability of so-called blocking techniques [61] for data de-duplication. This study first depicts a meaningful "high-level" view about the general data de-duplication process, and then a very comprehensive experimental analysis and evaluation of performance issues as well as scalability aspects of blocking techniques, which puts in emphasis benefits and limitations of state-of-the-art approaches, mostly focusing on main-memory management and time complexity results. As a secondary, yet useful, result of [16] the author concludes that blocking techniques are very sensitive to the ranging of parameter values, and, since finding the optimal setting for these parameters depends on the quality and the specific characteristics of the databases to be de-duplicated, it is not easy to apply these techniques in real-life database settings, where the ranging above can become arbitrary and completely-impredicative at all. The latter one is another significant lesson to be learned with respect to data de-duplication research principles.

Christen again proposes the *Febrl* (Freely Extensible Biomedical Record Linkage) system in [17]. *Febrl* is an open source tool, provided with a nice graphical user interface, which embeds a relevant number of state-of-the-art data de-duplication approaches in order to improve data cleaning and standardization in the domain of health databases of Australia. The most distinctive characteristic of *Febrl* relies in the fact that it is prone to house and integrate any novel arbitrary data de-duplication technique one would like to embed in the system, hence providing interesting extensible and cross-analysis facilities that allow researchers to boost the finding capabilities in this so-interesting scientific area. In [17], Christen provides us with a very detailed description of architecture and main functionalities of *Febrl*, plus discussion on data de-duplication tasks end-users are allowed to define and run in this open source environment.

Finally, [51] focuses the attention on studying the *quality* of de-duplication results due to *clustering-based approaches*, in order to discover limitations and potentialities of this family of data de-duplication methods. The analysis is conducted within the context of the *Stringer* system, a modular architecture that provides a reliable evaluation framework for stressing the effectiveness-on-arbitrary-domains and the scalability of clustering-based data de-duplication approaches. The result of this study makes us aware about some surprising evidences: some clustering algorithms that have never been considered for duplicate detection problems expose good performance as regards both accuracy and scalability of the data de-duplication phase. As a further result, in [51] authors conclude that it is not possible to obtain perfect duplicate clusters by means of actual methods available in literature, so that *quantitative approaches* (e.g., [52]) oriented to find accurate confidence scores for each duplicate cluster detected are mandatory.

## 3.2  Multidimensional Data De-duplication Approaches

In the context of Data Warehousing systems, [27] proposes an efficient *data reconciliation approach*, whose main benefit consists in introducing an *approximate matching method* that incorporates *Machine Learning* [74] tools and statistical techniques with the goal of sensitively reducing the spatio-temporal complexity due to the very large number of comparisons to be performed in order to detect matches across different data sources. The approach in [27], mainly proposed for data reconciliation purposes (which, in some sense, is more general than data de-duplication issues), can be straightforwardly adapted for data de-duplication purposes, yet preserving similar performance.

In [2], authors investigate the problem of detecting duplicate records in the relevant application scenario represented by Data Warehousing systems. To this end, authors propose and experimentally assess DELPHI, a duplicate detection algorithm tailored to the specific target of *dimensional tables* one can find in the data layer of Data Warehousing systems. DELPHI takes advantages from the semantics expressed by *dimensional hierarchies* available in such systems in order to improve the quality and the effectiveness of state-of-the-art approaches for duplicate record detection that, as authors correctly state, still expose a high number of false positives over conventional database tables.

[69] proposes a nice theoretical work focusing on general aspects of data cleaning for Data Warehousing. The idea carried out by [69] consists in a novel *knowledge-based approach* that exploits the degree of similarity of nearby records of large databases to detect duplicates within such databases. In more details, [69] trades-off the *recall* of the duplicate detection phase, i.e. the sensitivity of the target method in accepting as duplicates those records exposing a low degree of similarity with other records, and the *precision* of the duplicate detection phase, i.e. the sensitivity of the target method in accepting as duplicates only records having an high degree of similarity with other records. This conveys in the so-called *recall-precision dilemma* [69]. Authors propose a solution to this dilemma by means of a theoretical framework that makes use of *transitive-closure tools* over suitable graphs modeling the uncertainty of similarities among candidate records, and advocates for approximate solutions.

Finally, in [30] authors address the particular context represented by *large high-dimensional data sets*, and propose the innovative tasks of *"entity-name matching"* and *"entity-name clustering"*, which, as authors claim, play critical roles in data cleaning over high-dimensional data [30]. In more detail, *"entity-name matching"* deals with the problem of taking two lists of entity names from two different sources and determining which pairs of names are co-referent (i.e., refer to the same real-world entity). The term *"entity-name clustering"* instead refers to the task of taking a single list of entity names and assigning entity names to clusters such that all names in a cluster are co-referent. The major benefit due to [30] consists in proposing an *adaptive* approach, meaning that accuracy of the duplicate detection phase can be progressively improved by training.

## 3.3  Data-Mining Data/Results De-duplication Approaches

In the context of Data Mining, Winkler [99] studies the conceptual/theoretical over-lap between duplicate detection and *Bayesian networks*, which are well-understood tools of Machine Learning research [74]. In more detail, Winkler finds five spe-cial conditions such that the *Expectation Maximization* (EM) algorithm [34] can be used for parameter estimation within the core layer of Bayesian networks, in order to gain a significant speed-up during the duplicate detection phase performed by these networks. Hence, main results from [99] are of theoretical nature, and have been further exploited for subsequent research in the duplicate detection context. On the other hand, in [99] the author experimentally shows the performance gain due to applying the EM algorithm for making Bayesian networks faster on dupli-cate detection situations, and he also proves that this approach performs better than traditional approaches that are, generally, based on iterative refinement methods.

In [88], authors address the record de-duplication problem in the context of conventional *Knowledge Discovery in Databases* (KDD) [38] processes from a completely-novel perspective that, indeed, turns to be innovative with respect to previous research experiences. Here, the main idea consists in making use of an in-novative *multi-relational approach* able of performing simultaneous inference for all the pairs of candidate duplicate records, while allowing the model information to propagate from one candidate match to another via exploiting the set of (database) attributes these candidates share in common. Based on this theoretical model for rep-resenting and processing candidate duplicate records simultaneously, authors design the guidelines of a general framework for supporting the so-called *collective infer-ence* of possible duplicates. Analytical and theoretical results presented in [88] are fully-supported by a comprehensive campaign of experiments that truly demonstrate the benefits of the multi-relational de-duplication approach over state-of-the-art con-ventional ones on both syntectic and real-life data sets.

Finally, [44] deals with the problem of making data de-duplication methods *au-tomatic*, in order to provide a more reliable and effective support to Data Mining processes, hence avoiding tedious manual clerical reviews needed to (manually) check possible data links that may still exist. It should be noted that the situation above would be not even possible in some real-life critical application scenarios where real-time linkage of streaming data is required (e.g., credit card transactions, public health survey systems, and so forth). The main proposal from authors in [44] consists in a *decision-tree-based approach* that embeds some state-of-the-art *compression methods* aimed at improving the efficiency of the data de-duplication phase. Combining efficient previous methods makes the approach proposed in [44] particularly suitable to support record de-duplication over very large data sets, perhaps via invoking well-understood *parallel processing* paradigms.

## 3.4   Linked and XML Data De-duplication Approaches

The *identity uncertainty* problem related to the context of citing research papers, which fall in the context of linked data, is investigated in [84], as a practical data/object de-duplication issue arising in real-life information systems. This challenge is becoming relevant for a large spectrum of modern application scenarios, as actually enormous, massive amounts of bibliographic knowledge is made available to open research communities via citation tools and more-conventional digital libraries via the Web. Hence, multiple observations may easily correspond to the same (bibliographic) object, thus introducing severe flaws during the fruition of bibliographic knowledge. Starting from these motivations, authors in [84] attack uncertainty of identities via innovative *probabilistic models* over mappings defined between terms and objects, and infer (probabilistic) matches via well-understood *Markov-Chain Monte Carlo* [42] solving paradigms on top of such models. One of the significant and singular particularities of the approach [84] is represented by some specific optimizations introduced in the classical Markov-Chain Monte Carlo solving method in order to generate efficient object candidates when the target domain contains a large number of object duplicates for a certain term. Experimental results shown in [84] demonstrate the effectiveness of the proposed identity uncertainty management approach over the bibliographic data set underlying the well-known Web citation system *Citeseer* [68].

*Correlation clustering* [8] is an elegant clustering method that was originally proposed for clustering graphs with binary edge labels modeling the existence of correlation or un-correlation between the connected nodes. As highlighted in previous studies, correlation clustering can be adapted as a solving method for data de-duplication problems. In fact, (*i*) nodes can store candidate duplicate data/objects, (*ii*) labels can be assigned to edges on the basis of user-defined similarity scores between pairs of nodes, and (*iii*) unconstrained clustering algorithms can be used to cluster duplicate data/objects, based on a given threshold over similarity scores.

*Detecting duplicates in XML data*, which are relevant for modern Web applications and systems, is first investigated in [95]. [95] particularly considers the interesting scenario of detecting duplicates for all the kinds of parent/child relationships that can occur in an XML data set. Previous studies have focused on the problem of detecting duplicates for $1 : n$ parent/child (XML) relationships only. The most relevant contribution of [95] consists in the fact that an innovative *comparison order* of pairwise classification is introduced, being this order able of reducing the number of (re)classifications of the same pair of candidate duplicate XML objects. Two different algorithms that efficiently exploit this novel ordering solution are presented, and experimentally tested on real-life movie data sets.

[96, 55] instead introduces the *duplicate detection problem for graph data*. Similarly to the case of XML data [95], here authors make use of relationships among objects (of the underlying database) in order to build a suitable graph modeling such relationships, and then study how to improve the effectiveness and the efficiency of traditional duplicate detection approaches over this graph, by reducing spatial and time complexity. Here, the main idea consists of an *hybrid approach* that encompasses an initialization phase and an iterative phase, both aimed at gaining performance over traditional solutions. Furthermore, the proposed framework argues to achieve high scalability over large amount of data thanks to a proper RDBMS layer, which provides support for some specific elementary routines of the duplicate-detection-in-graphs phase by means of very efficient SQL statements implemented in forms of well-understood *stored procedures*. In [96, 55], a wide and pertinent experimental assessment of the proposed RDBMS-supported framework on both synthetic and real-life data sets clearly demonstrates the effectiveness and the efficiency of the framework, even in comparison with traditional data de-duplication approaches.

Finally, [70] moves the attention on *duplicate detection over large XML documents*, as a further extension of previous work [96, 55]. The proposed method is based on the well-known cryptographic hashing algorithm *Message Digest algorithm 5* (MD5) by Rivest *MD5 algorithm* [12], an efficient algorithm that takes as input messages of arbitrary length and returns as output message digests of fixed length, and it encompasses three different modules: (*i*) the *selector*, which retrieves candidate duplicate objects from the input XML document and a fixed set of candidate definitions; (*ii*) the *pre-processor*, which pre-processes candidate duplicate objects in order to code them into 512-bit padded messages; (*iii*) the *duplicate identifier*, which finally selects the duplicate objects based on the MD5 algorithm running on the padded messages generated by the pre-processor module.

## 3.5  *Streaming Data De-duplication Approaches*

Finally, in the context of streaming data, [100] addresses the relevant research challenge represented by *detecting duplicates in streaming data*, which has received considerable attention from the Database and Data Mining research communities (e.g., [87]). Indeed, in the context of data stream processing, renouncing to the uniqueness assumption on observed data items from an input stream is very demanding, as almost all the actual *aggregation approaches* over streaming data available in literature would need significant revision at both the theoretical and the implementation-wise level. Based on this key observation, authors propose a family of techniques able of computing *duplicate-insensitive order statistics* over data streams, with provable error guarantees. The proposed techniques are proven to be space- and time-efficient and suitable to support on-line computation of very high-speed data streams. Authors complete their nice analytical and theoretical contributions by means of a comprehensive set of experiments on both syntectic and real-life data stream sets, which further confirm the effectiveness and the efficiency of the proposed techniques.

# 4  Unsupervised Approaches to De-duplication

A major disadvantage of supervised approaches to duplicate detection is the requirement for appropriate amounts of labeled training data, which involves a considerable human effort in labeling pairs of training data as either duplicates or non-duplicates. This task becomes especially challenging when it comes to provide examples of ambiguous cases (e.g., apparently duplicate tuples that are really non duplicates and viceversa), from which to learn more effective models, that are capable to sharply discriminate duplicates from non-duplicates [58]. In practice, supervised approaches assume the availability of training data explaining the wide variety of possible errors for each targeted entity. However, such comprehensive collections of training data very rarely exist. Partially, this issue was addressed by resorting to active learning [85, 91]. Unfortunately, this still requires an interactive manual guidance. As a matter of fact, in many practical applications of supervised duplicate detection, it is not plain to obtain satisfying training data or interactive user guidance.

To avoid the limitations of supervised duplicate detection, a large body of unsupervised approaches to data de-duplication has been proposed in the literature. Such approaches essentially define suitable techniques for grouping duplicate tuples, so that to minimize two types of incorrect matchings: false-positives (i.e., tuples recognized as similar, that actually do not correspond to the same entity) and false-negatives (i.e., tuples corresponding to the same entity, that are not recognized as similar). The pursuit of such objectives has largely prompted the design of unsupervised classification methods, mostly based on clustering or nearest-neighbor classification. Therein, in order to meet the earlier requirements on effectiveness, efficiency and scalability, various categories of schemes for approaching de-duplication in terms of unsupervised classification have been developed. We focus on three major categories of unsupervised de-duplication schemes, which are discussed separately in the rest of this section. Precisely, subsection 4.1 covers the exploitation of consolidated clustering schemes for duplicate detection. Subsection 4.2 is devoted to de-duplication via (dis)similarity-search in metric spaces. Finally, subsection 4.3 deals with duplicate-detection through locality-sensitive hashing.

## 4.1  De-duplication Based on Clustering

Clustering methods [45, 59, 60] have been exploited for the de-duplication purpose to divide a set of tuples into various clusters. The individual clusters refer to corresponding real-world entities and meet the following two requirements: *homogeneity*, i.e. pairs of tuples within a same cluster are highly similar and, hence, expected to be duplicates, and *neatly separation*, i.e. pairs of tuples within distinct clusters are very dissimilar and, therefore, deemed to refer to distinct real-world entities.

A variety of clustering methods can be exploited in the context of duplicate detection. A very effective approach would be using a hierarchical clustering method[2], equipped with an accurate component-wise similarity metric, such as edit distance, affine gap distance, smith-waterman distance and Jaro distance (see [58] for a detailed survey) to match tuple tokens. Unfortunately, the quadratic complexity of hierarchical clustering in the number of available tuples, combined with the high computational cost of the schemes for matching tuple components (that becomes quadratic w.r.t. the length of tokens in the case of edit distance), would penalize the efficiency and scalability of the resulting de-duplication process, up to the point of making the latter impractical in the great majority of applicative domains, where even a small amount of data is available.

Apart from hierarchical methods, several consolidated clustering algorithms [37, 40, 48, 49, 53] are at the heart of various de-duplication techniques. However, although generally effective, these techniques do not generally guarantee an adequate level of scalability. As a matter of fact, these approaches would not work adequately in a scenario, where far too many clusters are expected to be found, as it does happen in a typical de-duplication scenario, where the actual number of clusters of duplicate tuples can be of the same order as the size of the database. The only suitable approaches appear to be the ones in [23, 72].

Precisely [72] avoids costly pairwise comparisons by grouping objects in *canopies*, i.e., subsets containing objects suspected to be similar according to some cheap (i.e. computationally inexpensive) similarity function and, then, computing actual pairwise similarities only within the discovered canopies. Since in a typical duplicate detection scenario there are several canopies, and an object is shared in a very few number of canopies, the main issue of the approach is the creation of canopies.

In [23], an efficient two-phase approach is proposed: first determine the nearest neighbors of every tuple in the database and, then, partition the original collection into groups of duplicates. The efficiency of the algorithm strictly relies on the nearest neighbors computation phase, where the availability of any disk-based index (i.e., inverted index associated with edit or fuzzy similarity functions) is assumed. Efficiency comes from the lookup order in which the input tuples are scanned, in order to retrieve nearest neighbors. The order corresponds to a breadth first traversal of a tree, where the children of any node are its nearest neighbors. The benefit consists in accessing, for consecutive tuples, the same portion of the index, thus improving the buffer hit ratio.

Despite their strengths, the approaches in [23, 72] are not meant for incremental de-duplication.

---

[2] Hierarchical clustering algorithms are well known in the literature for producing top quality results [60].

## 4.2 De-duplication Based on (dis)Similarity-Search in Metric Spaces

De-duplication can also be performed by grouping duplicates through neighbor-driven clustering. Given a collection of tuples equipped with a suitable (dis)similarity metric, the idea is that the cluster membership of a tuple should be established by looking at the clusters to which other similar tuples belong.

Neighbor-driven de-duplication consists of three basic steps. Initially, the pair-wise distance between tuples is computed. Then, a list of *neighbors* is retrieved for each query tuple, that is, for each tuple to be de-duplicated. A tuple is essentially a neighbor of the query tuple, when the former is similar to the latter according to the adopted similarity metric. Ultimately, the cluster membership for the query tuple is determined through a *voting* procedure, in which the retrieved neighbors, that are likely duplicates of the query tuple, vote for the cluster to which the latter should be assigned. The most basic voting scheme is the *majority* one, in which the query tuple is assigned to the cluster that is most common among its neighbors.

Similarity-search [24, 56] plays a major role in neighbor-driven de-duplication, since it allows the identification of all neighbors of a query tuple. However, the high dimensionality [94] of the space in which the search is typically performed is a major weakness of neighbor-driven de-duplication. Moreover, similarity-search requires setting an appropriate upper bound (or, also, a threshold) to the maximum distance from the query tuple, that actually identifies the neighborhood of the latter. Computing a distance threshold is problematic, since, as it is pointed out in [23], one absolute global distance threshold does not guarantee an effective retrieval of neighbors. This has undesirable effects. Indeed, the identification of too few neighbors may make duplicate clustering susceptible to overfitting. Instead, too many neighbors may lead to noisy duplicate detection, since far dissimilar tuples may be involved in the voting procedure. According to the results in [23], the distance threshold should be actually considered as a local property of the individual group of duplicates [23]. Therefore, each real-world entity in the data should require the computation of a suitable distance threshold, that differs from the thresholds associated with the other entities in the same data. Unfortunately, the tuning of several distance thresholds would make the resulting de-duplication process impractical. Additionally, the basic similarity-search for neighboring duplicates involves computing similarities between all pairs of tuples in the underlying data, thus being computationally quadratic in the overall number of available tuples. This exceedingly penalizes both the efficiency and the scalability of de-duplication.

In order to speed up the basic approach to neighbor-driven de-duplication, various refinements have been proposed [2, 22, 46], that exploit efficient indexing schemes. Unfortunately, these refinements are not specifically designed to approach neighbor-driven de-duplication from an incremental clustering perspective. Therein, neighbor-driven clustering would in principle benefit from an indexing scheme, that supports the execution of similarity queries and can even be incrementally updated with new tuples. Nonetheless, the syntactic heterogeneity of the tuples at hand is likely to heavily increase the size of the index, which would ultimately degrade the performance of the overall neighbor-driven de-duplication process. To further elaborate on this point, the exploitation of an indexing scheme for neighbor-driven de-duplication was empirically investigated in [32]. In particular, this study focused on the *M-Tree* index [25], a well-known, state-of-the-art index/storage structure, which looks like a *n*-ary tree.

The M-Tree allows to index and organize tuples, provided that a suitable distance metric *dist* is defined for pairwise tuple comparison. Precisely, tuples are arranged into a balanced tree structure, in which each node has a fixed size (related to the size of a page to be stored on disk). The individual entries of a non-leaf node store *routing* objects, i.e., summaries of the information about the contents of the subtrees rooted at the children of that node. In turn, each routing object $O_r$ is associated with two further elements: a *pointer* referencing the root of a sub-tree $T(O_r)$ (the so-called *covering tree* of $O_r$) and a *covering radius* $r(O_r)$, which guarantees that all objects in $T(O_r)$ are within the distance $r(O_r)$ from $O_r$. The search for the neighbors of a query tuple $t$ can be efficiently answered by simply traversing the M-Tree: at each non-leaf node storing a routing object $O_r$, the evaluation of both $dist(t, O_r)$ and $r(O_r)$ allows to decide whether the corresponding subtree $T(O_r)$ contains candidate neighbors and, hence, whether it has to be explored or not. In other words, querying the M-Tree for the neighbors of a query tuple involves traversing the M-Tree and ignoring those subtrees, that are reputed uninteresting for the search purpose.

A neighbor-driven approach to duplicate detection would strongly benefit from the exploitation of an index structure such as the M-Tree, since the latter would, in principle, answer similarity queries with minimal processing time and I/O cost. Additionally, the M-tree has three features, that are highly desirable in a de-duplication setting. First, it is a paged, balanced, and dynamic secondary-memory structure, capable to index data sets from generic metric spaces. Second, similarity range and nearest-neighbor queries can be performed and results can be ranked with respect to a given query tuple. Third, query execution is optimized to reduce both the number of pages read and the number of distance computations. However, the empirical analysis in [32] revealed that, when several tuples exhibit heterogeneous syntactic representations, the magnitude of the covering radii increases (especially at higher levels) and, hence, most of the internal nodes of the M-Tree tend to correspond to quite heterogeneous groups of tuples. Therefore, a high number of levels, nearly linear in the number of distinct entities in the data collection, is required to suitably index and organize the original collection of tuples. Thus, since in the typical de-duplication scenario the number of entities is likely to be of the same order as the number of available data tuples, the cost of similarity search actually tends to be nearly linear in the number of the original tuples. Clearly, this negatively affects the performance of the M-Tree and makes the overall neighbor-driven de-duplication process unable to scale for manipulating very large collections of data.

The problem of finding all tuples similar to a certain query tuple has been intensively studied within the database and information-retrieval communities with important achievements. Unfortunately, the incorporation of these results in neighbor-driven de-duplication would not make its computational cost independent on the size of the available data.

Some works from the database community, such as [86] and [5], focused on solving the problem exactly, by defining *set-similarity joins*, i.e. suitable operators for database management systems. Informally, a similarity join is an operation for recognizing different representation of a real-world entity. More precisely, given two relations, a similarity join finds all pairs tuples from the two relations, that are syntactically similar. Similarity is evaluated by means of a string-based similarity function: two tuples are considered similar if the value of the similarity function for these two tuples is greater than a certain threshold. An important drawback of the operators in [86] is that they scale quadratically with respect to the size of the data, which makes their exploitation impractical in the de-duplication of very large databases. The notion of set-similarity join was introduced in [5] as a primitive that takes two collections of sets as input and identifies all pairs of sets exhibiting a strong similarity. The latter is established through suitable predicates concerning the size and overlap of the sets. Set-similarity joins are performed through signature-based algorithms. These algorithms generate signatures for the input sets, with the desirable property that, if the similarity of two sets exceeds a certain threshold, then the two sets share a common signature. By exploiting this property, signature-based schemes find all pairs of sets with common features and, eventually, output all those pairs of sets whose pairwise similarity actually trespasses some preestablished threshold. The algorithms developed in [5] for performing set-similarity joins improve the basic performance of signature-based algorithms in two respects: the adoption of a different scheme for computing set signatures as well as the incorporation of a theoretical guarantee, according to which two highly dissimilar sets are not considered as duplicates with a high probability. The latter property of set-similarity joins considerably lowers the overall number of false-positive candidate pairs and, also, increases the efficiency of the resulting operators, that scale almost linearly in the size of the input set. However, linear scalability comes at expense of a non trivial parameter tuning, since no single parameter setting is appropriate for all computations. In practice, for a fixed parameter setting, the operators still scale quadratically and some properties of the input data must be analyzed so that to establish an optimal tuning, that ensures linear scalability.

Recently, an approach inspired from information retrieval methods [10] proposed to scale exact join-set methods to large volumes of real-valued vector data. This work refines the basic intuition in [86] of dynamically building an inverted list index of the input sets with some major indexing and optimization strategies, mainly concerning how the index is manipulated to evaluate the (cosine) similarity between the indexed records and the query one.

As a concluding remark, despite its effectiveness and the recent efforts for improving its scalability, similarity search for duplicates is not always a feasible approach to neighbor-driven de-duplication. It was shown in [3, 4] that either the space or the time required by the solutions devoted to expedite similarity search is exponential in the dimensionality of the data[3]. Moreover, it was also proven in [94] that similarity search based on space-partitioning indexing-schemes degenerates into a sequential scan of the data, even when dimensionality is moderately high. Therein, theoretical and empirical achievements lead to postulate in [94] that all approaches to nearest-neighbor search ultimately become linear in the size of the data, when dimensionality is sufficiently high. In the context of neighbor-driven de-duplication, the degeneration of the search for neighbors of the query tuple to a linear scan is undesirable, especially when processing very large volumes of data tuples.

## 4.3   De-duplication Based on Locality-Sensitive Hashing

De-duplication based on locality-sensitive hashing overcomes the limited scalability of neighbor-driven de-duplication. The premise is that in many de-duplication scenarios there exist several tuples, that are predominantly dissimilar from one another. Therefore, the number of groups of duplicates is likely to be of the same order as the overall number of tuples. In this context, finding few tuples mostly similar to the query tuple (i.e., to the generic tuple to be de-duplicated) is deemed to provide enough information for assigning the latter to the most appropriate group of duplicates through voting mechanisms.

The foregoing arguments motivate the exploitation of approximated similarity search for nearest neighbors of the query tuple, which can be performed much faster than exact similarity search by means of a suitable hash-based indexing method, that expedites the overall de-duplication process. To elaborate, an index structure is used to allow direct access to subsets of tuples with the same features.[4] The assignment of each individual tuple to the buckets of the index structure is managed by means of suitable hashing schemes. Precisely, the buckets associated to one tuple are the values of some hash function(s) on the features of the tuple. De-duplication benefits from a typical situation of hashing known as *collision*: the higher the similarity between two tuples, the likelier it is that these share the same features and, consequently, that both are assigned the same hash values, thus falling within the same buckets of the underlying index structure. This permits to narrow the search for neighbors of the query tuple to a focused linear search for nearest neighbors among those tuples falling within the buckets associated to (the individual features of) the query tuple. Therefore, in the de-duplication of the latter, nearest neighbors can be retrieved by issuing, against the index structure, similarity queries for neighboring

---

[3] The dimensionality of a structured tuple is simply the number of attributes in its schema. Instead, the dimensionality of a textual sequence can be viewed, in principle, as the number of distinguishing string tokens, chosen to represent the textual sequence as a point in a multidimensional space, according to the vector space model [7].

[4] Features are essentially distinguishing properties of a tuple, that are also commonly referred to as indexing keys (e.g., q-grams [46], i.e., contiguous substrings of length $q$, can be adopted to define suitable features for strings). Various key-generation schemes exist, that operate according to both the nature of available data and the specific applicative requirements.

tuples with the same features as the query tuple. Hash-based indexing guarantees that such queries can be efficiently answered. In particular, *locality-sensitive* hashing allows an indexing method for approximated similarity search, with a sub-linear dependence on the number of tuples. The idea behind locality-sensitive hashing is to bound the probability of collisions to the similarity between the tuples. In other words, a locality-sensitive hash function guarantees that the probability of collisions is much higher for similar tuples than it is for dissimilar ones. An important family of locality-sensitive hash functions, in which the similarity of two tuples is measured by their degree of overlap, can be naturally defined through the theory of *min-wise independent permutations* [19].

Various approaches to de-duplication based on locality-sensitive hashing have been proposed in the literature [20, 32, 43, 57].

Locality-sensitive hashing was originally developed in [57], as an efficient technique for accurately approximating the nearest-neighbor search problem.

The approach in [43] refines the basic proposal of [57] in several respects, including new theoretical guarantees on the worst-case time required for performing a nearest neighbor search as well as the generalization to the case of external memory. In particular, given a set of tuples and a family $\mathcal{H}$ of locality-sensitive hash functions, the generic tuple is associated with a corresponding bitstring signature, that identifies an index bucket. The bitstring is obtained from the values of $k$ locality-sensitive hash functions (randomly extracted from $\mathcal{H}$) over the tuple. Since the overall number of buckets identified by the resulting bitstrings can be huge, a second level of standard hashing is exploited to compress the foresaid buckets by mapping their contents into one hash table $\mathcal{T}$, whose buckets are directly mapped to blocks on disk. The size of $\mathcal{T}$ is proportional to the ratio of the number of tuples to the maximal bucket-size of $\mathcal{T}$. In general, it is possible to loose proximity relationships if a point and its nearest neighbor are hashed to distinct buckets. Therefore, in order to lower the probability of such an event, the same tuple is stored in [43] into $l$ hash tables $\mathcal{T}_1, \ldots, \mathcal{T}_l$, respectively indexed by as many independent bitstrings.

When it comes to de-duplicate a query tuple, its nearest neighbors are identified through the technique in [43] as follows. The query tuple is hashed to the buckets within the individual hash tables $\mathcal{T}_1, \ldots, \mathcal{T}_l$. All tuples previously hashed within these same buckets are gathered as candidate neighbors. A linear search is then carried out across these candidates, to find the neighbors actually closest to the query tuple. These are guaranteed to be at a distance from the latter within a small error factor of the corresponding optimal neighbors.

The drawback of the technique in [43] is the requirement for the identification of an optimal, data-specific tradeoff between two contrasting aspects of the index [9], namely accuracy and storage space. By increasing $l$, accuracy is guaranteed for the great majority of queries, through a correspondingly larger number of hash tables. However, this makes the storage requirement inversely proportional to the error factor. Also, it raises the number of potential neighbors and, hence, the overall response time. In such cases, one may act on $k$, since a high value of this parameter would sensibly lower the number of collisions and, hence, mitigate the increase in response time. Unfortunately, large values of $k$ augment the miss rate. By the converse, small values of parameter $l$ cannot guarantee accuracy for all queries.

An approach for identifying near duplicate Web pages is proposed in [20]. Here, each Web page is first tokenized and then represented as the set of its distinct, contiguous $n$-grams (referred to as shingles). The most frequent shingles are removed from the set to both improve performance and avoid potential causes of false resemblance. After preprocessing, near duplicates are identified via a clustering strategy that consists of the following four steps. A sketch is computed for each Web page, by applying a suitable min-wise independent permutation to its shingle representation. Sketches are then expanded to generate a list of all the individual shingles and the Web pages they appear in. Subsequently, this list is exploited to generate a new list of all the pairs of Web pages with common shingles, along with the number of shared shingles. Clustering is eventually achieved by examining the triplet elements of the latter list. If a certain pair of Web pages exceeds a pre-specified threshold for resemblance (estimated by the ratio of the number of shingles they have in common to the total number of shingles between them), the two Web pages are connected by a link in a union-find algorithm, that outputs final clusters in terms of connected components.

The algorithm in [20] requires a considerable amount of time and space on disk, especially due to the third phase, which makes it unscalable. Optimizations based on the notion of super-shingle addressed such an aspect, although these do not properly work with short Web pages. Yet, the de-duplication process strictly requires that the resemblance threshold is very high to effectively prune several candidate pairs of similar Web pages. Lower values of the threshold, corresponding to a typical setting for similarity search, cause several negative effects. False positive candidates are not appropriately filtered, which lowers precision. Very low values of the similarity threshold may also diminish false negatives, with a consequently moderate increase in recall. However, in such cases, the impact on the effectiveness of a small gain in recall would be vanished by the corresponding (much larger) loss in precision.

An incremental clustering technique for duplicate detection in very large databases of textual sequences is proposed in [32]. The techniques works by assigning each newly arrived tuple to an appropriate cluster of duplicates. More precisely, the de-duplication of a new tuple is accomplished by retrieving a set of neighboring tuples from a hash-based index structure. Neighbors are highly similar to the new tuple and, hence, their cluster membership provides useful information about the real-world entity corresponding to the new tuple. The latter is eventually assigned to the cluster of duplicates shared by the majority of neighbors.

The de-duplication process in [32] relies on a suitable hash-based index, that maps any tuple to a set of indexing keys and assigns syntactically similar tuples to the same buckets. In this manner, the neighbors of a query tuple can be efficiently identified by simply retrieving those tuples stored within the same buckets assigned to the query tuple itself, without either completely scanning the original database or using costly similarity metrics. Indexing keys are computed through a two-step key-generation procedure, in which locality-sensitive hash functions (based on a family of practically min-wise independent permutations [19, 43]) are hierarchically combined for a twofold purpose: reflecting the syntactic differences both among tuples and their components as well as enabling effective on-line matching. The first step of the key-generation scheme recognizes similar tuple components (i.e. string tokens) across tuples, despite some extent of syntactic heterogeneity. For this purpose, the individual tuples are purged into intermediate representations. These are obtained by encoding each component of the generic tuple via a min-wise hash function, that bounds the probability of collisions of two string tokens to the overlap between their respective sets of 1-grams (i.e., substrings of unit length) [46]. The choice of such a hash function guarantees that two similar but different tokens are, with high probability, assigned a same encoding. Therefore, any two tuples sharing syntactically similar tokens are purged into two intermediate representations, where such tokens converge towards a unique encoding. The second step of the key-generation scheme associates the intermediate representations of the original tuples with their indexing keys, through another min-wise hash function. The latter bounds the probability of collisions of two intermediate tuple representations to the overlap between their first-step token encodings. Again, this guarantees that two intermediate tuple representations sharing several first-step token encodings are associated with a same indexing key.

As a matter of fact, multiple min-wise hash functions are exploited both at the first and at the second step, for a twofold purpose, i.e., lowering the probability of false positives and false negatives (that essentially allows for a controlled level of approximation in the search for the nearest neighbors of the query tuple) as well as gaining a direct control over the number of keys used for indexing any tuple, which is necessary to guarantee the compactness of the overall storage space.

Interestingly, the approach in [32] has connections with several techniques from the literature.

Foremost, the two-step procedure for hashing tuples can be viewed as a smarter implementation of canopies (which are collected within the same buckets in the index). The main difference is that the properties of min-wise hashing functions allow to approximately detect such canopies incrementally.

Also, the key-generation scheme allows a constant (moderate) number of disk writes and reads to/from the index structure on secondary memory, which are two key aspects. Indeed, on one hand, the hash-based approach to de-duplication could cause continuous leaps in the disk-read operations, even when a small number of comparisons is needed for retrieving the neighbors of the query tuple. On the other hand, the number of disk pages written while updating the index structure is especially relevant in the incremental maintenance of the latter. Notably, the constant number of disk writes and reads is achieved in [32] along with a fixed (low) rate of false negatives. These are likely contrasting objectives in hash-based approaches to de-duplication, since few indexing keys must generally be produced to lower the amount of I/O operations, although this tends to increase the rate of false negatives.

Yet, from a methodological point of view, the approach in [32] exhibits analogies with the ones in [10, 86]. Indeed, a hash-based index is employed to maintain associations between a certain tuple feature and the subset of all available tuples that share that same feature. The actual difference with respect to the techniques in [10, 86] is that these approaches essentially pre-compute the nearest neighbors of each tuple, so that retrieving them becomes a simple lookup. Instead, the approach in [32] does not support neighbor pre-computation. Notwithstanding, it still enables neighbor search in a time that, on average, is independent on the number of database tuples.

Finally, the tradeoff between accuracy and storage space in [32] is much less challenging than in [43], since there exists a single hash-based index. Moreover, guaranteeing accuracy for all queries in [32] (i.e., lowering both the false-positive and false-negative rates) can be simply achieved by hashing a same tuple into as many buckets of the index as the number of first-step encodings of the tuple itself (for each such a representation, the concatenation of its second-step encodings yields the hash key associated to the tuple). Actually, a very limited number of distinct hash functions is used both at the first and at the second step to enforce de-duplication effectiveness, at the acceptable cost of a compact storage space.

# 5    Conclusions and Further Research

Duplicate detection is a necessary building block both for information integration and for the design of information systems, that allows more effective information handling and knowledge extraction, through a unified access and manipulation of consolidated and reconciled data. The current practices behind state-of-the-art approaches to duplicate detection have been discussed so far. Despite the considerable advances in the field, there are still various opportunities for further improvements [41, 47, 58, 98]. An overview of some major open problems and challenges is provided below.

## 5.1    Efficiency and Scalability

The detection of duplicates in very large volumes of data involves a huge number of pairwise tuple comparisons. In turn, each tuple comparison encompasses several matchings between the individual tuple components. A meaningful reduction of the number and the cost of both types of comparisons is key to devise solutions with which to improve the efficiency and scalability of de-duplication. Therein, various strategies have been proposed: blocking [58], sorted neighborhood [54], canopies [72], similarity joins [28] are some proposals for reducing comparisons between pairs of tuples without penalizing de-duplication effectiveness, whereas expedients such as [93] aim to speed up the individual comparison between tuple components. However, despite all such developments, efficiency and scalability often contrast with effectiveness. Indeed, the current state-of-the-art techniques can be divided into two broad classes [58], that highlight the foresaid contrast. Precisely, the techniques relying upon results from the field of databases privilege efficiency and scalability over effectiveness. On the contrary, other techniques exploiting contributions from the areas of machine learning and statistics reveal to be more effective. Unfortunately, their reduced efficiency and scalability makes the application of such techniques better suited for collections of data, whose size is orders of magnitude smaller than the collections processable through the techniques of the former type.

A worthwhile opportunity of further research is to devise new approaches combining the strengths of both types of de-duplication techniques.

## 5.2 Systematic Assessment of De-duplication Performance

Although several approaches to data de-duplication have been proposed in the literature, there is not yet a comprehensive and systematic knowledge of their relative performances in terms of de-duplication effectiveness as well as efficiency, across various data collections and domains. Recently, some efforts have been geared towards the design of evaluation frameworks [63, 64, 65], wherein to compare some de-duplication techniques. However, the process of gaining a systematic insight into the actual performances of de-duplication techniques is still in its infancy. The performed comparisons take into account only a limited number of approaches, that are compared on few collections of data from few real-world domains. Additionally, some collections of data were not effectively processed in [64]. This deserves further insights, that may be useful for the development of more sophisticated approaches.

Additionally, further research efforts for the identification of suitable performance metrics and state-of-the-art approaches across real-world domains should be fostered by the recent release of standard benchmarking data sets as the ones from [33]. Hopefully, the increasing availability of standard data sets will shed light on which metrics and techniques are best suited for the various applicative domains.

## 5.3 Ethical, Legal and Anonymity Aspects

Duplicate detection raises ethical, legal and privacy issues, whenever dealing with sensitive information about persons [62, 80]. Meeting the related national and international laws in force is thus a major requirement in the process of data de-duplication and subsequent management. A wide discussion concerning some major issues in data de-duplication and the mechanisms for the protection of individual privacy can be found in [90]. Anonymity preservation is one challenging issue, that is related to the release of large data collections for research and analytical purposes. The public release of any collections of data demands balancing analytic and research requirements with anonymity and confidentiality. Currently, a commonly adopted expedient to maintain the anonymity of persons in the underlying data is encrypting the identifying information. However, encryption may decrease the effectiveness of duplicate detection over time [78]. Another strategy for anonymity preservation would be the focused removal of the information necessary for tuple identification (and, hence, duplicate detection) from the available data sets. Unfortunately, this would still involve trust in some suitable preprocessing of the data [62].

One promising line of research for anonymity and confidentiality preservation is the development of ad-hoc models, with which to accurately assess identifying information leakage in the context of duplicate detection [41]. This would be useful to evaluate the leakage consequent to the release of identifying or sensible information and may be adopted as a criterion for carrying out the necessary precautions, e.g., global recoding and local suppression, additive noise and data perturbation as well as micro-aggregation (see [98] for a discussion of such precautions). Also, it would be very useful to incorporate models of analytic properties and re-identification risk in the process of de-duplication to answer two research questions, namely, understanding how the foresaid precautions compromise the analytic validity of the released data and estimating the re-identification rate if analytic validity is maintained.

## 5.4 Uncertainty

Uncertainty plays an important role in duplicate detection and is thus necessary to deal with it. Therein, a proposal is to attach confidences to the individual tuples. Confidences are essentially quantitative beliefs, whose actual meaning of confidences depends on the targeted applicative settings. In certain domains, confidences can be interpreted as the probability that the related tuples faithfully correspond to certain entities [41]. In other settings, confidences can be viewed as measures of the accuracy of the data.

While some efforts have been performed to efficiently compute the confidence of the de-duplicated tuples when confidences are explicitly attached to the tuples [73], the indirect estimation of some sort of de-duplication confidence, when uncertainty is not originally available, appears to be still unexplored.

# References

1. Agichtein, E., Ganti, V.: Mining Reference Tables for Automatic Text Segmentation. In: Proc. of ACM SIGKDD Int. Conf. On Knowledge Discovery and Data Mining, Seattle, Washington, USA, pp. 20–29 (2004)
2. Ananthakrishna, R., Chaudhuri, S., Ganti, V.: Eliminating Fuzzy Duplicates in Data Warehouses. In: Proc. of Int. Conf. on Very Large Databases, Hong Kong, China, pp. 586–597 (2002)
3. Andoni, A., Indyk, P.: Near-Optimal Hashing Algorithms for Approximate Nearest Neighbor in High Dimensions. In: Proc. of IEEE Symposium on Foundations of Computer Science, Las Vegas, Nevada, USA, pp. 459–468 (2006)
4. Andoni, A., Indyk, P.: Near-optimal Hashing Algorithms for Approximate Nearest Neighbor in High Dimensions. Communications of the ACM 51(1), 117–122 (2008)
5. Arasu, A., Ganti, V., Kaushik, R.: Efficient Exact Set-Similarity Joins. In: Proc. of Int. Conf. on Very Large Databases, Seoul, Korea, pp. 918–929 (2006)
6. Axford, S.J., Newcombe, H.B., Kennedy, J.M., James, A.P.: Automatic Linkage of Vital Records. Science 130, 954–959 (1959)
7. Baeza-Yates, R., Ribeiro-Neto, B.: Modern Information Retrieval. Addison-Wesley, Reading (1999)

8. Bansal, N., Blum, A., Chawla, S.: Correlation Clustering. Machine Learning 56(1-3), 89–113 (2004)

9. Bawa, M., Tyson Condie, S., Ganesan, P.: LSH Forest: Self-Tuning Indexes for Similarity Search. In: Proc. of Int. Conf. on World Wide Web, Chiba, Japan, pp. 651–660 (2005)

10. Bayardo, R.J., Srikant, R., Ma, Y.: Scaling Up All Pairs Similarity Search. In: Proc. of Int. Conf. on World Wide Web, Banff, Alberta, Canada, pp. 131–140 (2007)

11. Benjelloun, O., Garcia-Molina, H., Menestrina, D., Su, Q., Whang, S.E., Widom, J.: Swoosh: a generic approach to entity resolution. VLDB Journal 18(1), 255–276 (2009)

12. Berson, T.A.: Differential Cryptanalysis Mod 232 with Applications to MD5. In: Proc. of Ann. Conf. on Theory and Applications of Cryptographic Techniques, pp. 71–80 (1992)

13. Bhattacharya, I., Getoor, L.: Collective Entity Resolution in Relational Data. ACM Trans. Knowl. Discovery from Data 1(1), 1–35 (2007)

14. Bhattacharya, I., Getoor, L., Licamele, Louis: QueryTime Entity Resolution. In: Proc. of ACM SIGKDD Int. Conf. on Knowledge Discovery and Data Mining, Philadelphia, Pennsylvania, USA, pp. 529–534 (2006)

15. Bilenko, M., Mooney, R.J.: Adaptive Duplicate Detection Using Learnable String Similarity Measures. In: Proc. of ACM SIGKDD Int. Conf. on Knowledge Discovery and Data Mining, Washington, DC, USA, pp. 39–48 (2003)

16. Christen, P.: Towards Parameter-free Blocking for Scalable Record Linkage. Tech. Rep. TR-CS-07-03, Australian National University, Canberra, Australia (2007)

17. Christen, P.: Febrl - An Open Source Data Cleaning, Deduplication and Record Linkage System with a Graphical User Interface. In: Proc. of ACM Int. Conf. on Knowledge Discovery and Data Mining, pp. 1065–1068 (2008)

18. Borkar, V.R., Deshmukh, K., Sarawagi, S.: Automatic Segmentation of Text into Structured Records. In: Proc. of ACM SIGMOD Int. Conf. on Management of Data, Santa Barbara, California, USA, pp. 175–186 (2001)

19. Broder, A., Charikar, M., Frieze, A.M., Mitzenmacher, M.: Minwise Independent Permutations. In: Proc. of ACM Symposium on Theory of Computing, Dallas, Texas, USA, pp. 327–336 (1998)

20. Broder, A., Glassman, S., Manasse, M., Zweig, G.: Syntactic Clustering on the Web. In: Proc. of Int. Conf. on World Wide Web, Santa Clara, California, USA, pp. 1157–1166 (1997)

21. Cesario, E., Folino, F., Locane, A., Manco, G., Ortale, R.: Boosting Text Segmentation Via Progressive Classification. Knowl. and Inf. Syst. 15(3), 285–320 (2008)

22. Chaudhuri, S., Ganjam, K., Ganti, V., Motwani, R.: Robust and Efficient Fuzzy Match for Online Data Cleaning. In: Proc. of ACM SIGMOD Conf. on Management of Data, San Diego, California, USA, pp. 313–324 (2003)

23. Chaudhuri, S., Ganti, V., Motwani, R.: Robust Identification of Fuzzy Duplicates. In: Proc. of Int. Conf. on Data Engineering, Tokyo, Japan, pp. 865–876 (2005)

24. Chavez, E., Navarro, G., Baeza-Yates, R., Marroquin, J.L.: Searching in Metric Spaces. ACM Comput. Surv. 33(3), 273–321 (2001)

25. Ciaccia, P., Patella, M., Zezula, P.: M-Tree: An Efficient Access Method for Similarity Search in Metric Spaces. In: Proc. of Int. Conf. on Very Large Databases, Athens, Greece, pp. 426–435 (1997)

26. Cochinwala, M., Dalal, S., Elmagarmid, A.K., Verykios, V.S.: Record Matching: Past, Present and Future. Technical Report, number CSD-TR #01-013. Department of Computer Sciences, Purdue University (2001)

27. Cochinwala, M., Kurien, V., Lalk, G., Shasha, D.: Efficient Data Reconciliation. Information Sciences 137(1-4), 1–15 (2001)
28. Cohen, W.W.: Data Integration using Similarity Joins and a Word-based Information Representation Language. ACM Trans. on Inf. Syst. 18(3), 228–321 (2000)
29. Cohen, W.W., Ravikumar, P., Fienberg, S.E.: A Comparison of String Distance Metrics for Name-Matching Tasks. In: Proc. of IJCAI Workshop on Information Integration on the Web, Acapulco, Mexico, pp. 73–78 (2003)
30. Cohen, W.W., Richman, J.: Learning to Match and Cluster Large High-Dimensional Data Sets for Data Integration. In: Proc. of ACM SIGKDD Int. Conf. on Knowledge Discovery and Data Mining, Edmonton, Alberta, Canada, pp. 475–480 (2002)
31. Cohn, D.A., Atlas, L., Ladner, R.E.: Improving Generalization with Active Learning. Machine Learning 15(2), 201–221 (1994)
32. Costa, G., Manco, G., Ortale, R.: An Incremental Clustering Scheme for Data De-duplication. Data Min. and Knowl. Discovery 20(1), 152–187 (2010)
33. Database Group Leipzig. Benchmark datasets for entity resolution, http://dbs.uni-leipzig.de/en/research/projects/object_matching/fever/benchmark_datasets_for_entity_resolution
34. Dempster, A.P., Laird, N.M., Rubin, D.B.: Maximum Likelihood from Incomplete Data via the EM Algorithm. Journal of the Royal Statistical Society, Series B 39(1), 1–28 (2001)
35. Dittrich, J.-P., Seeger, B.: Data Redundancy and Duplicate Detection in Spatial Join Processing. In: Proc. of IEEE Int. Conf. on Data Engineering, pp. 535–546 (2000)
36. Elmagarmid, A.K., Ipeirotis, P.G., Verykios, V.S.: Duplicate Record Detection: A Survey. IEEE Transanctions on Knowledge and Data Engineering 19(1), 1–16 (2007)
37. Ester, M., Kriegel, H.P., Sander, J., Xu, X.: A Density-Based Algorithm for Discovering Clusters in Large Spatial Databases with Noise. In: Proc. of Int. Conf. on Knowledge Discovery and Data Mining, Portland, Oregon, USA, pp. 226–231 (1996)
38. Fayyad, U., Piatetsky-Shapiro, G., Smyth, P., Widener, T.: The KDD Process for Extracting Useful Knowledge from Volumes of Data. Communications of the ACM 39(11), 27–34 (1996)
39. Fellegi, I.P., Sunter, A.B.: A Theory for Record Linkage. Am. Stat. Assoc. 64, 1183–1210 (1969)
40. Ganti, V., Ramakrishnan, R., Gehrke, J., Powell, A.: Clustering Large Datasets in Arbitrary Metric Spaces. In: Proc. of Int. Conf. on Data Engineering, Sydney, Austrialia, pp. 502–511 (1999)
41. Garcia-Molina, H.: Entity resolution: Overview and challenges. In: Atzeni, P., Chu, W., Lu, H., Zhou, S., Ling, T.-W. (eds.) ER 2004. LNCS, vol. 3288, pp. 1–2. Springer, Heidelberg (2004)
42. Gilks, W.R., Richardson, S., Spiegelhalter, D.J.: Markov Chain Monte Carlo in Practice. Chapman and Hall, Boca Raton (1996)
43. Gionis, A., Indyk, P., Motwani, R.: Similarity Search in High Dimensions via Hashing. In: Proc. of Int. Conf. on Very Large Databases, Edinburgh, Scotland, pp. 518–529 (1999)
44. Goiser, K., Christen, P.: Towards Automated Record Linkage. In: Proc. of Australasian Data Mining Conf., pp. 23–31 (2006)
45. Grabmeier, J., Rudolph, A.: Techniques of Cluster Algorithms in Data Mining. Data Min. and Knowl. Discovery 6(4), 303–360 (2002)
46. Gravano, L., Ipeirotis, P.G., Jagadish, H.V., Koudas, N., Muthukrishnan, S., Srivastava, D.: Approximate String Joins in a Database (Almost) for Free. In: Proc of Int. Conf. on Very Large Databases, Rome, Italy, pp. 491–500 (2001)

47. Gu, L., Baxter, R.A., Vickers, D., Rainsford, C.: Record Linkage: Current Practice and Future Directions. Technical Report, number 03/83. CSIRO Mathematical and Information Sciences (2001)

48. Guha, S., Rastogi, R., Shim, K.: CURE: An Efficient Clustering Algorithm for Large Databases. In: Proc. of ACM SIGMOD Int. Conf. on Management of Data, Seattle, Washington, USA, pp. 73–84 (1998)

49. Guha, S., Rastogi, R., Shim, K.: ROCK: A Robust Clustering Algorithm for Categorical Attributes. Inf. Syst. 25(5), 345–366 (2001)

50. Gunsfield, D.: Algorithms on Strings, Trees and Sequences. Cambridge University Press, Davis (1997)

51. Hassanzadeh, O., Chiang, F., Lee, H.C., Miller, R.J.: Framework for Evaluating Clustering Algorithms in Duplicate Detection. Proceedings of VLDB 2(1), 1282–1293 (2009)

52. Hassanzadeh, O., Miller, R.J.: Creating Probabilistic Databases from Duplicated Data. The VLDB Journal 18(5), 1141–1166 (2009)

53. Hernández, M.A., Stolfo, S.J.: The Merge/Purge Problem for Large Databases. In: Proc. of ACM SIGMOD Int. Conf. on Management of Data, San Jose, California, USA, pp. 127–138 (1995)

54. Hernández, M.A., Stolfo, J.: Real-world Data is Dirty: Data Cleansing and the Merge/Purge Problem. Data Min. and Knowl. Discovery 2(1), 9–37 (1998)

55. Herschel, M., Naumann, N.: Scaling up Duplicate Detection in Graph Data. In: Proc. of ACM Int. Conf. on Information and Knowledge Management, pp. 1325–1326 (2008)

56. Hjatason, G.R., Samet, H.: Index-Driven Similarity Search in Metric Spaces. ACM Trans. on Database Syst. 28(4), 517–518 (2003)

57. Indyk, P., Motwani, R.: Approximate Nearest Neighbor - Towards Removing the Curse of Dimensionality. In: Proc. of Symposium on Theory of Computing, Dallas, Texas, USA, pp. 604–613 (1998)

58. Ipeirotis, P.G., Verykios, V.S., Elmagarmid, A.K.: Duplicate Record Detection: A Survey. IEEE Trans. Knowl. Data Eng. 19(1), 1–16 (2007)

59. Jain, A.K., Dubes, R.C.: Algorithms for Clustering Data. Prentice-Hall, Englewood Cliffs (1998)

60. Jain, A.K., Murty, M.N., Flynn, P.J.: Data Clustering: A Review. ACM Comput. Surv. 31(3), 264–323 (1999)

61. Jaro, M.A.: Advances in Record Linkage Methodology as Applied to Matching the 1985 Census of Tampa, Florida. Journal of the American Statistical Society 84, 420–424 (1989)

62. Kingsbury, N.R., et al.: Record Linkage and Privacy: Issues in Creating New Federal Research and Statistical Information. U.S. General Accounting Office (2001)

63. Kopcke, H., Rahm, E.: Frameworks for Entity Matching: A Comparison Data and Know. Engineering 69(2), 197–210 (2010)

64. Kopcke, H., Thor, A., Rahm, E.: Evaluation of entity resolution approaches on real-world match problems. Proc. of the VLDB Endowment 3(1), 484–493 (2010)

65. Kopcke, H., Thor, A., Rahm, E.: Evaluation of Learning-Based Approaches for Matching Web Data Entities. IEEE Internet Computing 14(4), 23–31 (2010)

66. McCallum, A.: MALLET: A Machine Learning for Language Toolkit, http://mallet.cs.umass.edu

67. Koudas, N., Sevcik, K.C.: Size Separation Spatial Join. In: Proc. of ACM Int. Conf. on Management of Data, pp. 324–335 (1997)

68. Lawrence, S., Bollacker, K., Giles, C.L.: Autonomous Citation Matching. In: Proc. of ACM Int. Conf. on Autonomous Agents, pp. 392–393 (1999)

69. Low, W.L., Lee, M.L., Ling, T.W.: A Knowledge-Based Approach for Duplicate Elimination in Data Cleaning. Information Systems 26(8), 585–606 (2001)

70. Lwin, T., Nyunt, T.T.S.: An Efficient Duplicate Detection System for XML Documents. In: Proc. of IEEE Int. Conf. on Computer Engineering and Applications, pp. 178–182 (2010)

71. McCallum, A., Freitag, D., Pereira, F.: Maximum Entropy Markov Models for Information Extraction and Segmentation. In: Proc. of Int. Conf. on Machine Learning, Standord, California, USA, pp. 591–598 (2000)

72. McCallum, A., Nigam, K., Ungar, L.: Efficient Clustering of High-Dimensional Data Sets with Application to Reference Matching. In: Proc. of ACM SIGKDD Int. Conf. on Knowledge Discovery and Data Mining, Boston, Massachusetts, USA, pp. 169–178 (2000)

73. Menestrina, D., Benjelloun, O., Garcia-Molina, H.: Generic Entity Resolution with Data Confidences. In: Int. VLDB Workshop on Clean Databases, Seoul, Korea (2006)

74. Mitchell, T.M.: Machine Learning. McGraw-Hill, New York (1997)

75. Monge, A.E., Elkan, C.P.: An Efficient Domain-Independent Algorithm For Detecting Approximately Duplicate Database Records. In: Proc. of SIGMOD Workshop on Research Issues on Data Mining and Knowledge Discovery, Tucson, Arizona, USA, pp. 23–29 (1997)

76. Monge, A.E., Elkan, C.P.: The Field Matching Problem: Algorithms and Applications. In: Proc. of Int. Conf. on Knowledge Discovery and Data Mining, Portland, Oregon, USA, pp. 267–270 (1996)

77. Mukherjee, S., Ramakrishnan, I.V.: Taming the Unstructured: Creating Structured Content from Partially Labeled Schematic Text Sequences. In: Proc. of CoopIS/DOA/ODBASE Int. Conf., Agia Napa, Cyprus, pp. 909–926 (2004)

78. Muse, A.G., Mikl, J., Smith, P.F.: Evaluating the quality of anonymous record linkage using deterministic procedures with the New York State AIDS registry and a hospital discharge file. Statistics in Medicine 14, 499–509 (1995)

79. Neiling, M., Jurk, S.: The Object Identification Framework. In: Proc. KDD Workshop on Data Cleaning, Record Linkage, and Object Consolidation, Washington, DC, USA, pp. 37–39 (2003)

80. Neutel, C.I.: Privacy Issues in Research Using Record Linkage. Pharmcoepidemiology and Drug Safety 6, 367–369 (1997)

81. Newcombe, H.B.: Record Linking: The Design of Efficient Systems for Linking Records into Individual and Family Histories. American Journal of Human Genetics 19, 335–359 (1967)

82. Newcombe, H.B., Kennedy, J.M., Axford, S.J., James, A.P.: Automatic Linkage of Vital Records. Science 130, 954–959 (1959)

83. Patel, J., DeWitt, D.J.: Partition Based Spatial-Merge Join. In: Proc. of ACM Int. Conf. on Management of Data, pp. 259–270 (1996)

84. Pasula, H., Marthi, B., Milch, B., Russell, S.J., Shpitser, I.: Identity Uncertainty and Citation Matching. In: Proc. of Ann. Conf. on Neural Information Processing Systems, pp. 1401–1408 (2002)

85. Sarawagi, S., Bhamidipaty, A.: Interactive Deduplication using Active Learning. In: Proc. of ACM SIGKDD Int. Conf. on Knowledge Discovery and Data Mining, Edmonton, Alberta, Canada, pp. 269–278 (2002)

86. Sarawagi, S., Kirpal, A.: Efficient set joins on similarity predicates. In: Proc. of SIGMOD Int. Conf. on Management of Data, Paris, France, pp. 743–754 (2004)

87. Shen, H., Zhang, Y.: Improved Approximate Detection of Duplicates for Data Streams over Sliding Windows. Journal of Computer Science and Technology 23(6), 973–987 (2008)

88. Singla, P., Domingos, P.: Multi-Relational Record Linkage. In: Proc. of ACM Int. Ws. on Multi-Relational Data Mining, pp. 31–38 (2004)

89. Smith, S., Waterman, M.S.: Identification of Common Molecular Subsequences. Journal of Molecular Biology 147(1), 195–197 (1981)

90. Statistical Linkage Key Working Group. Statistical Data Linkage in Community Services Data Collections (2002)

91. Tejada, S., Knoblock, C.A., Minton, S.: Learning Domain-Independent String Transformation Weights for High Accuracy Object Identification. In: Proc. of ACM SIGKDD Int. Conf. on Knowledge Discovery and Data Mining, Edmonton, Alberta, Canada, pp. 350–359 (2002)

92. Tepping, J.B.: A Model for Optimum Linkage of Records. Journal of the American Statistical Association 63, 1321–1332 (1968)

93. Verykios, V.S., Elmagarmid, A.K., Houstis, E.N.: Automating the approximate record-matching process. Inf. Sci. 126(1-4), 83–98 (2000)

94. Weber, R., Schek, H.J., Blott, S.: A Quantitative Analsysis and Performance Study for Similarity Search in High-Dimensional Spaces. In: Proc. of Int. Conf. on Very Large Databases, New York City, USA, pp. 194–205 (1998)

95. Weis, M., Naumann, N.: Detecting Duplicates in Complex XML Data. In: Proc. of IEEE Int. Conf. on Data Engineering, p. 109 (2006)

96. Weis, M., Naumann, N.: Space and Time Scalability of Duplicate Detection in Graph Data. Tech. Rep. 25, Hasso-Plattner Institut, Potsdam, Germany (2007)

97. Winkler, W.E.: String Comparator Metrics and Enhanced Decision Rules in the Fellegi-Sunter Model of Record Linkage. In: Proc. Section on Survey Research Methods, American Statistical Association, pp. 354–359 (1990)

98. Winkler, W.E.: Overview of Record Linkage and Current Research Directions. Technical Report. Statistical Research Division, U.S. Census Bureau (1999)

99. Winkler, W.E.: Methods for Record Linkage and Bayesian Networks. Tech. Rep. RRS2002/05, U.S. Bureau of the Census, Washington, D.C., USA (2002)

100. Zhang, Y., Lin, X., Yuan, Y., Kitsuregawa, M., Zhou, X., Yu, J.X.: Duplicate-insensitive Order Statistics Computation over Data Streams. IEEE Transanctions on Knowledge and Data Engineering 22(4), 493–507 (2010)

# A Survey on Integrating Data in Bioinformatics

Andrea Manconi and Patricia Rodriguez-Tomé

**Abstract.** Data integration is an open challenge in bioinformatics. Querying and retrieving data from remote and/or local sources and analyzing them are very time consuming tasks for biologists. Data integration allows biologists to combine knowledge from multiple disciplines. This has become a critical issue in biological research in recent years. Advances in technology have pathed the way to a huge and growing amount of available biological data. However, it is important to highlight that the distinctive feature in integrating biological data is not mainly concerned with the amount of data but with their complexity. Biological data sources are considered strongly heterogeneous in many aspects. Several approaches and systems based on different technologies and techniques, have been proposed in the literature to deal with the problem of integrating biological sources. Nevertheless it does not exist yet an approach able to solve all mentioned problems. This chapter provides a survey on data integration in the field of biological sources.

## 1 Introduction

In the last years, the steady growth of the Internet fostered the information explosion on the Web. This exponential growth makes it increasingly difficult to manage the information. The more the increase of heterogeneous sources explodes, the more combining the data sources under a single query interface becomes necessary.

Andrea Manconi
Institute for Biomedical Technologies, National Research Council, Via F.lli Cervi 93, 20090, Segrate (MI), Italy
e-mail: andrea.manconi@itb.cnr.it

Dept. of Electrical and Electronic Engineering, Univ. of Cagliari, P.zza D'Armi, 09123, Cagliari, Italy
e-mail: manconi@diee.unica.it

Patricia Rodriguez-Tomé
CRS4 - Center for Advanced Studies, Research and Development in Sardinia, Loc. Piscina Manna, Ed.1, 09010 Pula (CA), Italy
e-mail: prtome@crs4.it

M. Biba and F. Xhafa (Eds.): Learning Structure and Schemas from Documents, SCI 375, pp. 413–432.
springerlink.com                                                                      © Springer-Verlag Berlin Heidelberg 2011

Advances in technology (e.g., high throughput technologies) have given rise to a huge and growing amount of available biological data. This rapid growth has been accompanied by an equivalent increase of biological sources and databases. The number of resources listed in the Nucleic Acids Research database issues has increased over the years (see Fig. 1), and in the 2010 issue [1], 1230 databases are listed. Several types of data are exported by these biological sources. Genome, protein structure, protein model, protein protein, and microarray databases are some of the typical data types. It should be stressed that relationships among these biological data usually exist and are often difficult to identify.

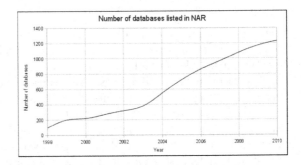

**Fig. 1** The number of resources listed from 1998 to 2010 in the Nucleic Acids Research database issues

In this perspective, data integration is a very important task for biologists. Data integration allows biologists to combine knowledge from multiple disciplines. This has become a critical issue in biological research in recent years. Querying and retrieving data from remote and/or local sources and analyzing them are very time consuming tasks.

Databases have become a common and everyday tool for biologists involved in genome analysis, protein studies, microarray and proteomics, as well as molecular entities. Databases in biology did not begin with sequencing. Before that, genetic analysis projects had already managed their data on computer systems, to share, compare, and analyze the information. Only to recall few examples, we can cite the CEPH database [2][3], aimed at collecting markers from an international collaboration, and OMIM [4][5] the Online Mendelian Inheritance in Man system, a compendium of information on genetic disorders and genes now hosted at NCBI[1] (National Center for Biotechnology Information). Subsequently, the increase of biological data promoted the design and the development of centralized archives of data. In particular, GenBank[2] [6] in the United States of America, EMBL[3] (European Molecular Biology Laboratory) [7] in Europe, and DDBJ[4] (DNA Data Bank of Japan) [8] in Japan have been developed. Independent at the beginning, the three databanks soon formed an international collaboration with daily exchange of sequence and annotation data.

---

[1] http://www.ncbi.nlm.nih.gov/omim/

[2] http://www.ncbi.nlm.nih.gov/genbank/

[3] http://www.ebi.ac.uk/embl/

[4] http://www.ddbj.nig.ac.jp/

However, this collaboration has been aimed at unifying the data collected neglecting any aspect on their representation. In fact, the data stored in these archives are presented differently to the user using a form of a textual representation called flat file. This heterogeneity in format obliged users to design and develop different and specialized tools aimed at wrapping each database. The same events were repeated for each new database created, being it protein sequence (e.g., PIR [9], SwissProt [10])-protein structure (e.g., PDB [11], MSD [12])-microarray (e.g., GEO [13], ArrayExpress [14]),- as well as proteomics identification (e.g. PRIDE [15])-databases. EMBOSS [16], the software suite provided by EBI (European Bioinformatics Institute), lists more than 20 different formats currently in use. The same approach for handling data has been adopted in designing specialized analysis software tools. Very often, a new tool will define its input and output format independently from those already in use. This entails that a user involved in performing analysis using both specialized data retrieved from biological sources and analysis software tools, spends part of her/his time in the task of processing data for migrating them.

It is worth noting that the nonexistence of a standard for representing biological data is not the only problem when analyzing them. Another known problem is related with the heterogeneity in naming. Each database uses its own way of naming unique identifiers. Then, two different entities could very well be identified with the same unique identifier on different systems. Without a convention aimed at tying a name with its original resource, data end up losing their origin and get confused between them. An effort to solve the problem has been made around 2005 to create a global identity naming mechanism for objects in life science[5]. Unfortunately, this proposal has not been widely accepted and very few databases have implemented this identification mechanism.

The heterogeneity in formats and identifiers extends also to the naming conventions. In fact, gene names and functions are not identified using the same conventional names in all resources. While format heterogeneity can be solved designing and developing specialized tools, it does not exist a software able to decide if two different named genes are identical and have the same function. GMOD (Genome Model Organism Database Project) has been the first that tackled the problem. In 1998, a consortium formed by researchers working on Drosophila (fruit fly), Mus musculus (mouse) and Saccharomyces cerevisiae (baker's yeast) began developing an ontology to describe their data. Many other groups have joined the consortium, extending the initial ontology to other species and creating new ones. This work resulted in a new ontology called Gene Ontology (GO) [17] that has become fundamental for genome research. All these considerations show that the distinctive feature in integrating biological data is not mainly concerned with the amount of data but with their complexity [18].

Another important aspect that needs to be considered when dealing with data integration is related to the autonomy of biological sources. Due to their autonomy, biological sources are free to modify their design or schema or remove data without

---

[5] http://lsids.sourceforge.net

any warning. Moreover, each source defines and provides different query interfaces and often restricts the access to their data. These restrictions force end-users and external systems to adapt and limit their queries to a certain form.

## 2   Mission

Integration of biological data is an open challenge for bionformatics. Several approaches and systems (i.e. WWW-query [19], k2/BioKleisli [20], and DiscoveryLink [21]), based on different technologies and techniques, have been proposed in the literature to deal with the problem of integrating biological sources. In particular, these approaches can be classified as data warehouse integration, mediator-based integration, and navigational integration [22]. Despite the fact that several approaches and systems have been proposed, it does not exist an approach able to solve all mentioned problems. This chapter is a survey on approaches, methods and systems for integrating data in the field of biological sources. We will discuss about the integration approaches and technologies adopted to deal with the specific problems of integrating biological data. Both the advantages and the disadvantages of the described approaches and systems will be shown. We will also present the new trends and challenges.

The remainder of the chapter is organized as follows. First, the main approaches adopted to deal with the problem of data integration will be described and discussed. It will be followed by a survey of the main systems devised for integrating biological data. After that, new trends and challenges in biological data integration will be discussed. Finally, conclusions end the chapter.

## 3   Background

Data integration is the problem of combining data residing at different sources, and of providing the user with a unified view of these data [23]. The systems devised for integrating biological sources may differ in several aspects including *i*) the integration performed, *ii*) the data model, *iii*) the source model, and *iv*) the level of transparency [24]. In the first point, it is possible to distinguish between browsing oriented and query oriented systems. Browsing oriented systems are aimed at supporting the user in an integrated browsing, whereas query oriented systems are aimed at providing an environment where users can pose queries on an integrated system. In the second point, it is possible to distinguish between systems that export structured data from those that export unstructured or semi-structured data. In the third point, it is possible to distinguish the systems on the basis of the existing relationships among the integrated sources. Data sources can export complementary or overlapping data. As for the fourth point, it is possible to distinguish between systems that enable the user to select the sources to involve in the query from systems that do not enable the user to do this task.

An accepted classification of these systems relies on their data model. In particular, it is possible to distinguish between systems that consider the sources as exporting linked sets of browsable records, and systems that consider the sources as exporting structured data. As for the former case, the integration involves the supporting effective navigation across sources (navigational or linked-based integration). As for the latter case, we can detect two approaches depending whether the integration brings all information retrieved by its sources in a centralized database (data warehousing integration) or it leaves the information in the sources (view or mediator-based integration and federated databases). In the following of this section we summarize these approaches.

## 3.1 Linked-Based Integration

Typically, biological data exported by a source are cross-referenced with other biological data exported by other sources. It means that to obtain the desired data, users browse across several Web pages by following hyperlinks. It is worth noting that some of the Web pages provided in this way are otherwise hard to discover browsing the Web. The linked-based is an approach aimed at facilitating the discovery of these linked sources. It cross-references data in a source with data in another source. When a query is asked, the answer detects all possible paths between two entities [25]. Modeling the data with a classical relational database does not permit to answer these queries. Therefore, data are represented using a model where sources are defined as sets of pages with their interconnections [26]. It is important to highlight that this approach performs a quick search of related links, rather than a real integration of sources.

Several problems have to be considered when the linked-based approach is used. In particular, we can observe that *i*) it is particularly vulnerable to naming ambiguities resulting in a lot of wrong connections among the sources, and *ii*) that to work well this approach requires the cooperation of the curators of the Web sources. Obviously, this is a wishful thinking.

## 3.2 Data Warehousing Integration

The data warehousing approach [27] is aimed at retrieving information from distributed sources and storing them in a centralized database (see Fig. 2). This approach relies on both an ad-hoc unified data model able to store the data from the sources, and a specialized software called ETL (Extract Transform Load). ETL is aimed at extracting the data from the sources, transforming (e.g. cleaning and filtering) them to fit the data model, and loading the processed data into the warehouse.

When a query is posed this is not evaluated on the sources but on the local schema. Then, a data warehouse can be seen as a set of materialized views defined over the sources [28].

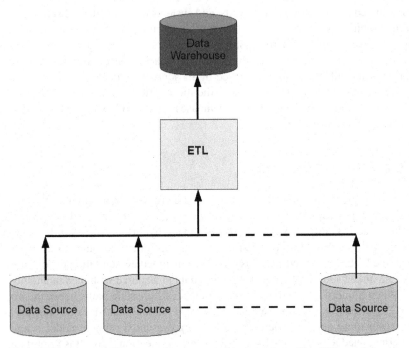

**Fig. 2** The macro-architecture of a data warehouse system. The ETL module extracts the information from the data sources, transforms and loads it into the data warehouse.

This approach presents some advantages and disadvantages. As for the advantages, it can be observed that being based on a local repository of the data, this approach allows to reduce the problems associated with the network. Problems related with a low response time, and temporary unavailability of sources are avoided. Another advantage is related with the possibility of optimizing the query using the materialized views. It is worth noting that, additional and very important benefits for bioinformatics result from the possibility to transform and validate the data before storing them in the warehouse [29]. As for the disadvantages, to maintain the data warehouse is very demanding and extremely complicated. In fact, to deal with the integration of new sources, and changes in the model of integrated sources, the data model requires continuous modifications. Moreover, it is very important that the warehouse be continuously updated according to changes on the data sources.

## 3.3  Mediator-Based Integration

Conversely to the data warehousing, in the mediator-based approach [30] the information is left in the sources. This approach relies on a mediated or global schema aimed at describing the sources and the relations among them (see Fig. 3). In particular, it is responsible for providing an interface - a virtual set of relations - between the sources [31].

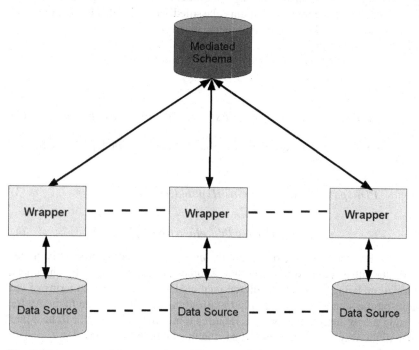

**Fig. 3** The macro-architecture of a mediator-based data integration system. The mediated schema provides an interface between the sources.

Users ask queries on the global schema as if this was the only one schema, without knowing the details of the data sources. To answer a query involves different tasks. In particular, when a query is posed on the mediated schema a suitable mediator will be responsible to reformulate the query on the local schema of the underlying sources. Then, this reformulated query is passed to a query optimizer which produces a query execution plan that specifies how to evaluate the query. Finally, the query execution plan obtains information from the sources. To this end, a set of suitable wrappers specific for each source work to translate data from the source to a form usable by the system.

Two approaches are mainly used to describe the relationships between the source relations and the mediated schema, the Global As View (GAV) [32][33] and the Local As View (LAV) [34][35][36].

### 3.3.1 Global as View

In the GAV approach, each relation in the mediated schema is expressed in terms of a query over the sources relations. To exemplify the concept, let us consider to integrate two biological sources: the structure protein database PDB and the database of structural classifications CATH [37]. The first database contains data of proteins with known structure (e.g. protein identifier, primary structure, atomic coordinates, chemical features, biochemical features, experimental details), whereas the second database contains data about the proteins classified with CATH. The relationships between the sources and the mediated schema can be expressed as follows:

$$PDB(pdbid, structure) \Rightarrow Protein(pdbid, structure) \tag{1}$$

$$CATH(pdbid, classification) \Rightarrow Classification(pdbid, classification) \tag{2}$$

The following expression describes how the CATH classified proteins can be obtained:

$$PDB(pdbid, structure) \wedge CATH(pdbid, classification) \Rightarrow \tag{3}$$

$$\Rightarrow Classified(pdbid, classification) \tag{4}$$

We can observe that if, on the one hand, the query reformulation process is immediate, on the other hand, to integrate new sources is non-trivial. As for the query reformulation, being the relations on the mediated schema defined in terms of the relations on the sources, we need only to unfold the definition of the firsts. As for integrating a new source, this process requires to modify the mediator to take into account all the ways the source can be used in the mediated schema.

### 3.3.2 Local as View

In the LAV approach, each relation in the sources is expressed in terms of a query over the mediated schema relations. To exemplify the concept, we consider to integrating two biological sources $S_1$ and $S_2$. The first containing protein structures experimentally determined using X-ray crystallography, and the second containing protein structures experimentally determined using the nuclear magnetic resonance (NMR). In this case, the sources can be described as follows.

$$S_1(pdbid, structure) \Rightarrow Protein(pdbid, structure) \wedge exp\_method = Xray \tag{5}$$

$$S_2(pdbid, structure) \Rightarrow Protein(pdbid, structure) \wedge exp\_method = NMR \tag{6}$$

Conversely of the GAV approach, we can observe that if on the one hand to integrate new sources is immediate, on the other hand the query reformulation process is non-trivial.

## 3.4 Federated Databases

As in the mediator based approach, also in federated databases [38] the information is left in the sources and a global schema materializes views on the local schemas. The differences between the two approaches are to be found in the building of the global schema. Federated databases systems adopt a bottom-up procedure, whereas mediator based systems adopt a top-down procedure to build the global schema. In fact, in mediator-based systems the global schema is built according to the local schemas, whereas in federated databases the local schemas are built or extended according to the global and federal schema.

# 4 Biological Data Integration Systems

## 4.1 Data Warehouse

A data warehouse is a physical collection of data taken from different sources and merged into one single and homogeneous database. Data warehouses are read-only. Updates are done by reprocessing all the different data sources. This process includes the extraction of semantics, format transformation, the identification of attributes correspondence, and finally data integration. These steps are very much time consuming. Furthermore, very often the original databases are independent and not in a formal collaboration with the data warehouse. Changes in the original schemas/formats are not reported in time to the data warehouse maintainers. This adds to the difficulty of maintenance of the system. We can summarize the pros and cons of a data warehouse system:

- *Pros* - All data are stored in one place. Users queries arc not dependent of network hazards between the different original data sources. Queries can be optimized.
- *Cons* - Huge maintenance costs. Updating data is very much time consuming. The acceleration rate of data production makes it very difficult to maintain data up to date according to changes in the original sources. Due to the proliferation of independent databases containing relevant data it is sometimes difficult to decide where and when to stop integrating.

In the following, systems built to integrate biological data exploiting the data warehouse approach are presented and discussed.

*SRS* - The Sequence Retrieval System [39], originally developed at EMBL is currently distributed by Biowisdom[6]. SRS is a scalable data integration platform which makes the data available on the Web. Any database distributed as a text file, XML file or even relational database can be integrated, regardless of its original format. SRS uses the ICARUS (Interpreter of Commands And Recursive Syntax) language for creating scripts that describe the data format, the links between the different data sources and other specifics for parsing the files. A parser script is specific to a database - SRS comes by default with more than 80 parsers for the common biological databases. The datafiles are then indexed. The SRS system provides the user with an intuitive Web form based interface to query simultaneously as many databases as are installed in the local system. It provides also a query language for advanced users and a programing API for developers. Furthermore, SRS includes access to some common bioinformatics tools from within its interface. SRS can be considered an efficient data warehouse system which can integrate many data source. Its drawback is the complexity of the system administration, which needs a dedicated full time administrator for medium to large installations. An example of a large installation can be found at the EBI[7].

*Entrez* - The Entrez system [40] is provided by the NCBI[8] as the access point to all databases provided by NCBI. The system has an in-built neighboring concept that allows linking different data sources without explicitly specifying the links - as it is done by SRS. This allows the user to see the whole system as one big database - a data warehouse. Queries will use those "links" to retrieve related data. A web services API allows developers to build their own tools. The NCBI system works only with the NCBI databases, and it is not possible to add other data sources, as it is possible with SRS.

*IGD* - The Integrated Genome Database [41] project has been discontinued. It is nevertheless the best example of a relational data integration. IGD has integrated more than 20 different databases, federating them in one huge relational database, a true data warehouse. Its own global schema was different from those of the underlying databases. For each different data source, the IGD team had to write specific data and schema parsers and transformers, to extract data from its original database and store it in the IGD database. The data was presented to the user by a graphical interface. The IGD system was also integrated with AceDB (see later on), offering a small portable database interface. The main drawback of IGD was the difficult and time consuming maintenance of the system. These difficulties became more acute with the advent of high throughput genomic projects.

---

[6] http://www.biowisdom.com/2009/12/srs/
[7] http://srs.ebi.ac.uk
[8] http://www.ncbi.nm.nih.gov

## 4.2   Database Federation

The idea behind a federation is that it is not necessary to create a central data warehouse. Instead, databases participating in a federation are distributed and locally maintained, but can cooperate. The end user will query the federation through a software that will make appear the multiple databases as only one. Depending on the application, the user might be able to choose which original source should participate in the query. There are two different types of federation: mediator-based and database-based.

### 4.2.1   Mediator-Based Federation

Databases to be integrated are independent from each other. A mediated schema provides an interface between the data sources and the client programs (software). The systems described below are not any more in development while still referenced.

*CORBA* - CORBA (Common Object Request Broker Architecture) is an architecture that facilitate data integration. It uses an interface definition language (IDL) to describe data and methods available to query the database/run software, independently of the database schema, database system and language. A server must be implemented using that IDL interface, to provide access to the clients through the Internet. Client and server do not need to be written in the same language. While it is better that the server is developed by the database itself, it is not mandatory. In 1995, at EBI was started a project[9] to develop new tools and methods for integrating biological databases. The aim of that project was to investigate the CORBA methodology and its possible applications for biological databases. Effort was made at EMBL-EBI to create public interfaces to databases (EMBL [42], RHdb [43]) and programs [44]. While not very complex, CORBA has a steep learning curve. It never took out for production purposes in bioinformatics. It has been abandoned at EMBL-EBI. International efforts were made within the CORBA OMG[10] to create some international standard for biomolecules [45] and maps [46].

*TAMBIS* - The TAMBIS (Transparent Access to Multiple Bioinformatics Information Sources) system [47] is based on the CPL (Collection Programming Language) a functional programming language that allows data to be described and manipulated. Data sources must each develop their own "CPL driver", which means develop the wrapper of their data in the CPL language, to be accessible by the system. TAMBIS was a proof of concept and has never been fully applied.

---

[9] EU FP4 Project BIO4-CT96-0346 (1996).
[10] http://www.omg.org

*ISYS* - The Integrated SYStem[11] [48] was developed to provide a dynamic and flexible platform for integration of molecular data sources and software. The system is a Java application that has to be installed locally. Then it will give access to distributed sources on the internet.

All these mediator based federations were developed around the same time (1995-2005) and did not took off within the bioinformatics community, because the community did not see the benefits of the system versus their complexity.

### 4.2.2    Database-Based Federation

In this case the databases must be active part of the federation. Each federation will share a common database schema and software. Each database part of the federation will provide a public version of its data in that schema, while keeping the possibility of having another internal or even public schema and format. Applications can be developed to query this "federation" schema, even allowing for multiple queries across databases.

*AceDB* - While no longer in full development, AceDB[12] [49] has been the first system to provide a common schema for multiple small genome projects. AceDB is a combined software and specifically built object-oriented database, presenting the user with a very useful graphical interface. The force of the system was speed and data management, its drawback the user management - there was no concurrent access possible. Later on a web access solved that problem. AceDB was first built for the C.Elegans studies, but has been very much used for all the organism specific projects at Sanger Centre (UK), Arabidopsis project (AAtDB) [50], for the UK Cropnet databases.

*GMOD* - GMOD[13] [51] (Generic Model Organism Database project) is a set of software tools for creating and managing genome-scale biological databases. These databases will share a common schema, which allows interoperability and the reuse of client interfaces and tools. Unlike AceDB, the GMOD databases are multi-user. GMOD tools are in use at many community databases, where it is important to be able to interact. A Genome viewer (GBrowse) offers a graphical and intuitive Web client. While developed at the beginning for the Model Organism, organisms of huge interest as models for others species, the GMOD system has been extended and now can be applied to almost every genome project. GMOD databases are read-write and serve as data management systems for the projects. A set of tools are provided to help the maintainers to update the databases to a new version of the database schema, which is updated when there is a need for a new data type. By keeping the schemas up to date in all databases, the system acts as a federation of databases.

---

[11] http://sourceforge.net/projects/isys2/

[12] http://www.acedb.org/

[13] http://www.gmod.org

*Ensembl* - The Ensembl[14] [52] genome annotation system, is being developed jointly by the EBI and the Wellcome Trust Sanger Institute since 2000. While first developed to present the completed data (chromosome per chromosome) from the Human Genome sequencing project, it has been since then used for every completed sequenced genome. Sequence data are analyzed, and automatically annotated. The system is composed of a relational database, a complete graphical Web interface for interactive use and a set of API and web services for programming query scripts. An Ensembl database is read-only. Updates to the database are made at regular intervals by the maintainers - new sequence data is merged and the whole data set is reanalyzed. Here again, since all databases share the same schema, it is possible for a program to query multiple instances. The source code of the Ensembl database and software is freely available and can be installed locally. The drawback is its complexity, which demands a trained computer expert for installation and maintenance. It is more convenient for users to access one of the official Ensembl installations.

*BioMart* - BioMart[15] [53] is a query-oriented data management system developed jointly by the OICR (Ontario Institute for Cancer Research) and the EBI. BioMart consists in a database (with schema) and a set of tools to query this database. Databases participating to the federation must extract data from the original data sources, and import them in a local BioMart implementation. A centralized service is available to keep track of all the databases in the federation. Various graphical clients exist like Martview[16] . A web services API is also available (MartService). BioMart can be seen as a complement to Ensembl and GMOD, offering a greater federation service. BioMart is not widely used even though it is not a difficult system to implement.

## 5   Discussion

In the previous sections we have described and discussed the main approaches and systems for integrating biological data. It appears clear that more needs to be done to provide a suitable solution for the problem. For the aim of this discussion it is important to highlight that when we refer to biological data we are considering raw data, methods, tools, as well as data derived by analysis. We deem that advances in integrating biological data can be obtained exploiting interoperability. Interoperability refers to the ability of diverse systems and organizations to work together. Several criteria can be used to figure out the interoperability between two systems. For our discussion, we concentrate on three well-defined criteria: software, syntactic, and semantic interoperability.

---

[14] http://www.ensembl.org

[15] http://www.biomart.org/

[16] http://www.biomart.org/biomart/martview/

- *Software interoperability* refers to the ability of two or more software components to communicate and interact requiring little or no knowledge among the softwares.
- *Syntactic interoperability* refers to the ability of two system to communicate and exchange data. This involves that to ensure the interoperability between two systems common data formats and communication protocols must be adopted. Syntactic interoperability is a pre-requisite to semantic interoperability.
- *Semantic interoperability* refers to the ability of two systems to properly and automatically interpret the exchanged data. Semantic interoperability is aimed at avoiding ambiguities in interpretation of data.

In our opinion major benefits in biological data integration can be obtained exploiting the advances in Web technologies. In fact, advances in Web technologies are aimed at developing these criteria of interoperability. The Web is undergoing a continuous evolutionary process that crosses four evolutionary eras, from Web 1.0 to Web 4.0. The Web 1.0 or World Wide Web has been characterized by the absence of interaction with the users (read-only), and by a keyword search based approach. The Web 2.0 is mainly characterized by a social and collaborative nature. The communication is many-to-many and not one-to-many as in the Web 1.0. The Web 3.0 will bring semantic to the Web where any published document will be described with additional information and data (i.e. metadata) so that to enable the query, the interpretation, and to perform tasks automatically on behalf of the user. Finally, the Web 4.0 which currently is the last planned evolution of the Web will integrate the Web 3.0 with artificial intelligence.

As for the Web 2.0, it contributes to exploit the software interoperability. The Web 2.0 defined two technologies useful to this goal: web services and mashups. A *web service*[17] is typically a Web API that is accessed via Hypertext Transfer Protocol and executed on a remote system, hosting the requested service. The World Wide Web Consortium defines a web service as "a software system designed to support interoperable machine-to-machine interaction over a network". For Stein [22] web services are a variant of linked-based integration. The heterogenous collection of linked data resources on the Web becomes a world of services that are linked by service names and definitions. We deem web services a very powerful technology for bioinformatics. Web services enable users to create specific on-demand data integration. With web services, users dispose of a programming environment to easily retrieve and analyze data. In our opinion, a very interesting and important benefit of exploiting web services technology is related with its collaborative aspects. Typically, bioinformaticians define specialized pipelines of web services to query, retrieve, and analyze biological data. These pipelines can be shared with the scientific community as new web services [54][55]. In this way, the concept of software reuse goes hand in hand with the concept of data integration reuse that refers to the ability of generating new pipelines for integrating data from existing pipelines

---

[17] http://www.w3.org/2002/ws/Activity.html

rather than from scratch. A *mashup* is a new application development approach that allows users to aggregate multiple services to create a service for a new purpose [56]. Mashup is aggregation oriented rather than integration oriented. Data taken from different resources on the Web with RESTful API or RSS-based syndication are used to make a new web application. If web services aim at composing different services to create new ones, mashups are aimed at aggregating different services. The main disadvantage in using mashups is related with the fact that they are not designed to access desktop data. This is a considerable disadvantage for bioinformaticians. Despite that, mashups are rapidly gaining importance in life sciences.

As for the Web 3.0, it contributes to exploit both syntactic and semantic interoperability. Several technologies (e.g. RDF [57], RDFS [58], OWL [59]) aimed at providing a formal description of data have been developed. The goal in the biological context is to transform the Web of biological data in a single repository of data and knowledge easy to query. Despite efforts in this direction, due to the specific knowledge domain semantic integration continues to be an open challenge. Data integration in Web 3.0 exploits ontologies. Ontology-based integration can be classified as LAV.

An ontology is a formal representation of knowledge. Ontologies represent both the concepts within a domain, and the relationships between these concepts. Several ontologies aimed at representing several knowledge domains have been defined. Typically, an ontology can be anchored to other more general ontologies with the aim of reusing the knowledge. With reference to biological field it can be observed that ontologies are defined independently of other ones preventing the knowledge reuse and making even harder the problem of integration [60]. In general, researchers do not work to define standard ontologies for a specific knowledge domain. Obviously, this way of proceeding is in contradiction with the philosophy of the Web 3.0. Another relevant problem is related with the ambiguous identification of resources. Often, ontologies that refer to a common concept use different URIs. In so way, a resource could have ambiguous descriptions [61]. Despite several proposals [62] [63] aimed at standardizing the identification of biological concepts there is no common accepted identification scheme.

## 6 Conclusions

In this chapter we have discussed the current approaches, methods, systems, and open issues of integrating biological data. First we provided a detailed analysis of the main problems related with this task. We have discussed on the complexity of the problem due to the heterogeneity of both data and sources. We followed, a description of the state of the art of the approaches adopted to deal with the general problem of data integration. According to these approaches we have described the systems that define the state of the art in integrating biological data. Finally, we discussed and stated our vision on new trends and challenges. In particular, we exposed how the specificity of the domain makes the problem of data integration even harder.

Despite existing efforts, biological data integration continues to be an open challenge. It appears clear that integration can not be achieved without the cooperation of
the data providers. The Biocuration Association, formed by the curators/annotators
of databases might be the first step towards that cooperation, which could be driven
"bottom-up" from the biologist/users to the developers.

**Acknowledgements.** This work has been supported by the MIUR FIRB ITALBIONET
(RBPR05ZK2Z), Bioinformatics analysis applied to Populations Genetics (RBIN064YAT_003)
and SHIWA Projects, and by the Center for Advanced Studies, Research and Development in
Sardinia jointly with the Autonomous Region of Sardinia, under the project UOMO.

# References

1. Cochrane, G.R., Galperin, M.Y.: The 2010 Nucleic Acids Research Database Issue and
   online Database Collection: a community of data resources. Nucleic Acids Research 38,
   D1–D4 (2009)
2. Dausset, J., Cann, H., Cohen, D., Lathrop, M., Lalouel, J.M., White, R.: Centre
   d'etude du polymorphisme humain (CEPH): collaborative genetic mapping of the human genome. Genomics 6(3), 575–577 (1990)
3. Murray, J.C., Buetow, K.H., Weber, J.L., Ludwigsen, S., Scherpbier-Heddema, T., Man-
   ion, F., Quillen, J., Sheffield, V.C., Sunden, S., Duyk, G.M., Weissenbach, J., Gyapay,
   G., Dib, C., Morrissette, J., Lathrop, G.M., Vignal, A., White, R., Matsunamic, N.,
   Gerken, S., Melis, R., Albertsen, H., Plaetke, R., Odelberg, S., Ward, D., Dausset, J.,
   Cohen, D., Cann, H.: A comprehensive human linkage map with centimorgan density.
   Science 265(5181), 2049–2054 (1994)
4. McKusick, V.A.: Mendelian Inheritance in Man. A Catalog of Human Genes and Genetic
   Disorders, 12th edn. Johns Hopkins University Press, Baltimore (1998)
5. Hamosh, A., Scott, A.F., Amberger, J.S., Bocchini, C.A., McKusick, V.A.: Online
   Mendelian Inheritance in Man (OMIM), a knowledgebase of human genes and genetic
   disorders. Nucleic Acids Research 33(database Issue), D514–D517 (2005)
6. Benson, D.A., Boguski, M.S., Lipman, D.J., Ostell, J., Ouellette, B.F.F.: GenBank. Nu-
   cleic Acids Research 26(1), 1–7 (1997)
7. Kulikova, T., Akhtar, R., Aldebert, P., Althorpe, N., Andersson, M., Baldwin, A., Bates,
   K., Bhattacharyya, S., Bower, L., Browne, P., Castro, M., Cochrane, G., Duggan, K.,
   Eberhardt, R., Faruque, N., Hoad, G., Kanz, C., Lee, C., Leinonen, R., Lin, Q., Lombard,
   V., Lopez, R., Lorenc, D., McWilliam, H., Mukherjee, G., Nardone, F., Pastor, M.P.G.,
   Plaister, S., Sobhany, S., Stoehr, P., Vaughan, R., Wu, D., Zhu, W., Apweiler, R.: EMBL
   Nucleotide Sequence Database in 2006. Nucleic Acids Research 35(1), D16–D20 (2006)
8. Kaminuma, E., Mashima, J., Kodama, Y., Gojobori, T., Ogasawara, O., Okubo, K.,
   Takagi, T., Nakamura, Y.: DDBJ launches a new archive database with analytical
   tools for next-generation sequence data. Nucleic Acids Research 38(database issue),
   D33–D38 (2010)
9. Barker, W.C., Garavelli, J.S., McGarvey, P.B., Marzec, C.R., Orcutt, B.C., Srinivasarao,
   G.Y., Yeh, L.S.L., Ledley, R.S., Mewes, H.W., Pfeiffer, F., Tsugita, A., Wu, C.: The PIR-
   International Protein Sequence Database. Nucleic Acids Research 27(1), 39–43 (1998)
10. Bairoch, A., Boeckmann, B.: The SWISS-PROT protein sequence data bank. Nucleic
    Acids Research 20, 2019–2022 (1992)

11. Berman, H.M., Westbrook, J., Feng, Z., Gilliland, G., Bhat, T.N., Weissig, H., Shindyalon, I.N., Bourne, P.E.: The Protein Data Bank. Nucleic Acids Research 28, 235–242 (2000)
12. Boutselakis, H., Dimitropoulos, D., Fillon, J., Golovin, A., Henrick, K., Hussain, A., Ionides, J., John, M., Keller, P.A., Krissinel, E., McNeil, P., Naim, A., Newman, R., Oldfield, T., Pineda, J., Rachedi, A., Copeland, J., Sitnov, A., Sobhany, S., Suarez-Uruena, A., Swaminathan, J., Tagari, M., Tate, J., Tromm, S., Velankar, S., Vranken, W.: E-MSD: the European Bioinformatics Institute Macromolecular Structure Database. Nucleic Acids Research 31(1), 458–462 (2002)
13. Barrett, T., Troup, D.B., Wilhite, S.E., Ledoux, P., Rudnev, D., Evangelista, C., Kim, I.F., Soboleva, A., Tomashevsky, M., Marshall, K.A., Phillippy, K.H., Sherman, P.M., Muertter, R.N., Edgar, R.: NCBI GEO: archive for high-throughput functional genomic data. Nucleic Acids Research 37(database issue), D5–D15 (2009)
14. Parkinson, H., Kapushesky, M., Kolesnikov, N., Rustici, G., Shojatalab, M., Abeygunawardena, N., Berube, H., Dylag, M., Emam, I., Farne, A., Holloway, E., Lukk, M., Malone, J., Mani, R., Pilicheva, E., Rayner, T.F., Rezwan, F., Sharma, A., Williams, E., Bradley, X.Z., Adamusiak, T., Brandizi, M., Burdett, T., Coulson, R., Krestyaninova, M., Kurnosov, P., Maguire, E., Neogi, S.G., Rocca-Serra, P., Sansone, S.A., Sklyar, N., Zhao, M., Sarkans, U., Brazma, A.: ArrayExpress update from an archive of functional genomics experiments to the atlas of gene expression. Nucleic Acids Research 37(database issue), D868–D872 (2009)
15. Vizcaíno, J.A., Côté, R., Reisinger, F., Foster, J.M., Mueller, M., Rameseder, J., Hermjakob, H., Martens, L.: A guide to the Proteomics Identifications Database proteomics data repository. Proteomics 9(18), 4276–4283 (2009)
16. Rice, P., Longden, I., Bleasby, A.: EMBOSS: The European Molecular Biology Open Software Suite. Trends in Genetics 16(6), 276–277 (2000)
17. Harris, M.A., et al.: The Gene Ontology (GO) database and informatics resource. Nucleic Acids Research 61, D258–D261 (2004)
18. Goble, C., Stevens, R.: State of the nation in data integration for bioinformatics. Journal of Biomedical Informatics 41(5), 687–693 (2008)
19. Perrière, G., Gouy, M.: WWW-query: An on-line retrieval system for biological sequence banks. Biochimie 78(5), 364–369 (1999)
20. Davidson, S.B., Crabtree, J., Brunk, B.P., Schug, J., Tannen, V., Overton, G.C., Stoeckert Jr., C.J.: K2/Kleisli and GUS: Experiments in integrated access to genomic data sources. IBM Systems Journal 40(2), 512–531 (2001)
21. Haas, L.M., Schwarz, P.M., Kodali, P., Kotlar, E., Rice, J.E., Swope, W.C.: DiscoveryLink: a system for integrated access to life sciences data sources. IBM Systems Journal 40(2), 489–511 (2001)
22. Stein, L.D.: Integrating biological databases. Nature Reviews Genetics 4, 337–345 (2003)
23. Lenzerini, M.: Data integration: a theoretical perspective. In: Proceedings of the Twenty-First ACM SIGMOD-SIGACT-SIGART Symposium on Principles of Database Systems, pp. 233–246 (2002)
24. Hernandez, T., Kambhampati, S.: Integration of Biological Sources: Current Systems and Challenges Ahead. Sigmod Record 33, 51–60 (2004)
25. Mork, P., Halevy, A., Tarczy-Hornoch, P.: A model for data integration systems of biomedical data applied to online genetic databases. In: Proceedings of the AMIA Symposium, pp. 473–477 (2001)
26. Friedman, M., Levy, A., Millstein, T.: Navigational Plans For Data Integration. In: Proceedings of the National Conference on Artificial Intelligence (AAAI), pp. 67–73 (1999)

27. Widom, J.: Research Problems in Data Warehousing. In: The Proceedings of the 4th International Conference Information and Knowledge Management, pp. 25–30 (1995)
28. Theodoratos, D., Sellis, T.: Data Warehouse Configuration. In: Proceedings of 23rd International Conference on Very Large Data Bases, pp. 126–135 (1997)
29. Davidson, S.B., Overton, G.C., Tannen, V., Wong, L.: BioKleisli: a digital library for biomedical researchers. International Journal on Digital Libraries 1(1), 36–53 (1997)
30. Wiederhold, G.: Mediators in the architecture of future information systems. Computer 25(3), 38–49 (1992)
31. Levy, A.Y.: Logic-based techniques in data integration. Logic-Based Artificial Intelligence, 575–595 (2000)
32. Garcia-Molina, H., Papakonstantinou, Y., Quass, D., Rajaraman, A., Sagiv, Y., Ullman, J., Vassalos, V., Widom, J.: The TSIMMIS Approach to Mediation: Data Models and Languages. Journal of Intelligent Information Systems 8(2), 117–132 (1997)
33. Adali, S., Candan, K.S., Papakonstantinou, Y., Subrahmanian, V.S.: Query caching and optimization in distributed mediator systems. ACM SIGMOD Record 25(2), 137–146 (1996)
34. Duschka, O.M., Genesereth, M.R., Levy, A.Y.: Recursive query plans for data integration. Journal of Logic Programming 43, 49–73 (2000)
35. Friedman, M., Weld, D.S.: Efficiently Executing Information-Gathering Plans. In: Proceeding of the International Joint Conference of Artificial Intelligence, pp. 785–791 (1997)
36. Levy, A.Y., Rajaraman, A., Ordille, J.J.: Query-answering algorithms for information agents. In: Proceedings of the 13th National Conference on Artificial Intelligence, pp. 40–47 (1996)
37. Cuff, A.L., Sillitoe, I., Lewis, T., Redfern, O.C., Garratt, R., Thornton, J., Orengo, C.A.: The CATH classification revisited – architectures reviewed and new ways to characterize structural divergence in superfamilies. Nucleic Acids Research 37, D310–D314(2009)
38. Sheth, A.P., Larson, J.A.: Federated database systems for managing distributed, heterogeneous, and autonomous databases. ACM Computing Surveys 22(3), 182–236 (1990)
39. Etzold, T., Argos, P.: SRS–an indexing and retrieval tool for flat file data libraries. Bioinformatics 9(1), 49–57 (1993)
40. Schuler, G.D., Epstein, J.A., Ohkawa, H., Kans, J.A.: Entrez: molecular biology database and retrieval system. Methods in Enzymology 266, 141–162 (1996)
41. Ritter, O.: The integrated genomic database (IGD). In: Suhai, S. (ed.) Computational Methods in Genome Research, pp. 57–73. Plenum Press, New York (1994)
42. Wang, L., Rodriguez-Tomé, P., Redaschi, N., McNeil, P., Robinson, A., Lijnzaad, P.: Accessing and distributing EMBL data using CORBA. Genome Biology 1(5) (2000)
43. Barrillot, E., Lesser, U., Lijnzaad, P., Cussat-Blanc, C., Jungfer, K., Guyon, F., Vaysseix, G., Helgesen, C., Rodriguez-Tomé, P.: A proposal for a standard CORBA interface for genome maps. Bioinformatics 15(2), 157–169 (1999)
44. Parsons, J.D., Rodriguez-Tomé, P.: JESAM: CORBA software components to create and publish EST alignments and clusters. Bioinformatics 16(4), 313–325 (2000)
45. Biomolecular Sequence Analysis RFP response Joint Revised Submission. Concept Five Technologies Inc., EMBL-EBI, Genome Informatics Corp., Millenium Pharm. Inc., Neomorphic Software Inc., NetGenics Inc. OMG Document lifesci. (August 1, 1999)
46. Genomic Maps RFP response Joint Second Revised Submission (with errata). EMBL-EBI, Millenium Pharm Inc., NetGenics Inc. OMG Document lifesci. (November 11, 1999)
47. Stevens, R., Baker, P., Bechhofer, S., Ng, G., Jacoby, A., Paton, N.W., Goble, C.A., Brass, A.: TAMBIS: Transparent Access to Multiple Bioinformatics Information Sources. Bioinformatics 16(2), 184–186 (2000)

48. Siepel, A., Farmer, A., Tolopko, A., Zhuang, M., Mendes, P., Beavis, W., Sobral, B.: ISYS: a decentralized, component-based approach to the integration of heterogeneous bioinformatics resources. Bioinformatics 17(1), 83–94 (2000)

49. Durbin, R., Mieg, J.T.: A C. elegans Database (1991) Documentation, code and data available from anonymous FTP servers at , `lirmm.lirmm.fr, cele.mrc-lmb.cam.ac.uk` and `ncbi.nlm.nih.gov`

50. Cherry, J.M., Cartinhour, S.W., Goodman, H.M.: AAtDB, an Arabidopsis thaliana database. Plant Molecular Biology Reporter 10, 308–309 (1992)

51. Stein, L.D., Mungall, C., Shu, S., Caudy, M., Mangone, M., Day, A., Nickerson, E., Stajich, J.E., Harris, T.W., Arva, A., Lewis, S.: The generic genome browser: a building block for a model organism system database. Genome Research 12(10), 1599–1610 (2002)

52. Hubbard, T., et al.: The Ensembl genome database project. Nucleic Acids Research 30(1), 38–41 (2001)

53. Smedley, D., Haider, S., Ballester, B., Holland, R., London, D., Thorisson, G., Kasprzyk, A.: BioMart - biological queries made easy. BMC Genomics (2009), doi:10.1186/1471-2164-10-22

54. Armano, G., Manconi, A.: ProDaMa: an open source Python library to generate protein structure datasets. BMC Research Notes 2, 202 (2009)

55. Armano, G., Manconi, A.: A Collaborative Web Application for Supporting Researchers in the Task of Generating Protein Datasets. In: Proceeding of DART 2010 - 4th International Workshop on Distributed Agent-Based Retrieval Tools (2010)

56. Di Lorenzo, G., Hacid, H., Paik, H.: Data Integration in Mashups. Services Computing 38(1), 59–66 (2009)

57. Mandola, F., Miller, E.: RDF Primer (2004),
`http://www.w3.org/TR/rdf-primer/`

58. Brickley, D., Guha, R.V.: RDF Vocabulary Description Language 1.0: RDF Schema (2004),
`http://www.w3.org/TR/rdf-schema/`

59. Smith, M.K., Welty, C., McGuiness, D.L.: OWL Web Ontology Language (2004),
`http://www.w3.org/TR/owl-guide/`

60. Soldatova, L.N., King, R.D.: Are the Current Ontologies used in Biology Good Ontologies? Nature Biotechnology 23, 1095–1098 (2005)

61. Kim, D.H., Sreenivasaiah, K.: Curren trends and new challenges of databses and web applications for system driven biological research. Frontiers in Physiology 1, 147 (2010), doi:10.3389/fphys.2010.00147.

62. Martin, S., Hohman, M.M., Liefeld, T.: The impact of Life Science Identifier on informatics data. Drug Discovery Today 10, 1566–1572 (2005)

63. Laibe, C., Le Novere, N.: MIRIAM Resources: tools to generate and resolve robust cross-references in Systems Biology. BMC Systems Biology 1, 58 (2007), doi:10.1186/1752-0509-1-58.

# Glossary

- API: Application Programming Interface
- CPL: Collection Programming Language
- ETL: Extract Transform Load
- GAV: Global As View
- IDL: Interface Definition Language
- LAV: Local As View
- NMR: Nuclear Magnetic Resonance
- OWL: Web Ontology Language
- RDF: Resource Description Framework
- RDFS: Resource Description Framework Schema
- REST: Representational Transfer State
- RSS: Really Simple Syndication
- URI: Uniform Resource Identifier.

# Author Index

# Subject Index

Printed in the United States
By Bookmasters